HAZARDOUS MATERIALS COMPLIANCE POCKETBOOK

©Copyright 1995, 1996

J. J. Keller & Associates, Inc.
3003 W. Breezewood Lane – P.O. Box 368
Neenah, Wisconsin 54957-0368
USA
Phone: (414) 722-2848

LIBRARY OF CONGRESS CATALOG CARD NUMBER 95-78579

ISBN 1-877798-41-X

Canadian Goods and Services Tax
(GST) Number: R123-317687

"INTERNATIONAL PUBLISHERS OF TRANSPORTATION GUIDES AND FORMS"

All rights reserved. Neither the Hazardous Materials Compliance Pocketbook nor any part thereof may be reproduced in any form or manner without the written permission of the publisher.

Due to the constantly changing nature of government regulations, it is impossible to guarantee absolute accuracy of the material contained herein. The Publisher and Editors, therefore, cannot assume any responsibility for omissions, errors, misprinting, or ambiguity contained within this publication and shall not be held liable in any degree for any loss or injury caused by such omission, error, misprinting or ambiguity presented in this publication.

This publication is designed to provide reasonably accurate and authoritative information in regard to the subject matter covered. It is sold with the understanding that the Publisher is not engaged in rendering legal, accounting, or other professional service. If legal advice or other expert assistance is required, the services of a competent professional person should be sought.

115-ORS

TABLE OF CONTENTS

The Law Affects You Directly 5
How to Comply with the Law 8
How to Identify a Hazardous
   Materials Shipment 12
Special Notes 20
Equipment You Must Have When
   Hauling Hazardous Materials 25
Loading and Unloading Rules for All
   Hazardous Materials Shipments 40
On the Road with Hazardous Materials 51
Loading, Unloading & on the Road with
   Specific Hazardous Materials 57
   Shipments of Class 1 (Explosives) 58
   Shipments of Class 3 (Flammable Liquids) ... 67
   Shipments of Class 2 (Gases) 73
   Shipments of Class 8
     (Corrosive Liquids) 79
   Shipments of Division 6.1 (Poisonous
     Materials) and Division 2.3
     (Poisonous Gases) 84
   Shipment of Class 4 (Flammable Solids)
     and Class 5 (Oxidizers) 88
   Shipments of Class 7
     (Radioactive Materials) 94
Hazardous Materials Table 99
Appendix A to §172.101 — List of Hazardous
   Substances 323
Appendix B to §172.101 — List of
   Marine Pollutants 410
Special Provisions 430

The Law Affects You Directly

The current Hazardous Materials Regulations can be traced to the Transportation Safety Act of 1974, the hazardous materials regulations "consolidation" of 1976, the Hazardous Materials Transportation Uniform Safety Act of 1990, and the numerous amendments issued to date. Carriers are subject to both the Hazardous Materials Regulations and Motor Carrier Safety Regulations (49 CFR Parts 107, 171-180 & 390-397).

In turn, carriers (employers) are responsible for training, testing and certifying drivers and other employees in proper hazardous materials handling, and for providing necessary documents and information to drivers hauling hazardous materials loads.

Drivers are responsible for following the procedures prescribed in the regulations. Drivers can be fined for not complying with parts of the regulations that relate to them. A person who violates a requirement in the regulations is liable for a penalty of not more than $25,000 and not less than $250 for each violation.

The regulations for transporting hazardous materials are based on good common sense, and are designed to protect the safety of all concerned. It makes good sense for you to follow them carefully!

Besides obeying requirements specifically aimed at them, responsible drivers can do a lot to ensure that hazardous materials shipments are handled properly by all concerned. The following checklist for drivers rounds up the ways you can help.

Checklist

1. Be able to recognize discrepancies in documents, packaging, labeling and compatibility.

2. Inspect all hazardous material shipments before loading and contact carrier management for instructions if there are any suspicious shipments offered.

3. Refuse to accept hazardous material freight from shippers or interline carriers, if the shipping papers are improperly prepared or don't check out with the freight. Also refuse leaking containers, those that might be damaged or any that appear improper.

4. Be sure that packages bearing POISON labels are not loaded with foodstuffs, animal feeds or any other edible cargo intended for consumption by humans or animals.

5. Be sure that all hazardous material is properly blocked and secured for transportation. Be sure containers won't be damaged by any other freight or by nails, rough floors, etc.

6. Be sure any required placards are in place before starting.

7. Have all necessary shipping papers (bills of lading, hazardous waste manifest, etc.) in good order, in your possession and available for immediate use in case of accident or inspection.

8. Know your responsibilities for attending your vehicle while carrying hazardous materials.

You must report leaking containers, defective valves, etc. or any accident that occurs during a trip. If you receive written instructions from the carrier on emergency procedures, etc., you are required to follow them carefully.

Pick-Up and Delivery Drivers

If you fall into this category, you also need to be acquainted with the regulations concerning transportation of hazardous materials. Knowing the basics will help you identify any discrepancies you might find in shipments.

You should be advised in advance that you will be picking up hazardous materials.

Before the vehicle is loaded, you must inspect the hazardous materials to be hauled. If you suspect that the shipment may not be in compliance with regulations, you should contact the terminal for instructions before proceeding.

You must comply with any special blocking and securing requirements for the type of cargo you will be hauling.

You should also know proper procedures to follow in accidents or emergencies during transit. Finally, you must comply with attendance requirements for vehicles hauling hazardous materials.

Passenger-Carrying Vehicles

In general, no hazardous materials **at all** should be hauled on any for-hire passenger vehicle, if any other practicable means of transportation is available. Extremely dangerous Division 6.1 (poisonous) or Division 2.3 (poisonous gas) liquids, or any paranitraniline may **not** be hauled in passenger vehicles. However, in instances where there is no other means of transportation available, limited quantities of some types of hazardous materials may be hauled in passenger vehicles. In general, any hazardous material (with the exception of small arms ammunition) hauled on a passenger-carrying vehicle must be hauled outside the passenger compartment. The following may be hauled in limited quantities:

- Up to 45 kg (99 lbs.) gross weight of Class 1 (explosives).

- Not more than 100 detonators, Division 1.4 (explosives) per vehicle.

- Not more than 2 lab samples of Class 1 (explosives) materials per vehicle.

- Not more than 225 kg (496 lbs.) total of other types of hazardous materials, with no more than 45 kg (99 lbs.) of any class.

- Not more than 45 kg (99 lbs.) total of non-liquid, less dangerous Division 6.1 (poisonous) materials.

- Radioactive materials with required labels, when no other practicable means of transportation is available, and then only in full compliance with specific loading requirements.
- Department of Defense materials under special circumstances.

How to Identify a Hazardous Materials Shipment

If you, as a driver, are to handle hazardous materials safely, you must be able to recognize them easily. HM shipments are recognizable by entries and the shipper's certification on shipping papers, HM warning labels and markings on the packages.

Hazardous materials fall into these categories:

Class 1 (explosives) — is is divided into six divisions as follows:

Division 1.1 — consists of explosives that have a mass explosion hazard.

Division 1.2 — consists of explosives that have a projection hazard.

Division 1.3 — consists of explosives that have a fire hazard and either a minor blast hazard or a minor projection hazard or both.

Division 1.4 — consists of explosive devices that present a minor blast hazard.

Division 1.5 — consists of very insensitive explosives.

Division 1.6 — consists of extremely insensitive detonating substances.

Class 2 (gases) — is divided into three divisions as follows:

Division 2.1 — consists of gases that are flammable.

Division 2.2 — consists of gases that are non-flammable and compressed.

Division 2.3 — consists of gases that are poisonous.

Class 3 (flammable and combustible liquids) — A flammable liquid has a flash point of not more than 60.5°C (141°F). A combustible liquid has a flash point above 60.5°C (141°F) and below 93°C (200°F).

Class 4 (flammable solids) — is divided into three divisions as follows:

Division 4.1 — consists of solids that are flammable.

Division 4.2 — consists of material that is spontaneously combustible.

Division 4.3 — consist of material that is dangerous when wet.

Class 5 (oxidizers and organic peroxides) — is divided into two divisions as follows:

Division 5.1 — oxidizer.

Division 5.2 — organic peroxide.

Class 6 (poisons) — is divided into two divisions as follows:

Division 6.1 — consists of material that is poisonous.

Division 6.2 — consists of material that is an infectious substance (etiologic agent).

Class 7 (radioactive materials) — consists of any material having a specific radioactive activity greater than 0.002 microcuries per gram.

Class 8 (corrosives) — consists of a material, liquid or solid, that causes visible destruction or irreversible alteration to human skin or a liquid that has a severe corrosion rate on steel or aluminum.

Class 9 (miscellaneous) — consists of a material which presents a hazard during transport, but which is not included in any other hazard class (such as a hazardous substance or a hazardous waste).

ORM-D (other regulated material) — consists of a material which although otherwise subject to the regulations, presents a limited hazard during transportation due to its form, quantity and packaging (consumer commodities).

Shipping Papers

A look at the shipping papers will help you identify the hazard or hazards you will be dealing with in a particular shipment. Hazardous materials are required to be indicated on shipping papers in one of three ways.

1. The hazardous materials may be listed **first**, before any non-hazardous materials, or

Number of packages	KIND OF PACKAGE, DESCRIPTION OF ARTICLES, SPECIAL MARKS, AND EXCEPTIONS	*Weight (Sub. to correction)	Class or Rate
10	drums, Ethyl methyl ether, 2.1, UN 1039	3400 lbs.	
3	skids, 100# Sorex paper 36" x 40"	3000 lbs.	

2. they may be listed in a **color** that clearly contrasts with entries for materials that are not subject to the hazardous materials regulations, or

	Number of packages	KIND OF PACKAGE, DESCRIPTION OF ARTICLES, SPECIAL MARKS, AND EXCEPTIONS	*Weight (Sub.to correction)	Class or Rate
Black →	3	skids, 100# Sorex paper 36" x 40"	3000 lbs.	
Red →	10	drums, Ethyl methyl ether, 2.1, UN 1039	3400 lbs.	
Black →	100	36" x 40" wooden skids	3000 lbs.	

3. by placing an "**X**" in a column captioned "HM".

No. Shipping Units	HM	Kind of Packages, Description of Articles (IF HAZARDOUS MATERIALS - PROPER SHIPPING NAME)	HAZARD CLASS	I.D. Number	PACKING GROUP	WEIGHT (subject to correction)	RATE
10	X	drums, Ethyl methyl ether	2.1	UN1039		3400 lbs.	
3		skids, 100# Sorex paper 36" x 40"				3000 lbs.	

Most n.o.s. and other generic proper shipping names must have the technical name of the hazardous material entered in parentheses in association with the basic description. If a hazardous material is a mixture or solution the technical names of at least two components most predominately contributing to the hazards of the mixture or solution must be entered.

No. Shipping Units	HM	Kind of Packages, Description of Articles (IF HAZARDOUS MATERIALS - PROPER SHIPPING NAME)	HAZARD CLASS	I.D. Number	PACKING GROUP	WEIGHT (subject to correction)	RATE
5	X	drums, Flammable liquids, n.o.s. (Xylene, Benzene)	3	UN1993	II	2250 lbs.	

Although there are a number of additional descriptions which may be required on shipping papers, there are several of special significance.

1. If the material is a hazardous substance, the letters "**RQ**" must appear either before or after the basic

entry. The letters "RQ" may be entered in the HM Column in place of the "X".

No. Shipping Units	HM	Kind of Packages, Description of Articles (IF HAZARDOUS MATERIALS - PROPER SHIPPING NAME)	HAZARD CLASS	I.D. Number	PACKING GROUP	WEIGHT (subject to correction)	RATE
10	RQ	drums, Aldrin, liquid	6.1	NA 2762	II	2000 lbs.	
3		skids, 100# Sorex paper				3000 lbs.	

2. For all materials meeting the poisonous by inhalation criteria, the words "Poison-Inhalation Hazard" and the words "Zone A", "Zone B", "Zone C", or "Zone D", for gases or "Zone A" or "Zone B" for liquids, as appropriate, must be added on the shipping paper immediately following the shipping description.

3. The fact that a material is an elevated temperature material must be indicated in the shipping name or the word "HOT" must immediately precede the proper shipping name.

4. If a material is a marine pollutant the words "Marine Pollutant" must be entered in association with the basic shipping description.

All hazardous material shipments (except those that do not require shipping papers) must have emergency response information on or in association with the shipping paper. If the information is in association with the shipping paper it may be in the form of the Emergency Response Guidebook, a Material Safety Data Sheet, or any other form that provides all the information required in 172.602.

A 24-hour emergency response phone number must be entered on the shipping paper. It can be immediately following the description of each hazardous material or if the number applies to every hazardous material entered on the shipping paper, entered once on the shipping paper in a clearly visible location.

Finally, check for the "Shipper's Certification" which states the shipment has been properly classified, described, packaged, marked and labeled/placarded.

<u>Labels</u>

When hazard warning labels are required on packages, the shipper is required to affix the labels before offering the shipment. The driver should not accept the shipment unless the required labels are affixed to the packages.

The Hazardous Materials Table at the back of this book provides a ready reference to the labels required for each proper shipping name.

If more than one label is indicated, the first one listed is the primary label and any others are subsidiary labels. Primary labels have a class or division number in their lower corner and subsidiary labels do not.

The hazard warning labels are diamond-shaped and should measure at least 100 mm (3.9 inches) on each side. The package pictured, has labels properly affixed.

PACKAGE MARKINGS

All non-bulk packagings (119 gallons or less) **must be marked with the proper shipping name** and the four digit UN/NA identification number as shown in the Hazardous Materials Table and illustrated above. Most n.o.s. and other generic proper shipping names must also have the technical name(s) entered in parentheses in association with the proper shipping name. In addition, other markings may be required, such as the package orientation arrows. However, two very important markings may also be required. First, if the package contains the reportable quantity of a hazardous substance, the letters "RQ" must be marked in association with the proper shipping name. And, second, for those materials meeting the poisonous by inhalation criteria, the words "Inhalation Hazard" must appear in association with the hazard warning labels or shipping name, when required.

Finally, packagings (except for cylinders, Class 7 packagings and bulk packagings) for most hazardous materials must meet the United Nations Performance-Oriented Packaging Standards (by October 1, 1996) and be marked by the manufacturer in a manner such as illustrated below for a fiberboard box with an inner packaging.

 **4G/Y145/S/95
USA/RA**

—SPECIAL NOTES—

Poison Inhalation Hazard

Because of the danger of transporting Poison Inhalation Hazard materials the Hazardous Materials Regulations have special requirements for shipments of materials meeting the definition of Poison Inhalation Hazard. If a material meets the criteria for inhalation hazard, whether listed in the hazardous materials table or not, it is subject to the requirements.

There are several things to be aware of when transporting Poison Inhalation Hazard materials.

1. **SHIPPING PAPERS** The words "Poison-Inhalation Hazard" and for gases "Zone A", "Zone B", "Zone C", or "Zone D", or for liquids "Zone A" or "Zone B", as appropriate, must be entered on the paper immediately following the shipping description. (For Anhydrous ammonia see 4 below.)

2. **MARKINGS** A package containing Poison Inhalation Hazard material must be marked "Inhalation Hazard" in association with any required labels, placards, or shipping name, when required. This marking must be at least 25 mm (1.0 inch) in height for portable tanks with capacities of less than 3,785 L (1,000 gallons) and at least 50 mm (2.0 inches) in height for cargo tanks and other bulk packages. Bulk packaging must be marked on two opposing sides with this marking. (For Anhydrous ammonia see 4 below.)

3. **PLACARDS** Any vehicle, tank or container that contains a Poison Inhalation Hazard material must be placarded on each side and each end with a POISON or POISON GAS placard, as appropriate, if it is not already placarded with these placards as required by the regulations. (For Anhydrous ammonia see 4 below.)

4. **EXCEPTION** Anhydrous ammonia transported within the U.S. is treated differently than other Poison Inhalation Hazard materials. Only the words "Inhalation Hazard" must be entered on the shipping paper in association with the shipping description. Bulk packages must be marked with "Inhalation Hazard" on two opposing sides. The additional placarding of POISON or POISON GAS is not required.

Elevated Temperature Materials

An elevated temperature material is a material offered for transportation in a **BULK** packaging and meets one of the following criteria:

1. Is a liquid at a temperature at or above 100°C (212°F);

2. Is a liquid with a flash point at or above 37.8°C (100°F) that is intentionally heated and transported at or above its flash point; or

3. Is in a solid phase and at a temperature at or above 240°C (464°F).

There are several things to be aware of when transporting elevated temperature materials.

1. **SHIPPING PAPERS** The fact that a material is an elevated temperature material must be indicated in the shipping name or the word "HOT" must immediately precede the proper shipping name.

2. **MARKINGS** Most bulk packages containing elevated temperature materials must be marked on two opposing sides with the word "HOT". However, if the package contains molten aluminum or molten sulfur the package must be marked "MOLTEN ALUMINUM" or "MOLTEN SULFUR" instead of the word "HOT".

   The "HOT" marking must be on the package itself or in black lettering on a white square-on-point configuration that is the same size as a placard. The lettering must be at least 50 mm (2 inches) high.

   The word "HOT" may be displayed in the upper corner of a white-square-on-point configuration that also displays the required identification number.

3. **PLACARDS** Bulk packages containing elevated temperature materials may require placards, such as Class 3 or Class 9 (not required for domestic transport).

4. **EXCEPTION** A package containing solid elevated temperature material is **not subject** to any of the Hazardous Materials Regulations, **except** that the word "HOT" **must** be on the package.

Marine Pollutants

A marine pollutant is a material that is listed in Appendix B to §172.101 and when in a solution or mixture of one or more marine pollutants, is packaged in a concentration which equals or exceeds:

1. Ten percent by weight of the solution or mixture; or

2. One percent by weight of the solution or mixture for materials that are identified as severe marine pollutants (PP) in Appendix B.

Marine pollutants transported by water, in any size packaging, are subject to all the applicable requirements. Marine pollutants transported by highway, rail or air are subject to the requirements when transported in **BULK** packages.

You should be aware of the following when transporting marine pollutants.

1. **SHIPPING PAPERS** If the proper shipping name for a material does not indicate that it is a marine pollutant, the words "Marine Pollutant" must appear in association with the basic description on the shipping papers.

2. **MARKINGS** Bulk packages must be marked, on all four sides, with the MARINE POLLUTANT marking, unless they are labeled or placarded.

Equipment You Must Have When Hauling Hazardous Materials

This chapter briefly details special requirements for vehicles transporting hazardous materials. Areas covered include **placards, identification numbers, shipping papers, cargo heaters and special documents**.

Placards

The federal Hazardous Materials Regulations require most vehicles hauling hazardous cargo to be placarded. **The shipper is responsible for providing the appropriate placards** to the motor carrier for a shipment. **The carrier is responsible for applying them correctly to the vehicle,** and for replacing them if they should be lost or damaged during a trip. In addition, carriers are responsible for any placarding necessitated by aggregate shipment which collect at their terminals. Large freight containers (640 cubic feet or more) must be placarded by the shipper.

Placards must be affixed on all four sides of the vehicle, trailer or cargo carrier. The front placard may be on the front of the tractor or the cargo body. Placards must be removed from any vehicle not carrying hazardous materials.

In general, some basic rules apply to placards:

- **Placards must be securely attached to the vehicle or placed in a proper placard holder.**

- **Placards must be located at least three inches from any other type of marking on the vehicle.**

- **Words on placards must be read horizontally.**

- **Placards must be clearly visible and maintained in a legible condition throughout the trip.**

As a rule, a driver may not move a vehicle that is not properly placarded, if placards are demanded by the cargo carried in the vehicle. However, in certain emergency situations, a vehicle not properly placarded may be moved if at least one of these three conditions is met:

- **Vehicle is escorted by a state or local government representative.**

- **Carrier has received permission from the Department of Transportation.**

- **Movement is necessary to protect life and property.**

Illustrations of all hazardous materials placards can be found at the back of this book.

Placards are chosen according to tables found in §172.504. Table 1 includes placard specifications for motor vehicles (as well as other types of equipment) carrying hazardous materials which require placarding no matter what amount of material is being hauled. Any quantity of hazardous material covered by Table 1 must be placarded as specified in Table 1.

Table 2 includes placard specifications for motor vehicles carrying hazardous materials, but, the amount of material being hauled determines when placards are required. When the gross weight of all hazardous materials covered by Table 2 is **less than 454 kg (1,001 pounds)**, no placard is required on a transport vehicle or freight container for the Table 2 materials. This paragraph does not apply to portable tanks, cargo tanks, or transport vehicles and freight containers subject to §172.505.

In the following tables the **bold face entry** describes the classification of the hazardous material, the light face entry indicates the placard required for that type of shipment.

TABLE 1	
HAZARD CLASS OR DIVISION	**PLACARD**
1.1	EXPLOSIVES 1.1
1.2	EXPLOSIVES 1.2
1.3	EXPLOSIVES 1.3
2.3	POISON GAS
4.3	DANGEROUS WHEN WET
6.1 (PG I, inhalation hazard only)	POISON
7 (Radioactive Yellow III label only)	RADIOACTIVE[1]

[1]RADIOACTIVE placard also required for exclusive use shipments of low specific activity material and surface contaminated objects transported in accordance with §173.427(b)(3) or (c) of this subchapter.

TABLE 2	
HAZARD CLASS OR DIVISION	PLACARD
1.4	EXPLOSIVES 1.4
1.5	EXPLOSIVES 1.5
1.6	EXPLOSIVES 1.6
2.1	FLAMMABLE GAS
2.2	NON-FLAMMABLE GAS
3	FLAMMABLE
Combustible liquid	COMBUSTIBLE
4.1	FLAMMABLE SOLID
4.2	SPONTANEOUSLY COMBUSTIBLE
5.1	OXIDIZER
5.2	ORGANIC PEROXIDE
6.1 (PG I or II, other than PG I inhalation hazard)	POISON
6.1 (PG III)	KEEP AWAY FROM FOOD
6.2	(None)
8	CORROSIVE
9	CLASS 9
ORM-D	(None)

Placarding exceptions.

1. When more than one division placard is required for Class 1 materials on a transport vehicle, rail car, freight container or unit load device, only the placard representing the lowest division number must be displayed.

2. A FLAMMABLE placard may be used in place of a COMBUSTIBLE placard on—

   (i) A cargo tank or portable tank.

   (ii) A compartmented tank car which contains both flammable and combustible liquids.

3. A NON-FLAMMABLE GAS placard is not required on a transport vehicle which contains non-flammable gas if the transport vehicle also contains flammable gas or oxygen and it is placarded with FLAMMABLE GAS or OXYGEN placards, as required.

4. OXIDIZER placards are not required for Division 5.1 materials on freight containers, unit load devices, transport vehicles or rail cars which also contain Division 1.1 or 1.2 materials and which are placarded with EXPLOSIVES 1.1 or 1.2 placards, as required.

5. For transportation by transport vehicle or rail car only, an OXIDIZER placard is not required for Division 5.1 materials on a transport vehicle, rail car or freight container which also contains Division 1.5 explosives and is placarded with EXPLOSIVES 1.5 placards, as required.

6. The EXPLOSIVE 1.4 placard is not required for those Division 1.4 Compatibility Group S (1.4S) materials that are not required to be labeled 1.4S.

7. For domestic transportation of oxygen, compressed or oxygen, refrigerated liquid, the OXYGEN placard in §172.530 of this subpart may be used in place of a NON-FLAMMABLE GAS placard.

8. Except for a material classed as a combustible liquid that also meets the definition of a Class 9 material, a COMBUSTIBLE placard is not required for a material classed as a combustible liquid when transported in a non-bulk packaging. For a material in a non-bulk packaging classed as a combustible liquid that also meets the definition of a Class 9 material, the CLASS 9 placard may be substituted for the COMBUSTIBLE placard.

9. For domestic transportation, a Class 9 placard is not required. A bulk packaging containing a Class 9 material must be marked with the appropriate identification number displayed on a Class 9 placard, orange panel or a white-square-on-point display configuration as required by subpart D of part 172.

10. For domestic transportation of Division 6.1, PG III materials, a POISON placard may be used in place of a KEEP AWAY FROM FOOD placard.

The compatibility group letter must be displayed on placards for shipments of Class 1 materials when transported by aircraft or vessel.

A transport vehicle or freight container which contains non-bulk packagings with two or more categories of materials, requiring different placards specified in Table 2, may be placarded DANGEROUS in place of the separate placarding specified for each of the materials in Table 2. However, when 2,268 kg (5,000 pounds) or more of **one category** of materials is loaded at **one loading facility**, the placard specified for that material in Table 2 must be applied.

Each transport vehicle, portable tank, freight container, or unit load device that contains a material subject to the "Poison-Inhalation Hazard" shipping paper description must be placarded "POISON" or "POISON GAS" on each side and each end if not so placarded under §172.504.

Placards for subsidiary hazards are required when certain hazardous materials are transported. The subsidiary placards that you will most likely encounter are DANGEROUS WHEN WET, POISON, POISON GAS and CORROSIVE. A subsidiary placard is easily recognized by the absence of a hazard class or division number in the bottom corner of the placard (see the placard illustrations at the back of the book).

SPECIAL PLACARDING TRANSITION PERIOD

The following placarding provision **applies only to highway transportation.** Until October 1, 2001, old (pre HM-181) placards, which conform to the specifications in effect on September 30, 1991, may be used in place of the current placards in accordance with the following table:

PLACARD SUBSTITUTION TABLE

Hazard class or division No.	Current placard name	Old (Sept. 30, 1991) placard name
DIVISION 1.1	EXPLOSIVES 1.1	EXPLOSIVES A.
DIVISION 1.2	EXPLOSIVES 1.2	EXPLOSIVES A.
DIVISION 1.3	EXPLOSIVES 1.3	EXPLOSIVES B.
DIVISION 1.4	EXPLOSIVES 1.4	DANGEROUS.
DIVISION 1.5	EXPLOSIVES 1.5	BLASTING AGENTS.
DIVISION 1.6	EXPLOSIVES 1.6	DANGEROUS.
DIVISION 2.1	FLAMMABLE GAS	FLAMMABLE GAS.
DIVISION 2.2	NONFLAMMABLE GAS	NONFLAMMABLE GAS
DIVISION 2.3	POISON GAS	POISON GAS.
CLASS 3	FLAMMABLE	FLAMMABLE.
COMBUSTIBLE LIQUID	COMBUSTIBLE	COMBUSTIBLE.
DIVISION 4.1	FLAMMABLE SOLID	FLAMMABLE SOLID.
DIVISION 4.2	SPONTANEOUSLY COMBUSTIBLE	FLAMMABLE SOLID.
DIVISION 4.3	DANGEROUS WHEN WET	FLAMMABLE SOLID W.
DIVISION 5.1	OXIDIZER	OXIDIZER.
DIVISION 5.2	ORGANIC PEROXIDE	ORGANIC PEROXIDE.
DIVISION 6.1, PG I and II	POISON	POISON.
DIVISION 6.1, PG III	KEEP AWAY FROM FOOD	(None required.)
CLASS 7	RADIOACTIVE	RADIOACTIVE.
CLASS 8	CORROSIVE	CORROSIVE.
CLASS 9	CLASS 9	(None required.)

Identification Numbers

Hazardous materials UN/NA identification numbers must be displayed on bulk packagings such as portable tanks, cargo tanks, and tank cars. The four digit ID numbers may be displayed on an "Orange Panel" (160 mm (6.3 inches) by 400 mm (15.7 inches)), across the center of the proper placard in place of the hazard wording, or in some cases on a white square-on-point display configuration having the same outside dimensions as a placard.

Identification numbers are required to be displayed on each side and each end of a bulk packaging which has a capacity of 3,785 L (1,000 gal.) or more. Bulk packagings that have a capacity of less than 3,785 L (1,000 gal.) only need the identification numbers displayed on two opposing sides.

Identification numbers are not required on the ends of portable tanks and cargo tanks having more than one compartment if hazardous materials having different identification numbers are transported in the compartments. The identification numbers on the sides of the tank must be displayed in the same sequence as the compartments containing the materials they identify.

Shipping Papers

Hazardous materials shipments must be accompanied by proper shipping papers, such as bills of lading, hazardous waste manifests, etc. During the course of the trip, the driver is responsible for maintaining the shipping papers according to requirements, so that they are easily accessible to authorities in case of accidents or inspection.

1. **If the hazardous material shipping paper is carried with any other papers, it must be clearly distinguished, either by tabbing it or having it appear first.**

2. **When the driver is at the controls, the shipping papers must be within his immediate reach when he is restrained by the seat belt.**

3. **The shipping papers must be readily visible to someone entering the driver's compartment, or in a holder mounted on the inside of the door on the driver's side.**

4. **If the driver is not in the vehicle, the shipping papers must be either in the holder on the door or on the driver's seat.**

The HM shipping paper requirements do not apply to materials, other than hazardous wastes or hazardous substances, identified by the letter "A" or "W" in column 1 of the 172.101 Table unless they are intended for transportation by air or water, and materials classed as ORM-D unless they are intended for transportation by air.

If the shipment will eventually interline with a rail carrier, the motor carrier must also mark on the shipping papers:

- **A clear description of the freight container or vehicle that will move into rail transportation.**
- **The kind of placard affixed to the container or vehicle.**

Cargo Heaters

If a cargo heater is located in a vehicle that will be used to transport hazardous materials, various restrictions apply.

<u>**When transporting Class 1 (explosives),**</u> any cargo heater must be made inoperable by draining or removing the heater's fuel tank and disconnecting its power source.

<u>**When transporting Class 3 (flammable liquid) or Division 2.1 (flammable gas)**</u>, any cargo heater in the vehicle must meet the following specifications:

- **It is a catalytic heater.**
- **Surface temperature on the heater cannot exceed 54°C (129°F) on a thermostatically controlled heater or on a heater without thermostat control when the outside temperature is 16°C (60°F) or less.**
- **The heater is not ignited in a loaded vehicle.**
- **There is no flame on the catalyst or anywhere else in the heater.**

- Heater is permanently marked by the manufacturer with "MEETS DOT REQUIREMENTS FOR CATALYTIC HEATERS USED WITH FLAMMABLE LIQUID AND GAS" and also "DO NOT LOAD INTO OR USE IN CARGO COMPARTMENTS CONTAINING FLAMMABLE LIQUID OR GAS IF FLAME IS VISIBLE ON CATALYST OR IN HEATER."

There are also special restrictions on the use of **automatic cargo space heating temperature control devices**. Such devices may be used in transporting Class 3 (flammable liquid) or Division 2.1 (flammable gas) materials only if all the following conditions are met:

- **Electrical apparatus in the cargo compartment is explosion-proof or non-sparking.**

- **No combustion apparatus is located in the cargo compartment.**

- **No air-return connection exists between the cargo compartment and the combustion apparatus.**

- **The system is not capable of heating any part of the cargo to more than 54°C (129°F).**

If the temperature control device in question does not meet the previous four specifications, the following steps must be taken before the vehicle can be used to transport flammable liquid or flammable gas:

- **Any cargo heater fuel tank (other than LPG) must be emptied or removed.**

- **LPG tanks must have all discharge valves closed and fuel feed lines disconnected.**

Special Documents

Certificate of Registration

Any Carrier who transports, in foreign, interstate or intrastate commerce, any of the following, is subject to the DOT Registration requirements.

- **Any highway route-controlled quantity of a Class 7 (radioactive) material;**

- **More than 25 kg (55 lbs.) of a Division 1.1, 1.2, 1.3 (explosive) material;**

- **More than one L (1.06 qts.) per package of a material extremely toxic by inhalation (i.e., "material poisonous by inhalation," as defined in §171.8, that meets a criteria for "hazard zone A," as specified in §§173.116(a) or 173.133(a));**

- **A hazardous material in a bulk packaging having a capacity equal to or greater than 13,248 L (3,500 gals.) for liquids or gases, or more than 13.24 cubic meters (468 cubic feet) for solids; or**

- **A shipment in other than a bulk packaging (being offered or loaded at one loading facility using one transport vehicle) of 2,268 kg (5,000 lbs.) gross weight or more of one class of hazardous materials for which placarding is required.**

Motor carriers subject to the registration requirements must carry, on board **each** vehicle that is transporting a hazardous material requiring registration:

- **A copy of the carriers' current Certificate of Registration; or**
- **Another document bearing the registration number identified as the "U.S. DOT Hazmat Reg. No.".**

Explosives Shipments

Some special documents are required to be carried by drivers of vehicles transporting Division 1.1, 1.2, or 1.3 (explosives) materials:

- **A copy of Part 397, Transportation of Hazardous Materials; Driving and Parking Rules, Federal Motor Carrier Safety Regulations.**
- **A document containing instructions on what to do in the event of an accident or delay in the shipment. This information must also include the name of the explosive hauled and names and phone numbers of all persons to be contacted in the event of accident or delay.**
- **Proper shipping papers.**
- **A written route plan for the movement of explosives that complies with the requirements.**

Hazardous Waste Shipment

A driver **must not** accept a shipment of hazardous waste unless it is accompanied by a properly prepared Uniform Hazardous Waste Manifest!

Delivery of hazardous wastes **must be** made ONLY to the facility or alternate facility designated on the hazardous waste manifest. If delivery can not be made, the driver should contact his dispatcher or other designated official, immediately.

Loading and Unloading Rules for All Hazardous Materials Shipments

There are some regulations for the safe and secure loading and unloading of hazardous materials shipments which apply to all hazardous commodities shipments. They are included in this chapter, and should be interpreted as general common sense rules to follow in the operation of your vehicle. Specifics on loading and unloading certain particular hazardous substances are contained in a later chapter.

Securing Packages

Tanks, barrels, drums, cylinders or other packaging not permanently attached to your vehicle and which contain any Class 3 (flammable liquid), Class 2 (gases), Class 8 (corrosive), Division 6.1 (poisonous), or Class 7 (radioactive) material must be secured against any movement within the vehicle during normal transportation.

Pole Trailers

No pole trailers may be used in the transportation of any hazardous material.

No Smoking

Smoking on or near any vehicle containing any Class 1 (explosive), Class 3 (flammable liquid), Class 4 (flammable solid), Class 5 (oxidizer), or Divison 2.1 (flammable gas) materials while loading or unloading is forbidden. Further, care should be taken to keep all fire sources — matches and smoking materials in particular — away from vehicles hauling any of the above materials.

Finally, Part 397 provides that no person may smoke or carry any lighted smoking materials on or within 25 feet of marked or placarded vehicles containing any Class 1 (explosive) materials, Class 5 (oxidizer) materials or flammable materials classified as Division 2.1, Class 3, Divisions 4.1 and 4.2, or any empty tank vehicle that has been used to transport Class 3 (flammable) liquid or Division 2.1 (flammable gas) materials.

Set Handbrake

During the loading and unloading of any hazardous materials shipment the handbrake on the vehicle must be set, and all precautions taken to prevent movement of the vehicle.

Tools

Any tools used in loading or unloading hazardous material must be used with care not to damage the closures on any packages or containers, or to harm packages of Class 1 (explosive) material and other hazardous materials in anyway.

Prevent Motion between Containers

Containers with valves or other similar fittings must be loaded so that there is minimum likelihood of any damage to them during transportation. Containers of Class 1 (explosives), Class 3 (flammable liquids), Class 4 (flammable solids), Class 5 (oxidizers), Class 8 (corrosives), Class 2 (gases) and Division 6.1 (poisonous) materials, must be blocked and braced to prevent motion of the containers relative to each other and to the vehicle during transit.

Attending Vehicles

There are attendance requirements for **cargo tanks** that are being loaded and unloaded with hazardous materials. Such a tank must be attended at all times during loading and unloading by a qualified person. The person who is responsible for loading the cargo is also responsible for seeing that the vehicle is attended. However, the carrier's obligation to oversee unloading ceases when all these conditions are met:

- **The carrier's transportation obligation is completed.**
- **The cargo tank is placed on the consignee's premises.**
- **Motive power is removed from the cargo tank and the premises.**

A person qualified to attend the vehicle during loading and/or unloading must be:

- Awake.
- Within 7.62 meters (25 feet) of the vehicle with an unobstructed view of it.
- Aware of the nature of the hazardous material in question.
- Instructed in emergency procedures.
- Authorized to and capable of moving the cargo tank if necessary.

Forbidden Articles

Common carriers may not haul nitroglycerin, desensitized liquid nitroglycerin or diethylene glycol dinitrate, except in certain special instances where these substances meet safety requirements of the regulations. Other carriers may transport these commodities.

No motor carrier may accept for transportation any article classed as "Forbidden" in the 172.101 Hazardous Materials Table.

Carrier personnel are also responsible for rejecting packages of hazardous materials which show signs of leakage or other damage.

Commodity Compatibility

The Hazardous Materials Regulations contain Segregation Tables which indicate which hazardous materials may not be loaded, transported or stored together.

Materials which are in packages that require labels, in a compartment within a multi-compartmented cargo tank or in a portable tank loaded in a transport vehicle or freight container are subject to the Segregation Tables. In addition to the following Tables, cyanides or cyanide mixtures may not be loaded or stored with acids.

Hazardous materials may not be loaded, transported, or stored together, except as provided in the following Table:

SEGREGATION TABLE FOR HAZARDOUS MATERIALS

Class or division	Notes	1.1 1.2	1.3	1.4	1.5	1.6	2.1	2.2	2.3 gas Zone A	2.3 gas Zone B	3	4.1	4.2	4.3	5.1	5.2	6.1 liquids PG 1 Zone A	7	8 liquids only
Explosives ..... 1.1 and 1.2	A	*	*	*	*	*	X	X	X	X	X	X	X	X	X	X	X	X	X
Explosives 1.3		*	*	*	*	*	X	X	X	X	X	X	X	X	X	X	X		X
Explosives 1.4		*	*	*	*	*	X		X	X	O	O	O	O	X	X	O		O
Very insensitive explosives 1.5	A	*	*	*	*	*	X	X	X	X	X	X	X	X	X	X	X	X	X
Extremely insensitive explosives 1.6		*	*	*	*	*													
Flammable gases 2.1		X	X	X	X						X	O			X		X		O
Non-toxic, non-flammable gases. 2.2		X	X															O	
Poisonous gas Zone A 2.3		X	X								X	X	X	X	X	X			X
Poisonous gas Zone B 2.3		X	X								O	O	O	O	O	O			O
Flammable liquids 3		X	X	O	X		X	O	X	O			X	O	X	X	X		O
Flammable solids 4.1		X	X		X				X	O	O				X	X	X		
Spontaneously combustible materials. 4.2		X	X	O	X				X	O	X				X	X	X	X	X
Dangerous when wet materials. 4.3		X	X		X				X	O	O				X	O	X		X
Oxidizers 5.1	A	X	X	O	X		X		X	O	O		X	X			X	X	O
Organic peroxides 5.2		X	X		X				X	O	O		X	O			X		O
Poisonous liquids PG 1 6.1 Zone A		X	X		X				X	O	X	X	X	X	X	X			X
Radioactive materials 7		X											X		X				
Corrosive liquids 8		X	X	O	X		O		X	O	O	O	X	O	O	O	X		

45

Instructions for using the segregation table for hazardous materials are as follows:

1. The absence of any hazard class or division or a blank space in the Table indicates that no restrictions apply.

2. The letter "X" in the Table indicates that these materials may not be loaded, transported, or stored together in the same transport vehicle or storage facility during the course of transportation.

3. The letter "O" in the Table indicates that these materials may not be loaded, transported, or stored together in the same transport vehicle or storage facility during the course of transportation unless separated in a manner that, in the event of leakage from packages under conditions normally incident to transportation, commingling of hazardous materials would not occur. Notwithstanding the methods of separation employed, Class 8 (corrosive) liquids may not be loaded above or adjacent to Class 4 (flammable) or Class 5 (oxidizing) materials; except that shippers may load truckload shipments of such materials together when it is known that the mixture of contents would not cause a fire or a dangerous evolution of heat or gas.

4. The "*" in the Table indicates that segregation among different Class 1 (explosive) materials is governed by the compatibility table in paragraph (f) of this section.

5. The note "A" in the second column of the Table means that, notwithstanding the requirements of the letter "X", ammonium nitrate fertilizer may be loaded or stored with Division 1.1 (Class A explosive) or Division 1.5 (blasting agents) materials.

6. When the §172.101 Table or §172.402 of this subchapter requires a package to bear a subsidiary hazard label, segregation appropriate to the subsidiary hazard must be applied when that segregation is more restrictive than that required by the primary hazard. However, hazardous materials of the same class may be stowed together without regard to segregation required any secondary hazard if the materials are not capable of reacting dangerously with each other and causing combustion or dangerous evolution of heat, evolution of flammable, poisonous, or asphyxiant gases, or formation of corrosive or unstable materials.

Class 1 (explosive) materials shall not be loaded, transported, or stored together, except as provided in this section, and in accordance with the following Table:

COMPATIBILITY TABLE FOR CLASS 1 (EXPLOSIVE) MATERIALS

Compatibility Group	A	B	C	D	E	F	G	H	J	K	L	N	S
A	X										X		
B	X	X									X	X	4/5
C	X		X	X	X						X	3	4/5
D	X	4	2	X	X						X	3	4/5
E	X	X	X	X	X						X	3	4/5
F	X	X	X	X	X	X					X		4/5
G	X	X	X	X	X		X				X	X	4/5
H	X	X	X	X	X			X			X	X	4/5
J	X	X	X	X	X				X		X	X	4/5
K	X	X	X	X	X					X	X	X	4/5
L	X	X	X	X	X	X	X	X	X	X	1	X	X
N	X	X	3	3	3						X		4/5
S	X	4/5	4/5	4/5	4/5	4/5	4/5	4/5	4/5	4/5	X	4/5	

Instructions for using the compatibility table for Class 1 (explosive) materials are as follows:

1. A blank space in the Table indicates that no restrictions apply.

2. The letter "X" in the Table indicates that explosives of different compatibility groups may not be carried on the same transport vehicle.

3. The numbers in the Table mean the following:

    - "1" means an explosive from compatibility group L shall only be carried on the same transport vehicle with an identical explosive.

    - "2" means any combination of explosives from compatibility groups C, D, or E is assigned to compatibility group E.

    - "3" means any combination of explosives from compatibility groups C, D, or E with those in compatibility group N is assigned to compatibility group D.

    - "4" means §177.835(g) when transporting detonators.

    - "5" means Division 1.4S fireworks may not be loaded on the same transport vehicle with Division 1.1 or 1.2 (Class A explosive) materials.

Except as provided in paragraph (i) of this section, explosives of the same compatibility group but of different divisions may be transported together provided that the whole shipment is transported as though its entire contents were of the lower numerical division (i.e., Division 1.1 being lower than Division 1.2). For example, a mixed shipment of Division 1.2 (Class A explosive) materials and Division 1.4 (Class C explosive) materials, both of compatibility group D, must be transported as Division 1.2 (Class A explosive) materials.

When Division 1.5 (blasting agent) materials, compatibility group D, are transported in the same freight container as Division 1.2 (Class A explosive) materials, compatibility group D, the shipment must be transported as Division 1.1 (Class A explosive) materials, compatibility group D.

On the Road with Hazardous Materials

Some general rules apply to drivers hauling hazardous materials during the actual time on the road. The following sections cover these driving rules for all hazardous materials shipments.

Railroad Crossings

Any marked or placarded vehicle (except Divisions 1.6, 4.2 and Class 9), or one carrying any amount of chlorine, must stop at railroad crossings.

Stops must be made within 50 feet of the crossing, but no closer than 15 feet. When you determine it is safe to cross the tracks, you may do so, but do not shift gears while on the tracks.

Stops need not be made at:

1. **Streetcar crossings or industrial switching tracks within a business district.**

2. **Crossings where a police officer or flagman is directing traffic.**

3. **Crossings which are marked by a stop-and-go traffic light which is green.**

4. **Abandoned rail lines and industrial or spur line crossings clearly marked "exempt".**

Tunnels

Unless there is no practicable other route, shipments of hazardous materials should not be driven through tunnels. Operating convenience cannot be used as a determining factor in such decisions.

In addition, the provisions of the Hazardous Materials Regulations do not supersede state or local laws and ordinances which may be more restrictive concerning hazardous materials and urban vehicular tunnels used for mass transportation.

Routing

A motor carrier transporting hazardous materials required to be marked or placarded shall operate the vehicle over routes which do not go through or near heavily populated areas, places where crowds are assembled, tunnels, narrow streets, or alleys, except when:

- There is no practicable alternative;
- It is necessary to reach a terminal, points of loading or unloading, facilities for food, fuel, rest, repairs, or a safe haven; or
- A deviation is required by emergency conditions;

Operating convenience is not a basis for determining if a route can be used. A motor carrier shall also comply with the routing designations of States or Indian tribes as authorized by federal regulations.

Attending Vehicles

Any marked or placarded vehicle containing hazardous materials which is on a public street or highway or the shoulder of any such road must be attended by the driver, except for times when the driver is performing duties necessary to the operation of the vehicle.

What exactly does "attended" mean? In terms of the regulations a motor vehicle is attended when the person in charge is on the vehicle and awake (cannot be in the sleeper berth) or is within 100 feet of the vehicle and has an unobstructed view of it.

Parking

Marked or placarded vehicles containing hazardous materials should not be parked on or within five feet of the traveled portion of any roadway, except for brief periods when operational necessity requires you to park the vehicle, and it would be impractical to stop elsewhere. Further restrictions apply to vehicles hauling Class 1.1, 1.2, or 1.3 (explosive) materials.

Standard warning devices are to be set out as required by law when a vehicle is stopped along a roadway.

Tire Checks

Any marked or placarded vehicle which contains hazardous materials and is equipped with dual tires on any axle must have tire checks performed every two hours or 100 miles, whichever comes first.

In addition, drivers must examine the tires on their vehicles at the beginning and ending of each hazardous materials trip, and each time the vehicle is parked.

If any defect is found in a tire, it should be repaired or replaced immediately. The vehicle may, however, be driven a short distance to the nearest safe place for repair.

If a hot tire is found, it must be removed from the vehicle immediately and taken to a safe distance. Such a vehicle may not be operated until the cause of the overheating is corrected.

Fires

A marked or placarded vehicle containing hazardous materials should not be driven near an open fire, unless careful precautions have been taken to be sure the vehicle can completely pass the fire without stopping.

In addition, a marked or placarded vehicle containing hazardous materials should not be parked within 300 feet of any open fire.

Accident Reporting

Any accident involving a vehicle hauling hazardous materials, in which there is any unintentional release of hazardous materials (a hazardous materials incident), must be reported to your carrier and within 30 days to the DOT on Form F 5800.1 (Hazardous Materials Incident Report).

In some cases, immediate notification of the DOT is required by the Hazardous Materials Regulations. In these cases, contact your carrier by phone as soon as possible after the accident, so the telephone report can be made to the Department of Transportation. Incidents which require this immediate notification include those in which:

- **As a direct result of hazardous materials**
    1. **A person is killed.**
    2. **A person receives injuries which require hospitalization.**
    3. **Property damage estimates exceed $50,000.**
    4. **An evacuation of the general public occurs lasting one or more hours.**
    5. **One or more major transportation arteries or facilities are closed or shut down for one hour or more.**
    6. **The operational flight pattern or routine of an aircraft is altered.**
- **Fire, breakage, spillage or suspected radioactive contamination occurs involving radioactive material.**
- **Fire, breakage, spillage or suspected contamination occurs involving infectious substances (etiologic agents).**
- **There has been a release of a marine pollutant in a quantity exceeding 450 L (119 gallons) for liquids or 400 kg (882 pounds) for solids.**

- **Some other situation occurs which the carrier considers important enough to warrant immediate reporting.**

As a driver, you should be prepared to supply complete and accurate information on any accident your vehicle is involved in on the road, to any proper authorities who require information for the record.

Loading, Unloading & on the Road
WITH SPECIFIC HAZARDOUS MATERIALS

Following are sections on specific classes of hazardous materials, with special emphasis on the role **YOU, the driver,** play in their safe and lawful transportation.

- **Class 1 (Explosives)**
- **Class 3 (Flammable Liquids)**
- **Class 2 (Gases)**
- **Class 8 (Corrosive Liquids)**
- **Division 6.1 (Poisonous Materials) and Division 2.3 (Poisonous Gases)**
- **Class 4 (Flammable Solids) and Class 5 (Oxidizers)**
- **Class 7 (Radioactive Materials)**

Another helpful feature is the reminder checklist which closes each section.

Shipments of CLASS 1 (Explosives)

General loading, transportation and unloading requirements that apply to all hazardous materials shipments can be found in the two preceding chapters. The following information is specific to Class 1 materials.

Combination Vehicles

Division 1.1 or 1.2 (explosives) materials may not be loaded on any vehicle of a combination of vehicles if:

1. More than two cargo carrying vehicles are in the combination;

2. Any full trailer in the combination has a wheel base of less than 184 inches;

3. Any vehicle in the combination is a cargo tank which is required to be marked or placarded; or

4. The other vehicle in the combination contains any:

    - Substances, explosive, n.o.s., Division 1.1A (explosive) material (initiating explosive),

    - Packages of Class 7 (radioactive) materials bearing "Yellow III" labels,

    - Division 2.3 (poison gas) or Division 6.1 (poison) materials, or

    - Hazardous materials in a portable tank or a DOT Spec. 106A or 110A tank.

Vehicle Condition

Vehicles transporting any Class 1 materials must be free of sharp projections into the body which could damage packages of explosives.

Vehicles carrying Division 1.1, 1.2, or 1.3 materials must have tight floors. They must be lined with non-metallic material or non-ferrous material in any portion that comes in contact with the load.

Motor vehicles carrying Class 1 materials should have either a closed body or have the body covered with a tarpaulin, so that the load is protected from moisture and sparks. However, explosives other than black powder may be hauled on flat-bed vehicles if the explosive portions of the load are packed in fire and water-resistant containers or are covered by a fire and water-resistant tarpaulin.

If the vehicle assigned to haul Class 1 materials is equipped with any kind of cargo heater, **it must be inoperable for the duration of that shipment.** Check to be sure the heater's fuel tank has been completely drained and the power source disconnected.

Loading & Unloading

Rules for loading Class 1 (explosive) materials are based on good common sense. They include:

1. **Be sure the engine of your vehicle is stopped before loading or unloading any Class 1 materials.**

2. **Don't use bale hooks or any other metal tool to handle Class 1 materials.**

3. Don't roll any Class 1 materials packages, except barrels or kegs.

4. Be careful not to throw or drop packages of Class 1 materials.

5. Keep Class 1 materials well clear of any vehicle exhaust pipes at all times.

6. Be sure any tarpaulins are well-secured with ropes or wire tie-downs.

7. Loads of Class 1 materials must be contained entirely within the body of the vehicle, with no projection or overhang.

8. If there is a tailgate or tailboard, it should be securely in place throughout the trip.

9. Don't load Class 1 materials with any other materials that might damage packages during transit. In an allowable mixed load, bulkheads or other effective means must be used to separate cargo.

Delivery

Shipments of Class 1 materials must be delivered only to specifically authorized persons. The only exception is if a Class 1 shipment is delivered to a magazine which is locked immediately after the shipment is unloaded.

Emergency Transfers

Division 1.1, 1.2, or 1.3 materials may not be transferred (from container to container, motor vehicle to motor vehicle, another vehicle to a motor vehicle) on any street, road or highway, except in emergencies.

In such cases all possible precautions must be taken to alert other drivers of the hazards involved in the transfer. Specifically, red electric lanterns, emergency reflectors or flags must be set out as prescribed for stopped or disabled vehicles.

Parking

Marked or placarded vehicles carrying Division 1.1, 1.2, or 1.3 materials must not be parked in the following locations:

- **On a public street or highway, or within five feet of such a roadway.**

- **On private property (including truck stops and restaurants) without the knowledge and consent of the person in charge of the property and who is aware of the nature of the materials.**

- **Within 300 feet of any bridge, tunnel, dwelling, building or place where people work or assemble, except for brief periods necessary to the operation of the vehicle when parking elsewhere is not within practical limits.**

Marked or placarded vehicles containing Division 1.4, 1.5, or 1.6 materials should not be parked on or within five feet of the traveled portion of any public roadway, except for brief periods necessary to vehicle operation.

Attending Vehicles

Marked or placarded vehicles containing Division 1.1, 1.2, or 1.3 materials must be attended at all times by their drivers or some qualified representative of the motor carrier, except in those situations where all the following conditions are met:

1. **Vehicle is located on the property of a carrier, shipper or consignee; in a safe haven; or on a construction or survey site (if the vehicle contains 50 pounds or less of Division 1.1, 1.2, or 1.3 materials).**

2. **A lawful bailee knows the nature of the vehicle's cargo and has been instructed in emergency procedures.**

3. **Vehicle is within the unobstructed field of vision of the bailee, or is in a safe haven.**

Marked or placarded vehicles containing Division 1.4, 1.5, or 1.6 materials, located on a public street or highway or the shoulder of the roadway, must be attended by the driver, except when the driver is performing necessary duties as operator of the vehicle.

No Smoking

An extra reminder! Don't smoke or carry a lighted cigarette, cigar or pipe on or within 25 feet of any vehicle containing Class 1 materials.

Required Documents

Motor carriers transporting Division 1.1, 1.2, or 1.3 materials must furnish drivers with the following required documents:

- **A copy of Part 397 — Transportation of Hazardous Materials; Driving and Parking Rules (Federal Motor Carrier Safety Regulations).**

- **Instructions on procedures to be followed in case of accident or delay, a full description of the load being hauled, along with names and phone numbers of persons to be contacted in emergencies.**

- **A written route plan for the Class 1 materials shipment. (In some cases, this plan may be prepared by the driver, if the trip begins at some other point than the carrier's terminal.)**

Drivers are required to sign a receipt for these items. In addition, don't forget that you must also have shipping papers for the load as described earlier in this book.

Accidents

If a vehicle transporting Class 1 materials is involved in an accident, you should take the following steps immediately:

- Arrange to keep unauthorized persons away from the area, and discourage smoking or the presence of any flame.

- Arrange to warn other drivers of the hazard.

- If the vehicle carrying Class 1 materials is entangled with another vehicle or some structure, get the entire load moved at least 200 feet from the vehicle (and preferably from any building) before beginning to move the vehicle.

- If there is fire involved in the accident, extra warnings of the danger of explosion are necessary for people in the area and other drivers.

- When clearing the wreck, be careful to avoid sparks from tools, etc. which might cause fire or explosion.

Damaged Packages

There are some special provisions for packages that are damaged in accidents. Packages may be repaired if it's practical and not dangerous. Packages that are beyond repair should be handled as follows:

Boxes: Reinforce with heavy wrapping paper and twine; place in strong box, surround with dry fine sawdust, dry clean cotton waste, or elastic dry newspaper wads; securely attach cover of outside box.

Can or Keg: Ruptured containers should be placed in good-quality grain bags, then boxed.

If these precautions have been taken, and the packages are clearly marked with their contents and consignee's name and address, the damaged packages may be delivered to the original destination. If these protective measures are not taken, damaged containers may only be moved to the nearest safe disposal site.

If the accident results in spillage of liquid nitroglycerin, desensitized liquid nitroglycerin or diethylene glycol dinitrate anywhere on a motor vehicle, the area must be washed with enough of the following solution to completely neutralize the danger:

30 g (1.1 ounce) 60% commercial sodium sulfide

222 ml (8 fluid ounces) denatured alcohol

60 ml (2 fluid ounces) acetone

90 ml (3 fluid ounces) water

Spillage of any other Class 1 material from damaged containers must be swept up thoroughly, and any portion of the vehicle floor where Class 1 materials have been absorbed must be removed.

Class 1 Do's

1. Check vehicle thoroughly to be sure it meets specifications.

2. Stop engine while loading or unloading.

3. Handle cargo as specified.

4. Deliver only to authorized persons.

5. Know the vehicle attendance requirements.

6. Have all required documents in order before starting.

7. Do know the correct procedure to follow in case of accident.

Class 1 Don'ts

1. Don't transfer Class 1 materials, except in emergencies.

2. Don't park a Class 1 materials vehicle in prohibited areas.

3. Don't smoke on or within 25 feet of a Class 1 materials vehicle.

4. Don't move vehicle involved in accident, until complete cargo has been moved to safety.

Shipments of CLASS 3 (Flammable Liquids)

General loading, transportation and unloading requirements that apply to all hazardous materials shipments can be found in the two preceding chapters. The following information is specific to Class 3 materials.

Loading & Unloading

There are some common sense rules to follow when loading and unloading Class 3 materials:

- Engine should be stopped. The only possible exception — if the engine is used to operate a pump used in the loading or unloading process.

- Tanks, barrels, drums, cylinders or other types of packaging containing Class 3 materials (and not permanently attached to the vehicle) must be secured to prevent movement during transportation.

- Keep fires (of all kinds and sizes) away!

- All manholes and liquid discharge valves must be closed and leak-free before a vehicle containing Class 3 materials can be driven.

Bonding & Grounding

Containers other than cargo tanks not in metallic contact with each other must have metallic bonds or ground conductors attached to avoid the possibility of static charge. Attach bonding first to the vehicle to be filled, then to the container from which the flammable liquid will be loaded. **Attachment must be done in this order!** To prevent ignition of vapors, connection should be made at some distance from the opening where Class 3 material is discharged.

For cargo tanks, similar provisions apply, but they vary depending on loading procedure.

When loading through an open filling hole, attach one end of a bond wire to the stationary system piping or integral steel framing, the other end to the cargo tank shell. Make the connection before any filling hole is opened, and leave it in place until the last hole has been closed.

If nonmetallic flexible connections exist in the stationary system piping, additional bond wires are required.

When unloading from open filling holes using a suction piping system, electrical connection must be maintained from cargo tank to receiving tank.

When loading through a vaportight top or bottom connection, and where there is no release of vapor at any point where a spark could occur, no bonding or grounding is required. However, contact of the closed connection must be made before flow starts and must not be broken until flow is complete.

When unloading through a non-vaportight connection, into a stationary tank, no bonding or grounding is required if the metallic filling connection is in contact with the filling hole at all times.

No Smoking

Smoking on or within 25 feet of any vehicle containing Class 3 materials is forbidden during the loading and unloading process, as well as the rest of the trip. Keep anyone with any kind of smoking materials away from the vehicle! The rule also applies to empty tank vehicles used to transport Class 3 materials.

Cargo Tank Attendance

A cargo tank must be attended by a qualified person at all times while it is being loaded or unloaded. The person responsible for loading the tank is also responsible for seeing that the tank is attended. The person attending must be awake, have an unobstructed view and be within 7.62 meters (25 feet) of the tank. The person must also know the nature of the hazard, be instructed in emergency procedures and be able to move the vehicle, if necessary.

Cargo Heaters

Vehicles equipped with combustion cargo heaters may transport Class 3 materials only if:

- **It is a catalytic heater.**
- **Heater surface cannot exceed 54°C (129°F) when the outside temperature is 16°C (61°F) or less.**

- Heater is not ignited in a loaded vehicle.
- No flame is present anywhere in or on the heater.
- Heater must bear manufacturer's certification that it meets DOT specifications.

There are also restrictions on the use of automatic cargo-space-heating temperature control devices. Such devices may be used in vehicles transporting Class 3 materials if the following conditions are met:

- Any electrical aparatus in the cargo area must be non-sparking or explosive-proof.
- No combustion apparatus is located in the cargo area, and there is no connection for air return between the cargo area and the combustion apparatus.
- The heating system will not heat any part of the cargo to more than 54°C (129°F).

If the above conditions are not met, the vehicle may still be used to transport Class 3 materials, if the device is rendered inoperable.

Parking & Attending Vehicles

Marked or placarded vehicles containing Class 3 materials should not be parked on or within five feet of the traveled portion of any public highway, except for brief periods necessary to vehicle operation. While such a vehicle is on the highway, it must be attended by the driver, except when the driver is performing necessary duties as operator of the vehicle.

Accidents

In case of an accident involving a vehicle containing Class 3 materials, the following steps should be taken:

- Keep all unauthorized persons away from the accident scene.

- Keep flame or fire of any sort out of the area.

- Take steps to warn other drivers of the hazard.

- If the Class 3 materials cargo is leaking, do whatever is necessary to keep the leakage from spreading over a large area.

Damaged Tanks

If a Class 3 materials cargo tank is damaged so badly that transportation of the vehicle any farther would be unsafe, the following emergency measures should be taken:

- Remove vehicle from traveled portion of the highway.

- Dig trenches, or do whatever is necessary to keep the spillage confined to a small area and away from sewers, streams, etc.

- As a general rule, it is not permissible to transfer Class 3 materials from one vehicle to another on a roadway. However, in cases of emergency, where transfer is possible, it may be done. Adequate warning devices must be set out to advise other drivers of the hazard. Extreme care should be taken to clear the area and to keep any possibility of fire or spark away.

- A leaking cargo tank vehicle may be moved, but only the minimum distance necessary to reach a spot where the cargo can be disposed of safely.

Class 3 Do's

1. Stop engine while loading and unloading.
2. Know the bonding and grounding procedures.
3. Check smaller cargo packages to be sure they are secure.
4. Check manholes and valves to be sure they are closed.
5. Know accident procedures thoroughly.

Class 3 Don'ts

1. Don't smoke or allow others to smoke while handling flammable liquids.
2. Don't leave vehicle unattended while unloading.
3. Don't use any cargo heater without knowing the restrictions.

Shipments of CLASS 2 (Gases)

General loading, transportation and unloading requirements that apply to all hazardous materials shipments can be found in the two preceding chapters. The following information is specific to Class 2 materials.

Vehicle Condition

Before any cylinders of Class 2 materials are loaded onto a motor vehicle, that vehicle should be inspected to determine that the floor or loading platform is essentially flat. If a solid floor or platform is not provided in the vehicle, there must be securely-fastened racks which will secure the cargo.

Loading & Unloading

Cylinders — To prevent overturning, Class 2 materials cylinders must be handled in one of four ways:

1. **Lashed securely in an upright position.**
2. **Loaded into racks securely fastened to the vehicle.**
3. **Packed in boxes or crates that would prevent overturn.**
4. **Loaded horizontially. (Specification DOT-4L cylinders must be loaded upright and securely braced, however.)**

Bulk packagings — Portable tank containers may be loaded on the flat floor or platform of a vehicle or onto a suitable frame. In either case, the containers must be blocked or held down in some way to prevent movement during transportation, including sudden starts or stops and changes in direction of the vehicle. Containers may even be stacked, if they are secured to prevent any motion.

If a Division 2.1 (flammable gas) material is being loaded or unloaded from a tank motor vehicle, the engine must be stopped. The only exception occurs if the engine is used to operate the transfer pump. The engine must also be off while filling or discharge connections are disconnected, unless the delivery hose is equipped with a shut-off valve.

Liquid discharge valves on cargo tanks must be closed securely after loading or unloading, before any movement of the vehicle.

Special handling provisions are specified for certain kinds of Class 2 materials. Information on these follows.

Liquefied Hydrogen

Specification DOT-4L cylinders filled with liquefied hydrogen, cryogenic liquid must be carried on vehicles with open bodies equipped with racks or supports having clamps or bands to hold the cylinders securely upright. These arrangements must be capable of withstanding acceleration of 2 "g" in any horizontal direction.

Cylinders of liquefied hydrogen are marked with venting rates. The combined total venting rates on all cylinders on a single motor vehicle must not exceed 60 standard cubic feet per hour.

Only private and contract carriers may haul liquefied hydrogen on the highway and the transportation must be a direct movement from origin to destination.

Vehicles hauling liquefied hydrogen may not be driven through tunnels.

Chlorine

Any shipment of chlorine in a cargo tank must be accompanied by a gas mask (approved for the purpose by U.S. Bureau of Mines) and an emergency kit to control leaks in the fittings on the dome cover plate.

Do not move, couple or uncouple any chlorine tank vehicle while loading or unloading connections are attached.

Do not leave any trailer or semitrailer without a power unit, unless it has been chocked or some adequate means provided to prevent it from moving.

No Smoking

Smoking on or near the vehicle while you are loading or unloading Division 2.1 material is forbidden.

Attending Vehicles

Marked or placarded vehicles located on a public street or highway or the shoulder of the roadway, must be attended by the driver, except when the driver is performing necessary duties as operator of the vehicle.

In addition, cargo tanks must be attended by a qualified person at all times during loading and unloading.

Parking

Marked or placarded vehicles containing Class 2 materials should not be parked on or within five feet of the traveled portion of any public roadway, except for brief periods necessary to vehicle operation.

Accidents

In case of an accident involving a vehicle containing Class 2 materials which would cause a hazard to other drivers if released, the following steps should be taken immediately:

- **Keep unauthorized personnel well clear of the accident area.**
- **Notify the shipper immediately.**
- **Place appropriate warning devices to warn other drivers of the hazard.**
- **If the gas is Division 2.1, keep all matches and flame-producing objects out of the area.**

Transfer of Cargo

Division 2.1 material may not be transferred from container to container or vehicle to vehicle on the highway except in emergency situations. If such a transfer is required, it must be carried out under the following conditions:

- **All precautions to prevent escaping gas must be taken.**
- **Warning devices must be set out to advise other drivers of the hazard.**
- **Cargo tanks must be grounded.**
- **Transfer must be made in daylight hours, except in extreme night-time emergencies where waiting until daylight poses a greater hazard than making the transfer in the dark.**
- **Extra care in warning of the hazard is necessary in night-time transfers of flammable compressed gas.**
- **Keep unauthorized persons out of the area.**
- **Take all precautions to keep any flame or fire away from the scene.**
- **Special care should be taken in operating any engine which might ignite the flammable gas.**

Class 2 Do's

1. Be sure vehicle is adequately equipped to haul Class 2 materials safely.

2. Stop engine while loading or unloading Division 2.1 materials.

3. Be sure liquid discharge valves are closed before moving vehicle.

4. Know special provisions for handling liquefied hydrogen and chlorine.

5. Know procedures to follow in case of an accident.

Class 2 Don'ts

1. Don't drive a vehicle hauling liquefied hydrogen through a tunnel.

2. Don't move chlorine tanks with loading or unloading connections attached.

3. Don't smoke on or near vehicles containing Division 2.1 materials.

Shipments of
CLASS 8 (Corrosive Liquids)

General loading, transportation and unloading requirements that apply to all hazardous materials shipments can be found in the two preceding chapters. The following information is specific to Class 8 materials.

Vehicle Condition

Vehicles assigned to carry Class 8 materials must have an even floor surface. A false floor or platform, which is well secured to prevent relative motion inside the vehicle's body, can be considered a floor.

Loading & Unloading

If individual carboys and breakable (frangible) containers of Class 8 materials, including charged electric storage batteries, are loaded or unloaded by hand, they must be individually handled. Care must be taken not to drop such containers, for obvious reasons.

If carboys or breakable containers of Class 8 materials are to be loaded more than one tier high, certain conditions must be met:

- **Carboys or containers must be boxed or crated, or be in barrels or kegs that meet required standards.**

- **The weight of each tier above the first must be entirely supported by the boxes, crates, barrels, kegs, etc.**

- **The number of tiers permissible will be judged by the amount of weight which can be supported without danger of crushing or breaking the base.**

Means must be taken to keep Class 8 containers or batteries from shifting during transit. Cleats or any suitable means may be used to keep cargo stationary.

There are some special provisions for particular types of cargo and vehicles. They follow here:

Nitric Acid

If you are hauling nitric acid as part of a mixed load, you should know that carboys or other containers of nitric acid may not be loaded above containers of any other material. Further, containers of nitric acid may only be loaded two tiers high.

Storage Batteries

Storage batteries containing any electrolyte and hauled in a mixed load must be positioned so that they are protected from the possibility of other cargo falling on or against them. Battery terminals must be insulated and protected from the possibility of short circuits.

Cargo Tanks

If you are hauling Class 8 materials (regardless of quantity) in a cargo tank, you must check the following before you move the vehicle:

- **All manhole closures are closed and secured.**

- All valves and closures in the liquid discharge systems are closed and leak-free.

Parking & Attending Vehicles

A marked or placarded vehicle containing Class 8 materials should not be parked on or within five feet of the traveled portion of any public roadway, except for brief periods ncessary for vehicle operation. While such a vehicle is on the highway, it must be attended by the driver, except when the driver is performing necessary duties as operator of the vehicle.

Accidents

If a vehicle hauling Class 8 materials is involved in an accident in which any of the Class 8 materials containers are left broken, spilled or leaking, take the following steps immediately.

- Be careful in handling other portions of the cargo which may have been damaged by the spillage.

- The interior of the vehicle where corrosives have spilled should be thoroughly washed with water as soon as possible after unloading, and absolutely before the vehicle is reloaded.

If leakage occurs from any part of a cargo tank hauling Class 8 materials during the course of a trip, take these steps immediately:

- Use whatever measures necessary to stop further leakage on the highway.

- **Drive the minimum distance possible to a location where the leaking cargo may be disposed of safely.**

However, if the leak is so severe that any further movement of the cargo tank is unsafe:

- **Get off the highway and take steps to prevent the spread of the leaking cargo over a wide area (dig trenches to lead the Class 8 materials to a depression or hole and away from sewers, etc.; absorb spillage with ashes, sand, etc.; catch liquid in containers if practical).**

- **Keep spectators away from the area where the Class 8 material or its fumes are present.**

- **Take necessary steps to minimize any hazard to other drivers or to livestock which might be caused by unloading Class 8 material.**

Class 8 Do's

1. Be sure vehicle condition meets requirements.

2. Know container handling requirements for loading by hand.

3. Be sure cargo is loaded in accordance with specifications.

4. If you are hauling batteries, be sure they are adequately protected during transit.

5. Know procedures to follow in case of an accident.

Class 8 Don'ts

1. Don't load nitric acid above other commodities, and don't load it more than two tiers high.

2. Don't park on or near the roadway, or leave your vehicle unattended for more than very brief periods necessary to your duties.

Shipments of
DIVISION 6.1 (Poisonous Materials) &
DIVISION 2.3 (Poisonous Gases)

General loading, transportation and unloading requirements that apply to all hazardous materials shipments can be found in the two preceding chapters. The following information is specific to Division 6.1 or Division 2.3 materials.

Loading & Unloading

Some types of Division 6.1 or Division 2.3 materials may be hauled in cargo tanks. However, before you begin transportation, all manhole closures on the tank must be closed and secured. In addition, any valves and closures in the liquid discharge system must be tightly closed and free of leaks. **NOTE: No Division 6.1 or 2.3 materials may be hauled if there is any interconnection between packaging.**

No package bearing a poison label can be loaded on a vehicle with any foodstuff, feed or edible material that might be consumed by humans or animals.

Special provisions apply to arsenical materials.

Arsenical Materials

Bulk arsenical compounds loaded and unloaded under certain conditions demand extra caution. These materials include:

- **arsenical dust**

- **arsenic trioxide**
- **sodium arsenate**

These materials may be loaded in siftproof steel hopper or dump vehicle bodies, if they are equipped with waterproof and dustproof covers secured on all openings. They must be loaded carefully, using every means to minimize the spread of the compounds into the atmosphere. Loading and unloading may not be done in any area where there might be persons not connected with the transportation. This includes all highways and other public places.

After a vehicle is used for transporting arsenicals, and before it may be used again for any other materials, the vehicle must be flushed with water or use other appropriate means to remove all traces of arsenical material.

Parking & Attending Vehicles

A marked or placarded vehicle containing Division 6.1 or 2.3 materials should not be parked on or within five feet of the traveled portion of any public roadway, except for brief periods necessary to vehicle operation. While such a vehicle is on the highway, it must be attended by the driver, except when the driver is performing necessary duties as an operator of the vehicle.

Accidents

If a vehicle carrying Division 6.1 or 2.3 materials which are flammable, noxious or toxic is involved in an accident, take the following steps immediately to protect persons and property in the vicinity:

- Set out appropriate warning devices.
- Keep unauthorized persons away.
- Take steps to prevent any type of smoking in the area, and keep all sources of flame away.
- Do whatever is necessary to stop any leaking poisons from reaching streams or sewers in the area, or to keep powdered poisons from being scattered over the area by wind.

Leaking Packages

Vehicles transporting Division 6.1 or 2.3 materials in packages which may develop leakage as a result of an accident must be thoroughly inspected for contamination before they are returned to use. If contamination is discovered, it must be completely removed before the vehicle can return to use.

If cargo tanks or compartments of cargo tanks develop leakage while carrying poisons which also fall into one of these classes:

- Class 3
- Class 2

You must follow the same procedures as for Class 3 and Class 2 materials.

Warn any bystanders and other drivers of the danger present in the Division 6.1 or 2.3 materials and/or inhaling its fumes.

Division 6.1 & Division 2.3 Do's

1. Know provisions for hauling Division 6.1 or 2.3 materials in cargo tanks, if your vehicle falls in that category.

2. Know special provisions for hauling arsenicals.

3. Be aware of vehicle parking and attendance rules.

4. Know what to do in case of an accident.

Division 6.1 & Division 2.3 Don'ts

1. Don't load any package marked poison with any type of edible material.

2. Don't haul any arsenical material unless vehicle is properly loaded and marked.

Shipments of
CLASS 4 (Flammable Solids)
& CLASS 5 (Oxidizers)

General loading, transportation and unloading requirements that apply to all hazardous materials shipments can be found in the two preceding chapters. The following information is specific to Class 4 and Class 5 materials.

Loading & Unloading

Logic dictates the rules that apply to the loading and unloading of Class 4 and Class 5 materials:

- **The entire load must be contained within the body of the vehicle.**

- **The load must be covered either by the vehicle body or tarpaulins or other similar means.**

- **If there is a tailgate or board, it must be closed securely.**

- **Care must be taken to load cargo while dry and to keep it dry.**

- **Provide adequate ventilation for cargos which are known to be subject to heating and/or spontaneous combustion, in order to prevent fires.**

- **Shipments of water-tight bulk containers do not have to be covered by a tarpaulin or other means.**

Pick-up & Delivery

Provisions concerning Class 4 and Class 5 loads being completely contained inside the body of the vehicle with the tailgate closed do not apply to pick-up and delivery vehicles used entirely for that purpose in and around cities.

Special provisions are necessary for some particular commodities. The following sections detail those requirements.

Charcoal

Charcoal screenings and ground, crushed, pulverized or lump charcoal become more hazardous when wet. Extra care should be taken to keep packages completely dry during loading and carriage.

In addition, these commodities must be loaded so that bags lay horizontally in the vehicle and piled so that there are spaces at least 10 cm (3.9 inches) wide for effective air circulation. These spaces must be maintained between rows of bags. No bags may be piled closer than 15 cm (5.9 inches) from the top of a vehicle with a closed body.

Nitrate of Soda Bags

Empty unwashed bags which contained nitrate of soda may be hauled loose or baled in truckload lots only. Vehicles hauling the bags must have closed or covered bodies and be lined with paper.

Such materials must be loaded by the shipper and unloaded by the consignee.

Matches

Unless strike-anywhere matches are packed in wooden outside boxes, careful inspection of the vehicle in which they are to be hauled is necessary. Either the interior surface must be smooth and without protrusions (bolts, nuts, sharp edges, corners) or a smooth wooden inner liner must be provided before matches may be loaded.

Special care also should be taken to prevent the shifting or jamming of packages of strike-anywhere matches during transit. Packages should be loaded as compactly as possible with the strongest dimension of the packages loaded lengthwise of the vehicle.

Matches must not be loaded next to any package bearing a Class 3 materials label.

Smokeless Powder

Smokeless powder for small arms in quantities not exceeding 45 kg (100 pounds) net weight per vehicle may be classed as a Class 4 material, if approved by the Bureau of Explosives or the Bureau of Mines and the Associate Administrator for Hazardous Materials Safety.

Each inside package must be 3.6 kg (7.9 pounds) or less in size. These packages must be protected and arranged to prevent simultaneous ignition.

The overall package must be of a type approved by the Bureau of Explosives or the Bureau of Mines and the Associate Administrator for Hazardous Materials Safety and each outside package must be labeled as a Division 4.1 (flammable solid).

Nitrates

All nitrates (except ammonium nitrate with organic coating) must be loaded in vehicles that have been swept clean and are free of projections which could damage the bags. The vehicles may be closed or open-type. If open, the cargo must be covered securely.

Ammonium nitrate with organic coating should not be loaded in all-metal closed vehicles, except those made of aluminum or aluminum alloys.

No Smoking

Don't smoke or carry a lighted cigarette, cigar or pipe on or within 25 feet of a vehicle containing Class 4 or Class 5.

Parking & Attending Vehicles

A marked or placarded vehicle containing Class 4 and/or Class 5 materials should not be parked on or within five feet of the traveled portion of any public roadway, except for brief periods necessary to vehicle operation. While such a vehicle is on the highway, it must be attended by the driver, except when the driver is performing necessary duties as operator of the vehicle.

Accidents

If a vehicle carrying Class 4 or Class 5 materials is involved in an accident, and cargo is spilled, you must take immediate steps to warn drivers and other persons in the vicinity of the danger of fire. Setting out standard hazard warning devices, verbal warnings, etc. should be employed.

If fire is already present, take any necessary steps to prevent it from spreading farther. As much as possible, portions of any broken packages should be gathered and removed to a safe place. In addition, if removing the undamaged portion of the load from the vehicle is possible and would decrease the risk of fire or any other hazard, it should be done.

In accidents involving **shipments of matches**, where packaging may have been damaged by fire or water, these steps should be taken:

1. **Examine outside packages and repair damage as much as possible.**

2. **Loose matches should be destroyed.**

3. **Smoking boxes of matches should be removed to a safe distance from the vehicle (do not open) and then destroyed by burning. Don't leave them while they're still smoking or burning.**

4. **Load Damaged packages of matches according to the same requirements given for the initial load.**

In accidents involving **shipments of dry calcium hypochlorite compounds** packed in metal drums, the drums must be held at least five days if any fire was involved in the accident. Any drum which shows signs of spontaneous heating or internal pressure or stress after that time may not be reshipped.

Class 4 & Class 5 Do's

1. Know loading and covering requirements for cargo.
2. Be sure ventilation is adequate for the cargo you're hauling.
3. Know the special handling requirements for your commodity, if any.
4. Know accident procedures.

Class 4 & Class 5 Don'ts

1. Don't leave vehicle unattended for long periods.
2. Don't reload damaged cargo without following correct safety procedures to determine if all hazards have been eliminated.
3. Don't smoke on or near vehicle.

Shipments of CLASS 7 (Radioactive Materials)

General loading, transportation and unloading requirements that apply to all hazardous materials shipments can be found in the two preceding chapters. The following information is specific to Class 7 materials.

Loading & Unloading

The number of packages of Class 7 material which may be carried in a transport vehicle is limited by the total transport index number. This number is determined by adding together the transport index numbers given on the labels of individual packages. The total may not exceed 50, except in certain special "exclusive use" shipments that the regulations describe.

Class 7 material in a package labeled "RADIOACTIVE YELLOW-II" or "RADIOACTIVE YELLOW-III" may not be positioned closer to any area which may be continuously occupied by passengers, employees, animals, or packages of undeveloped film than indicated in the following table. If several packages are involved, distances are determined by the total transport index number.

Total transport index	Minimum separation distance in meters (feet) to nearest undeveloped film for various times of transit					Minimum distance in meters (feet) to area of persons, or minimum distance in meters (feet) from dividing partition of cargo compartments
	Up to 2 hours	2-4 hours	4-8 hours	8-12 hours	Over 12 hours	
None	0.0 (0)	0.0 (0)	0.0 (0)	0.0 (0)	0.0 (0)	0.0 (0)
0.1 to 1.0 ...	0.3 (1)	0.6 (2)	0.9 (3)	1.2 (4)	1.5 (5)	0.3 (1)
1.1 to 5.0 ...	0.9 (3)	1.2 (4)	1.8 (6)	2.4 (8)	3.4 (11)	0.6 (2)
5.1 to 10.0 ..	1.2 (4)	1.8 (6)	2.7 (9)	3.4 (11)	4.6 (15)	0.9 (3)
10.1 to 20.0 .	1.5 (5)	2.4 (8)	3.7 (12)	4.9 (16)	6.7 (22)	1.2 (4)
20.1 to 30.0 .	2.1 (7)	3.0 (10)	4.6 (15)	6.1 (20)	8.8 (29)	1.5 (5)
30.1 to 40.0 .	2.4 (8)	3.4 (11)	5.2 (17)	6.7 (22)	10.1 (33)	1.8 (6)
40.1 to 50.0 .	2.7 (9)	3.7 (12)	5.8 (19)	7.3 (24)	11.0 (36)	2.1 (7)

Note: The distance in this table must be measured from the nearest point on the nearest packages of Class 7 (radioactive) material.

Shipments of low specific activity materials and surface contaminated objects must be loaded to avoid spilling and scattering loose materials.

Be sure packages are blocked and braced so they cannot change position during normal transportation conditions.

Fissile material, controlled shipments must be shipped as described in Section 173.457 of the hazardous materials regulations. No fissile material, controlled shipments may be loaded in the same vehicle with any other fissile Class 7 material. Fissile material, controlled shipments must be separated by at least 6 m (20 feet) from any other package or shipment labeled "Radioactive."

Keep the time spent in a vehicle carrying radioactive materials at an absolute minimum.

Vehicle Contamination

Vehicles used to transport Class 7 materials under exclusive use conditions must be checked after each use with radiation detection instruments. These vehicles may not be returned to use until the radiation dose at every accessible surface is 0.005 mSv per hour (0.5 mrem per hour) or less, and the removable radioactive surface contamination is not greater than prescribed in §173.443.

These contamination limitations do not apply to vehicles used only for transporting Class 7 materials, provided certain conditions are met:

- **Interior surface shows a radiation dose of no more than 0.1 mSv per hour (10 mrem per hour).**

- **Dose is no more than 0.02 mSv per hour (2 mrem per hour) at a distance of 1 m (3.3 feet) from any interior surface.**

- **The vehicle must be stenciled with "For Radioactive Materials Use Only" in letters at least 7.6 cm (3 inches) high in a conspicuous place, on both sides of the exterior of the vehicle.**

- **Vehicles must be kept closed at all times except during loading and unloading.**

Parking & Attending Vehicles

A marked or placarded vehicle containing Class 7 materials should not be parked on or within five feet of the traveled portion of any public roadway except for brief periods necessary to vehicle operation. While such a vehicle is on the highway, it must be attended by the driver, except when the driver is performing necessary duties as operator of the vehicle.

Accidents

In case of any accident involving Class 7 materials, in which there has been breakage, spillage or suspected radioactive contamination, the offeror shall be notified at the earliest practicable moment. This requirement is in addition to the incident reporting rules for any hazardous materials incident.

Vehicles, buildings, equipment or areas in which Class 7 materials have been spilled must be inspected and meet certain standards before being returned to use. The radioactive dose at every accessible surface must be less than 0.005 mSv per hour (0.5 mrem per hour). In addition, there should be no significant removable radioactive surface contamination.

Suspicious packages or materials should be kept away from people as much as possible. If there is obvious leakage, avoid inhaling or coming in contact with the Class 7 material in any way. Loose Class 7 material should be left in a well-segregated area until instructions for its disposal are received.

Class 7 Do's

1. Know how the transport index system works in limiting Class 7 shipments.

2. Be sure packages are blocked and braced correctly for shipment.

3. If your vehicle transports Class 7 materials only, know the special provisions for marking and acceptable radiation dose.

4. Know procedures for accidents involving Class 7 materials.

5. Be sure shipper is notified immediately in case of accident.

Class 7 Don'ts

1. Don't spend more time than absolutely necessary in a vehicle used for transporting Class 7 materials.

2. Don't leave vehicle unattended for long periods of time.

3. Don't handle damaged cargo unless you have been instructed in disposal procedures by authorized personnel.

Hazardous Materials Table

PUBLISHER'S NOTE: Space limitations preclude our using the entire §172.101 Hazardous Materials Table. We are including the first seven columns of the table, which directly relate to drivers of vehicles transporting hazardous materials.

§172.101 Purpose and use of hazardous materials table.

(a) The Hazardous Materials Table (Table) in this section designates the materials listed therein as hazardous materials for the purpose of transportation of those materials. For each listed material, the Table identifies the hazard class or specifies that the material is forbidden in transportation, and gives the proper shipping name or directs the user to the preferred proper shipping name. In addition, the Table specifies or references requirements in this subchapter pertaining to labeling, packaging, quantity limits aboard aircraft and stowage of hazardous materials aboard vessels.

(b) *Column 1: Symbols.* Column 1 of the Table contains five symbols ("+", "A", "D", "I", and "W"), as follows:

(1) The plus (+) fixes the proper shipping name, hazard class and packing group for that entry without regard to whether the material meets the definition of that class or packing group or meets any other hazard class definition. An appropriate alternate proper shipping name and hazard class may be authorized by the Associate Administrator for Hazardous Materials Safety.

(2) The letter "A" restricts the application of requirements of this subchapter to materials offered or intended for transportation by aircraft, unless the material is a hazardous substance or a hazardous waste.

(3) The letter "D" identifies proper shipping names which are appropriate for describing materials for domestic transportation but may be inappropriate for international transportation under the provisions of international regulations (e.g., IMO, ICAO). An alternate proper shipping name may be selected when either domestic or international transportation is involved.

(4) The letter "I" identifies proper shipping names which are appropriate for describing materials in international transportation. An alternate proper shipping name may be selected when only domestic transportation is involved.

(5) The letter "W" restricts the application of requirements of this subchapter to materials offered or intended for transportation by vessel, unless the material is a hazardous substance or a hazardous waste.

(c) *Column 2: Hazardous materials descriptions and proper shipping names*. Column 2 lists the hazardous materials descriptions and proper shipping names of materials designated as hazardous materials. Modification of a proper shipping name may otherwise be required or authorized by this section. Proper shipping names are limited to those shown in Roman type (not italics).

(1) Proper shipping names may be used in the singular or plural and in either capital or lower case letters. Words may be alternatively spelled in the same manner as they appear in the ICAO Technical Instructions or the IMDG Code. For example "aluminum" may be spelled "aluminium" and "sulfur" may be spelled "sul-

phur". However, the word "inflammable" may not be used in place of the word "flammable".

(2) Punctuation marks and words in italics are not part of the proper shipping name, but may be used in addition to the proper shipping name. The word "or" in italics indicates that terms in the sequence may be used as the proper shipping name, as appropriate.

(3) The word "poison" or "poisonous" may be used interchangeably with the word "toxic" when only domestic transportation is involved. The abbreviation "n.o.i." or "n.o.i.b.n." may be used interchangeably with "n.o.s.".

(4) Except for hazardous wastes, when qualifying words are used as part of the proper shipping name, their sequence in the package markings and shipping paper description is optional. However, the entry in the Table reflects the preferred sequence.

(5) When one entry references another entry by use of the word "see", if both names are in Roman type, either name may be used as the proper shipping name (e.g., Ethyl alcohol, *see* Ethanol).

(6) When a proper shipping name includes a concentration range as part of the shipping description, the actual concentration, if it is within the range stated, may be used in place of the concentration range. For example, an aqueous solution of hydrogen peroxide containing 30 percent peroxide may be described as "Hydrogen peroxide, aqueous solution *with not less than 20 percent but not more than 40 percent hydrogen peroxide*" or "Hydrogen peroxide, aqueous solution *with 30 percent hydrogen peroxide*".

(7) Use of the prefix "mono" is optional in any shipping name, when appropriate. Thus, Iodine monochloride may be used interchangeably with Iodine chloride. In "Glycerol alphamonochlorohydrin" the term "mono" is considered a prefix to the term "chlorohydrin" and may be deleted.

(8) Hazardous substances. Appendix A to this section lists materials which are listed or designated as hazardous substances under section 101(14) of the Comprehensive Environmental Response, Compensation, and Liability Act (CERCLA). Proper shipping names for hazardous substances (see the appendix to this section and §171.8 of this subchapter) shall be determined as follows:

(i) If the hazardous substance appears in the Table by technical name, then the technical name is the proper shipping name.

(ii) If the hazardous substance does not appear in the Table and is not a forbidden material, then an appropriate generic, or "n.o.s", shipping name shall be selected corresponding to the hazard class (and packing group, if any) of the material as determined by the defining criteria of this subchapter (see §§173.2 and 173.2a of this subchapter). For example, a hazardous substance which is listed in Appendix A but not in the Table and which meets the definition of flammable liquid might be described as "Flammable liquid, n.o.s." or other appropriate shipping name corresponding to the flammable liquid hazard class.

(9) Hazardous wastes. If the word "waste" is not included in the hazardous material description in Column 2 of the Table, the proper shipping name for a hazardous waste (as defined in §171.8 of this subchapter), shall include the word "Waste" preceding the

proper shipping name of the material. For example: Waste acetone.

(10) Mixtures and solutions. (i) A mixture or solution not identified specifically by name, comprised of a hazardous material identified in the Table by technical name and non-hazardous material, shall be described using the proper shipping name of the hazardous material and the qualifying word "mixture" or "solution", as appropriate, unless—

(A) Except as provided in §172.101(i)(4) the packaging specified in Column 8 is inappropriate to the physical state of the material;

(B) The shipping description indicates that the proper shipping name applies only to the pure or technically pure hazardous material;

(C) The hazard class, packing group, or subsidiary hazard of the mixture or solution is different from that specified for the entry;

(D) There is a significant change in the measures to be taken in emergencies;

(E) The material is identified by special provision in Column 7 of the §172.101 Table as a material poisonous by inhalation; however, it no longer meets the definition of poisonous by inhalation or it falls within a different hazard zone than that specified in the special provision; or

(F) The material can be appropriately described by a shipping name that describes its intended application, such as "Coating solution", "Extracts, flavoring" or "Compound, cleaning liquid".

(ii) If one or more of the conditions specified in paragraph (c)(10)(i) of this section is satisfied then a proper shipping name shall be selected as prescribed in paragraph (c)(12)(ii) of this section.

(11) Except for a material subject to or prohibited by §§173.21, 173.51, 173.56(d), 173.56(e)(1), 173.124(a)(2)(iii) or 173.128(e) of this subchapter, a material for which the hazard class is uncertain and must be determined by testing or a material that is a hazardous waste may be assigned a tentative shipping name, hazard class, identification number, and packing group, based on the shipper's tentative determination according to-

(i) Defining criteria in this subchapter;

(ii) The hazard precedence prescribed in §173.2a of this subchapter; and

(iii) The shipper's knowledge of the material.

(12) Except when the proper shipping name in the Table is preceded by a plus (+)—

(i) If it is specifically determined that a material meets the definition of a hazard class or packing group, other than the class or packing group shown in association with the proper shipping name, or does not meet the defining criteria for a subsidiary hazard shown in Column 6 of the Table, the material shall be described by an appropriate proper shipping name listed in association with the correct hazard class, packing group, or subsidiary hazard for the material.

(ii) Generic or n.o.s. descriptions. If an appropriate technical name is not shown in the Table, selection of a proper shipping name shall be made from the generic or n.o.s. descriptions corresponding to the specific hazard class, packing group, or subsidiary hazard, if any, for the material. The name that most appropriately describes the material shall be used; e.g, an alcohol not listed by its technical name in the Table shall be described as "Alcohol, n.o.s." rather than "Flammable liquid, n.o.s.". Some mixtures may be more appropriately described according to their application, such as "Coating solution" or "Extracts, flavoring, liquid", rather than by an n.o.s. entry, such as "Flammable liquid, n.o.s." It should be noted, however, that an n.o.s. description as a proper shipping name may not provide sufficient information for shipping papers and package marking. Under the provisions of subparts C and D of this part, the technical name of the constituent which makes the product a hazardous material may be required in association with the proper shipping name.

(iii) *Multiple hazard materials.* If a material meets the definition of more than one hazard class, and is not identified in the Table by a specific description, the hazard class of the material shall be determined by using the precedence specified in §173.2a of this subchapter, and an appropriate shipping description (e.g., "Flammable liquid, corrosive n.o.s.") shall be selected as described in paragraph (c)(12)(ii) of this section.

(iv) If it is specifically determined that a material is not a forbidden material and does not meet the definition of any hazard class, the material is not a hazardous material.

(13) *Self-reactive materials and organic peroxides.* A generic proper shipping name for a self-reactive material or an organic peroxide, as listed in Column 2 of the Table, must be selected based on the material's technical name and concentration, in accordance with the provisions of §§173.224 or 173.225 of this subchapter, respectively.

(d) ***Column 3: Hazard class or Division.*** Column 3 contains a designation of the hazard class or division corresponding to each proper shipping name, or the word "Forbidden".

(1) A material for which the entry in this column is "Forbidden" may not be offered for transportation or transported. This prohibition does not apply if the material is diluted, stabilized or incorporated in a device and it is classed in accordance with the definitions of hazardous materials contained in part 173 of this subchapter.

(2) When a reevaluation of test data or new data indicates a need to modify the "Forbidden" designation or the hazard class or packing group specified for a material specifically indentified in the Table, this data should be submitted to the Associate Administrator for Hazardous Materials Safety.

(3) A basic description of each hazard class and the section reference for class definitions appear in §173.2 of this subchapter.

(4) Each reference to a Class 3 material is modified to read "Combustible liquid" when that material is reclassified in accordance with §173.150(e) or (f) of this subchapter or has a flash point above 60.5°C (141°F) but below 93°C (200°F).

(e) *Column 4: Identification number.* Column 4 lists the identification number assigned to each proper shipping name. Those preceded by the letter "UN" are associated with proper shipping names considered appropriate for international transportation as well as domestic transportation. Those preceded by the letters "NA" are associated with proper shipping names not recognized for international transportation, except to and from Canada. Identification numbers in the "NA9000" series are associated with proper shipping names not appropriately covered by international hazardous materials (dangerous goods) transportation standards, or not appropriately addressed by the international transportation standards for emergency response information purposes, except for transportation between the United States and Canada.

(f) *Column 5: Packing group.* Column 5 specifies one or more packing groups assigned to a material corresponding to the proper shipping name and hazard class for that material. Classes 2 and 7 materials and ORM-D materials do not have packing groups, Packing Groups I, II, and III indicate the degree of danger presented by the material is either great, medium or minor, respectively. If more than one packing group is indicated for an entry, the packing group for the hazardous material is determined using the criteria for assignment of packing groups specified in subpart D of part 173. When a reevaluation of test data or new data indicates a need to modify the specified packing group(s), the data should be submitted to the Associate Administrator for Hazardous Materials Safety. Each reference in this column to a material which is a hazardous waste or a hazardous substance, and whose proper shipping name is preceded in Column 1 of the Table by the letter "A" or "W", is modified to read "III" on those occasions when the material is offered for transportation or transported by a mode in which its transportation is not otherwise subject to requirements of this subchapter.

(g) *Column 6: Labels.* Column 6 specifies the hazard warning label(s) required for a package filled with a material conforming to the associated hazard class and proper shipping name, unless the package is otherwise excepted from labeling by provisions in subpart E of part 172, or part 173 of this subchapter. The first label shown for each entry is indicative of the primary hazard of the material, additional labels are indicative of subsidiary hazards. Provisions in §172.402 of this part may require that a label other than that specified in Column 6 be affixed to the package in addition to that specified in Column 6. No label is required for a material classed as a combustible liquid or for a Class 3 material that is reclassed as a combustible liquid.

(h) ***Column 7: Special provisions.*** Column 7 specifies codes for special provisions applicable to hazardous materials. When Column 7 refers to a special provision for a hazardous material, the meaning and requirements of that special provision are as set forth in §172.102 of this subpart.

HAZARDOUS MATERIALS TABLE

§172.101 HAZARDOUS MATERIALS TABLE

Symbols (1)	Hazardous materials descriptions and proper shipping names (2)	Hazard class or Division (3)	Identification Numbers (4)	Packing group (5)	Label(s) required (if not excepted) (6)	Special provisions (7)
	Accellerene, see p-Nitrosodimethylaniline					
	Accumulators, electric, see Batteries, wet etc.					
D	Accumulators, pressurized, pneumatic or hydraulic *(containing non-flammable gas)*	2.2	NA1956		NONFLAMMABLE GAS	
	Acetal	3	UN1088	II	FLAMMABLE LIQUID	T7
	Acetaldehyde	3	UN1089	I	FLAMMABLE LIQUID	A3, B16, T20, T26, T29
A	Acetaldehyde ammonia	9	UN1841	III	CLASS 9	
	Acetaldehyde oxime	3	UN2332	III	FLAMMABLE LIQUID	B1, T8
	Acetic acid, glacial or Acetic acid solution, *with more than 80 percent acid, by mass*	8	UN2789	II	CORROSIVE, FLAMMABLE LIQUID	A3, A6, A7, A10 B2, T8
	Acetic acid solution, *with more than 10 percent but not more than 80 percent acid, by mass*	8	UN2790	II	CORROSIVE	A3, A6, A7, A10 B2, T8
	Acetic anhydride	8	UN1715	II	CORROSIVE, FLAMMABLE LIQUID	A3, A6, A7, A10, B2, T8
	Acetone	3	UN1090	II	FLAMMABLE LIQUID	T8
	Acetone cyanohydrin, stabilized	6.1	UN1541	I	POISON	2, A3, B9, B14, B32, B76, B77, N34, T38, T43, T45

HAZARDOUS MATERIALS TABLE

Sym.	Descriptions and shipping names	Hazard	ID No.	PG	Label(s)	Special provisions
	Acetone oils	3	UN1091	II	FLAMMABLE LIQUID	T7, T30
	Acetonitrile	3	UN1648	II	FLAMMABLE LIQUID	T14
	Acetyl acetone peroxide with more than 9 percent by mass active oxygen	Forbidden				
	Acetyl benzoyl peroxide, solid, or with more than 40 percent in solution	Forbidden				
	Acetyl bromide	8	UN1716	II	CORROSIVE	B2, T12, T26
	Acetyl chloride	3	UN1717	II	FLAMMABLE LIQUID, CORROSIVE	A3, A6, A7, B100, N34, T18, T26
	Acetyl cyclohexanesulfonyl peroxide, with more than 82 percent wetted with less than 12 percent water	Forbidden				
	Acetyl iodide	8	UN1898	II	CORROSIVE	B2, B101, T9
	Acetyl methyl carbinol	3	UN2621	III	FLAMMABLE LIQUID	B1, T1
	Acetyl peroxide, solid, or with more than 25 percent in solution	Forbidden				
	Acetylene, dissolved	2.1	UN1001		FLAMMABLE GAS	
	Acetylene (liquefied)	Forbidden				
	Acetylene silver nitrate	Forbidden				
	Acetylene tetrabromide, see **Tetrabromoethane**					
	Acid butyl phosphate, see **Butyl acid phosphate**					
	Acid, sludge, see **Sludge acid**					

HAZARDOUS MATERIALS TABLE

Sym.	Descriptions and shipping names	Hazard	ID No.	PG	Label(s)	Special provisions
	Acridine	6.1	UN2713	III	KEEP AWAY FROM FOOD	B1, T1
	Acrolein dimer, stabilized	3	UN2607	III	FLAMMABLE LIQUID	1, B9, B12, B14, B30, B42, B72, B77, T38, T43, T44
	Acrolein, inhibited	6.1	UN1092	I	POISON, FLAMMABLE LIQUID	T8
	Acrylamide	6.1	UN2074	III	KEEP AWAY FROM FOOD	T8
	Acrylic acid, inhibited	8	UN2218	II	CORROSIVE, FLAMMABLE LIQUID	B2, T8
	Acrylonitrile, inhibited	3	UN1093	I	FLAMMABLE LIQUID, POISON	B9, T18, T26
	Actuating cartridge, explosive, see **Cartridges, power device**					
	Adhesives, containing a flammable liquid	3	UN1133	II	FLAMMABLE LIQUID	B52, T7, T30
		3		III	FLAMMABLE LIQUID	B1, B52, T7, T30
	Adiponitrile	6.1	UN2205	III	KEEP AWAY FROM FOOD	T1
	Aerosols, corrosive, Packing Group II or III, (each not exceeding 1 L capacity)	2.2	UN1950		NONFLAMMABLE GAS, CORROSIVE	A34
	Aerosols, flammable, (each not exceeding 1 L capacity)	2.1	UN1950		FLAMMABLE GAS	
	Aerosols, non-flammable, (each not exceeding 1 L capacity)	2.2	UN1950		NONFLAMMABLE GAS	N82
	Aerosols, poison, each not exceeding 1 L capacity	2.2	UN1950		NONFLAMMABLE GAS	
	Air bag inflators or **Air bag modules** or **Seat-belt pre-tensioners** or **Seat-belt modules**	9	UN3268	III	CLASS 9	
	Air, compressed	2.2	UN1002		NONFLAMMABLE GAS	

HAZARDOUS MATERIALS TABLE

Sym.	Descriptions and shipping names	Hazard	ID No.	PG	Label(s)	Special provisions
	Air, refrigerated liquid, *(cryogenic liquid)*	2.2	UN1003		NONFLAMMABLE GAS, OXIDIZER	
	Air, refrigerated liquid, *(cryogenic liquid) non-pressurized*	2.2	UN1003		NONFLAMMABLE GAS, OXIDIZER	
	Aircraft evacuation slides, see **Life saving appliances** *etc.*					
	Aircraft hydraulic power unit fuel tank *(containing a mixture of anhydrous hydrazine and monomethyl hydrazine) (M86 fuel)*	3	UN3165	I	FLAMMABLE LIQUID, POISON, CORROSIVE	
	Aircraft survival kits, see **Life saving appliances** *etc.*					
	Alcoholates solution, n.o.s., *in alcohol*	3	UN3274	II	FLAMMABLE LIQUID, CORROSIVE	
	Alcoholic beverages	3	UN3065	II	FLAMMABLE LIQUID	24, B1, T1
				III	FLAMMABLE LIQUID	24, B1, N11, T1
	Alcohols, n.o.s	3	UN1987	I	FLAMMABLE LIQUID	T8, T31
				II	FLAMMABLE LIQUID	T8, T31
				III	FLAMMABLE LIQUID	B1, T7, T30
	Alcohols, toxic, n.o.s.	3	UN1986	I	FLAMMABLE LIQUID, POISON	T8, T31
				II	FLAMMABLE LIQUID, POISON	T8, T31
				III	FLAMMABLE LIQUID, KEEP AWAY FROM FOOD	B1, T8, T31

HAZARDOUS MATERIALS TABLE

Sym.	Descriptions and shipping names	Hazard	ID No.	PG	Label(s)	Special provisions
	Aldehydes, n.o.s.	3	UN1989	I	FLAMMABLE LIQUID	T8, T31
				II	FLAMMABLE LIQUID	T8, T31
				III	FLAMMABLE LIQUID	B1, T7, T30
	Aldehydes, toxic, n.o.s.	3	UN1988	I	FLAMMABLE LIQUID, POISON	T8, T31
				II	FLAMMABLE LIQUID, POISON	T8, T31
				III	FLAMMABLE LIQUID, KEEP AWAY FROM FOOD	B1, T8, T31
	Aldol	6.1	UN2839	II	POISON	T8
D	Aldrin, *liquid*	6.1	NA2762	II	POISON	
D	Aldrin, *solid*	6.1	NA2761	II	POISON	
	Alkali metal alcoholates, self-heating, corrosive, n.o.s.	4.2	UN3206	II	SPONTANEOUSLY COMBUSTIBLE, CORROSIVE	
				III	SPONTANEOUSLY COMBUSTIBLE, CORROSIVE	
	Alkali metal alloys, liquid, n.o.s.	4.3	UN1421	I	DANGEROUS WHEN WET	A2, A3, B48, N34
	Alkali metal amalgams	4.3	UN1389	I	DANGEROUS WHEN WET	A2, A3, N34
	Alkali metal amides	4.3	UN1390	II	DANGEROUS WHEN WET	A6, A7, A8, A19, A20, B106
	Alkali metal dispersions, or Alkaline earth metal dispersions	4.3	UN1391	I	DANGEROUS WHEN WET	A2, A3
	Alkaline corrosive liquids, n.o.s., see **Caustic alkali liquids, n.o.s.**					

HAZARDOUS MATERIALS TABLE

Sym.	Descriptions and shipping names	Hazard	ID No.	PG	Label(s)	Special provisions
	Alkaline earth metal alcoholates, n.o.s.	4.2	UN3205	II	SPONTANEOUSLY COMBUSTIBLE	
				III	SPONTANEOUSLY COMBUSTIBLE	
	Alkaline earth metal alloys, n.o.s.	4.3	UN1393	II	DANGEROUS WHEN WET	A19, B101, B106
	Alkaline earth metal amalgams	4.3	UN1392	I	DANGEROUS WHEN WET	A19, B101, B106, N34, N40
	Alkaloids, liquid, n.o.s., or Alkaloid salts, liquid, n.o.s.	6.1	UN3140	I	POISON	A4, T42
				II	POISON	T14
				III	KEEP AWAY FROM FOOD	T7
	Alkaloids, solid, n.o.s. or Alkaloid salts, solid, n.o.s. *poisonous*	6.1	UN1544	I	POISON	
				II	POISON	
				III	KEEP AWAY FROM FOOD	
	Alkyl sulfonic acids, liquid or Aryl sulfonic acids, liquid *with more than 5 percent free sulfuric acid*	8	UN2584	II	CORROSIVE	B2, T8, T27
	Alkyl sulfonic acids, liquid or Aryl sulfonic acids, liquid *with not more than 5 percent free sulfuric acid*	8	UN2586	III	CORROSIVE	T8
	Alkyl sulfonic acids, solid or Aryl sulfonic acids, solid, *with more than 5 percent free sulfuric acid*	8	UN2583	II	CORROSIVE	
	Alkyl sulfonic acids, solid or Aryl sulfonic acids, solid *with not more than 5 percent free sulfuric acid*	8	UN2585	III	CORROSIVE	

HAZARDOUS MATERIALS TABLE

Sym.	Descriptions and shipping names	Hazard	ID No.	PG	Label(s)	Special provisions
	Alkylphenols, liquid, n.o.s. *(including C2-C12 homologues)*	8	UN3145	I	CORROSIVE	T8
				II	CORROSIVE	T8
				III	CORROSIVE	T7
	Alkylphenols, solid, n.o.s. *(including C2-C12 homologues)*	8	UN2430	I	CORROSIVE	T8
				II	CORROSIVE	T8
				III	CORROSIVE	T8
	Alkylsulfuric acids	8	UN2571	II	CORROSIVE	B2, T9, T27
	Allethrin, see Pesticides, liquid, toxic, n.o.s.					
	Allyl acetate	3	UN2333	II	FLAMMABLE LIQUID, POISON	T8
	Allyl alcohol	6.1	UN1098	I	POISON, FLAMMABLE LIQUID	2, B9, B14, B32, B74, B77, T38, T43, T45
	Allyl bromide	3	UN1099	I	FLAMMABLE LIQUID, POISON	T18
	Allyl chloride	3	UN1100	I	FLAMMABLE LIQUID, POISON	T18, T26
	Allyl chlorocarbonate, see Allyl chloroformate					
	Allyl chloroformate	6.1	UN1722	I	POISON, FLAMMABLE LIQUID, CORROSIVE	2, A3, B9, B14, B32, B74, N41, T38, T43, T45
	Allyl ethyl ether	3	UN2335	I	FLAMMABLE LIQUID, POISON	T8
	Allyl formate	3	UN2336	I	FLAMMABLE LIQUID, POISON	T18, T26

HAZARDOUS MATERIALS TABLE

Sym.	Descriptions and shipping names	Hazard	ID No.	PG	Label(s)	Special provisions
	Allyl glycidyl ether	3	UN2219	III	FLAMMABLE LIQUID	B1, T7
	Allyl iodide	3	UN1723	II	FLAMMABLE LIQUID, CORROSIVE	A3, A6, B100, N34, T18
	Allyl isothiocyanate, stabilized	6.1	UN1545	II	POISON	A3, A7
	Allylamine	6.1	UN2334	I	POISON, FLAMMABLE LIQUID	2, B9, B14, B32, B74, T38, T43, T45
	Allyltrichlorosilane, stabilized	8	UN1724	II	CORROSIVE, FLAMMABLE LIQUID	A7, B2, B6, N34, T8, T26
	Aluminum alkyl halides	4.2	UN3052	I	SPONTANEOUSLY COMBUSTIBLE	B9, B11, T28, T29, T40
	Aluminum alkyl hydrides	4.2	UN3076	I	SPONTANEOUSLY COMBUSTIBLE	B9, B11, T28, T29, T40
	Aluminum alkyls	4.2	UN3051	I	SPONTANEOUSLY COMBUSTIBLE	B9, B11, T28, T29, T40
	Aluminum borohydride or Aluminum borohydride in devices	4.2	UN2870	I	SPONTANEOUSLY COMBUSTIBLE, DANGEROUS WHEN WET	B11
	Aluminum bromide, anhydrous	8	UN1725	II	CORROSIVE	B106
	Aluminum bromide, solution	8	UN2580	III	CORROSIVE	T8
	Aluminum carbide	4.3	UN1394	II	DANGEROUS WHEN WET	A20, B101, B106, N41
	Aluminum chloride, anhydrous	8	UN1726	II	CORROSIVE	B106
	Aluminum chloride, solution	8	UN2581	III	CORROSIVE	T8
	Aluminum dross, wet or hot	Forbidden				

HAZARDOUS MATERIALS TABLE

Sym.	Descriptions and shipping names	Hazard	ID No.	PG	Label(s)	Special provisions
	Aluminum ferrosilicon powder	4.3	UN1395	II	DANGEROUS WHEN WET, POISON	A19, B106, B108
	Aluminum hydride	4.3	UN2463	I	DANGEROUS WHEN WET	A19, A20, B106, B108
D	Aluminum, molten	9	NA9260	III	DANGEROUS WHEN WET, KEEP AWAY FROM FOOD	A19, B100, N40
	Aluminum nitrate	5.1	UN1438	III	OXIDIZER	A1, A29
	Aluminum phosphate solution, see Corrosive liquids, etc.					
	Aluminum phosphide	4.3	UN1397	I	DANGEROUS WHEN WET, POISON	A8, A19, B100, N40
	Aluminum phosphide pesticides	6.1	UN3048	I	POISON	A8
	Aluminum powder, coated	4.1	UN1309	II	FLAMMABLE SOLID	
				III	FLAMMABLE SOLID	
	Aluminum powder, uncoated	4.3	UN1396	II	DANGEROUS WHEN WET	A19, A20, B106, B108
				III	DANGEROUS WHEN WET	A19, A20, B106, B108
	Aluminum processing by-products	4.3	UN3170	II	DANGEROUS WHEN WET	B106
				III	DANGEROUS WHEN WET	B106
	Aluminum resinate	4.1	UN2715	III	FLAMMABLE SOLID	
	Aluminum silicon powder, uncoated	4.3	UN1398	III	DANGEROUS WHEN WET	A1, A19, B108
	Amatols, see Explosives, blasting, type B					

HAZARDOUS MATERIALS TABLE

Sym.	Descriptions and shipping names	Hazard	ID No.	PG	Label(s)	Special provisions
	Amines, flammable, corrosive, n.o.s. or **Polyamines, flammable, corrosive, n.o.s.**	3	UN2733	I	FLAMMABLE LIQUID, CORROSIVE	T42
				II	FLAMMABLE LIQUID, CORROSIVE	T8, T31
				III	FLAMMABLE LIQUID, CORROSIVE	B1, T8, T31
	Amines, liquid, corrosive, flammable, n.o.s. or **Polyamines, liquid, corrosive, flammable, n.o.s.**	8	UN2734	I	CORROSIVE, FLAMMABLE LIQUID	A3, A6, N34, T8, T31
				II	CORROSIVE, FLAMMABLE LIQUID	T8, T31
	Amines, liquid, corrosive, n.o.s or **Polyamines, liquid, corrosive, n.o.s.**	8	UN2735	I	CORROSIVE	A3, A6, B10, N34, T42
				II	CORROSIVE	B2, T8
				III	CORROSIVE	T8
	Amines, solid, corrosive, n.o.s., or **Polyamines, solid, corrosive n.o.s.**	8	UN3259	I	CORROSIVE	
				II	CORROSIVE	
				III	CORROSIVE	
	2-Amino-4-chlorophenol	6.1	UN2673	II	POISON	
	2-Amino-5-diethylaminopentane	6.1	UN2946	III	KEEP AWAY FROM FOOD	T1
	2-(2-Aminoethoxy) ethanol	8	UN3055	III	CORROSIVE	T2
	N-Aminoethylpiperazine	8	UN2815	III	CORROSIVE	T7
	Aminophenols (o-; m-; p-)	6.1	UN2512	III	KEEP AWAY FROM FOOD	T1
	Aminopropyldiethanolamine, see Amines, etc.					

HAZARDOUS MATERIALS TABLE

Sym.	Descriptions and shipping names	Hazard	ID No.	PG	Label(s)	Special provisions
	n-Aminopropylmorpholine, see Amines, etc.					
	Aminopyridines (o-; m-; p-)	6.1	UN2671	II	POISON	T7
	Ammonia, anhydrous, liquefied or **Ammonia solutions**, relative density less than 0.880 at 15 degrees C in water, with more than 50 percent ammonia	2.3	UN1005		POISON GAS, CORROSIVE	4
D	**Ammonia, anhydrous, liquefied** or **Ammonia solutions**, relative density less than 0.880 at 15 degrees C in water, with more than 50 percent ammonia	2.2	UN1005		NONFLAMMABLE GAS	13
	Ammonia solutions, relative density between 0.880 and 0.957 at 15 degrees C in water, with more than 10 percent but not more than 35 percent ammonia	8	UN2672	III	CORROSIVE	T14
	Ammonia solutions, relative density less than 0.880 at 15 degrees C in water, with more than 35 percent but not more than 50 percent ammonia	2.2	UN2073		NONFLAMMABLE GAS	
	Ammonium arsenate	6.1	UN1546	II	POISON	
	Ammonium azide	Forbidden				
	Ammonium bifluoride, solid, see Ammonium hydrogen difluoride, solid					
	Ammonium bifluoride solution, see Ammonium hydrogen difluoride, solution					
	Ammonium bromate	Forbidden				

HAZARDOUS MATERIALS TABLE

Sym.	Descriptions and shipping names	Hazard	ID No.	PG	Label(s)	Special provisions
	Ammonium chlorate	Forbidden				
	Ammonium dichromate	5.1	UN1439	II	OXIDIZER	
	Ammonium dinitro-o-cresolate	6.1	UN1843	II	POISON	
	Ammonium fluoride	6.1	UN2505	III	KEEP AWAY FROM FOOD	T8
	Ammonium fluorosilicate	6.1	UN2854	III	KEEP AWAY FROM FOOD	
	Ammonium fulminate	Forbidden				
	Ammonium hydrogen sulfate	8	UN2506	II	CORROSIVE	
	Ammonium hydrogendifluoride, solid	8	UN1727	II	CORROSIVE	B106, N34
	Ammonium hydrogendifluoride, solution	8	UN2817	II	CORROSIVE, POISON	N34, T15
				III	CORROSIVE, KEEP AWAY FROM FOOD	T8
	Ammonium hydrosulfide, solution, see Ammonium sulfide solution					
D	*Ammonium hydroxide, see Ammonia solutions, etc.*					
	Ammonium metavanadate	6.1	UN2859	II	POISON	
D	**Ammonium nitrate fertilizers**	5.1	NA2072	III	OXIDIZER	7

HAZARDOUS MATERIALS TABLE

Sym.	Descriptions and shipping names	Hazard	ID No.	PG	Label(s)	Special provisions
	Ammonium nitrate fertilizers: *uniform non-segregating mixtures of ammonium nitrate with added matter which is inorganic and chemically inert towards ammonium nitrate, with not less than 90 percent ammonium nitrate and not more than 0.2 percent combustible material (including organic material calculated as carbon), or with more than 70 percent but less than 90 percent ammonium nitrate and not more than 0.4 percent total combustible material.*	5.1	UN2067	III	OXIDIZER	52
AW	**Ammonium nitrate fertilizers:** *uniform non-segregating mixtures of nitrogen/phosphate or nitrogen/potash types or complete fertilizers of nitrogen/phosphate/potash type, with not more than 70 percent ammonium nitrate and not more than 0.4 percent total added combustible material or with not more than 45 percent ammonium nitrate with unrestricted combustible material*	9	UN2071	III	CLASS 9	
D	Ammonium nitrate-fuel oil mixture containing only prilled ammonium nitrate and fuel oil	1.5D	NA0331	II	EXPLOSIVE 1.5D	
	Ammonium nitrate, liquid *(hot concentrated solution)*	5.1	UN2426		OXIDIZER	B5, B100, T25
D	Ammonium nitrate mixed fertilizers	5.1	NA2069	III	OXIDIZER	10
	Ammonium nitrate, with more than 0.2 percent combustible substances, *including any organic substance calculated as carbon, to the exclusion of any other added substance*	1.1D	UN0222	II	EXPLOSIVE 1.1D	

HAZARDOUS MATERIALS TABLE

Sym.	Descriptions and shipping names	Hazard	ID No.	PG	Label(s)	Special provisions
	Ammonium nitrate, *with not more than 0.2 percent of combustible substances, including any organic substance calculated as carbon, to the exclusion of any other added substance*	5.1	UN1942	III	OXIDIZER	A1, A29
	Ammonium nitrite	Forbidden				
	Ammonium perchlorate	1.1D	UN0402	II	EXPLOSIVE 1.1D	107
	Ammonium perchlorate	5.1	UN1442	II	OXIDIZER	107, A9
	Ammonium permanganate	Forbidden				
	Ammonium persulfate	5.1	UN1444	III	OXIDIZER	A1, A29
	Ammonium picrate, *dry or wetted with less than 10 percent water, by mass*	1.1D	UN0004	II	EXPLOSIVE 1.1D	
	Ammonium picrate, *wetted with not less than 10 percent water, by mass*	4.1	UN1310	I	FLAMMABLE SOLID	23, A2, N41
	Ammonium polysulfide, solution	8	UN2818	II	CORROSIVE, POISON	T14
				III	CORROSIVE, KEEP AWAY FROM FOOD	T7
	Ammonium polyvanadate	6.1	UN2861	II	POISON	
	Ammonium silicofluoride, see **Ammonium fluorosilicate**					
	Ammonium sulfide solution	8	UN2683	II	CORROSIVE, POISON, FLAMMABLE LIQUID	T14
	Ammunition, blank, see **Cartridges for weapons, blank**					

HAZARDOUS MATERIALS TABLE

Sym.	Descriptions and shipping names	Hazard	ID No.	PG	Label(s)	Special provisions
	Ammunition, illuminating with or without burster, expelling charge or propelling charge	1.2G	UN0171	II	EXPLOSIVE 1.2G	
	Ammunition, illuminating with or without burster, expelling charge or propelling charge	1.3G	UN0254	II	EXPLOSIVE 1.3G	
	Ammunition, illuminating with or without burster, expelling charge or propelling charge	1.4G	UN0297	II	EXPLOSIVE 1.4G	
	Ammunition, Incendiary liquid or gel, with burster, expelling charge or propelling charge	1.3J	UN0247	II	EXPLOSIVE 1.3J	
	Ammunition, Incendiary (water-activated contrivances) with burster, expelling charge or propelling charge, see **Contrivances, water-activated, etc.**					
	Ammunition, Incendiary, white phosphorus, with burster, expelling charge or propelling charge	1.2H	UN0243	II	EXPLOSIVE 1.2H	
	Ammunition, Incendiary, white phosphorus, with burster, expelling charge or propelling charge	1.3H	UN0244	II	EXPLOSIVE 1.3H	
	Ammunition, Incendiary with or without burster, expelling charge, or propelling charge	1.2G	UN0009	II	EXPLOSIVE 1.2G	
	Ammunition, Incendiary with or without burster, expelling charge, or propelling charge	1.3G	UN0010	II	EXPLOSIVE 1.3G	
	Ammunition, Incendiary with or without burster, expelling charge or propelling charge	1.4G	UN0300	II	EXPLOSIVE 1.4G	
	Ammunition, practice	1.4G	UN0362	II	EXPLOSIVE 1.4G	
	Ammunition, practice	1.3G	UN0488	II	EXPLOSIVE 1.3G	
	Ammunition, proof	1.4G	UN0363	II	EXPLOSIVE 1.4G	

HAZARDOUS MATERIALS TABLE

Sym.	Descriptions and shipping names	Hazard	ID No.	PG	Label(s)	Special provisions
	Ammunition, rocket, see **Warheads, rocket** *etc.*					
	Ammunition, SA (small arms), see **Cartridges for weapons,** *etc.*					
	Ammunition, smoke (water-activated contrivances), white phosphorus, with burster, expelling charge or propelling charge, see **Contrivances, water-activated,** *etc. (UN 0248)*					
	Ammunition, smoke (water-activated contrivances), without white phosphorus or phosphides, with burster, expelling charge or propelling charge, see **Contrivances, water-activated,** *etc. (UN 0249)*					
	Ammunition smoke, white phosphorus with burster, expelling charge, or propelling charge	1.2H	UN0245	II	EXPLOSIVE 1.2H	
	Ammunition, smoke, white phosphorus with burster, expelling charge, or propelling charge	1.3H	UN0246	II	EXPLOSIVE 1.3H	
	Ammunition, smoke with or without burster, expelling charge or propelling charge	1.2G	UN0015	II	EXPLOSIVE 1.2G, CORROSIVE	
	Ammunition, smoke with or without burster, expelling charge or propelling charge	1.3G	UN0016	II	EXPLOSIVE 1.3G, CORROSIVE	
	Ammunition, smoke with or without burster, expelling charge or propelling charge	1.4G	UN0303	II	EXPLOSIVE 1.4G, CORROSIVE	
	Ammunition, sporting, see **Cartridges for weapons,** *etc. (UN 0012; UN 0328; UN 0339)*					
	Ammunition, tear-producing, non-explosive, without burster or expelling charge, non-fuzed	6.1	UN2017	II	POISON, CORROSIVE	

HAZARDOUS MATERIALS TABLE

Sym.	Descriptions and shipping names	Hazard	ID No.	PG	Label(s)	Special provisions
	Ammunition, tear-producing with burster, expelling charge or propelling charge	1.2G	UN0018	II	EXPLOSIVE 1.2G, CORROSIVE, POISON	
	Ammunition, tear-producing with burster, expelling charge or propelling charge	1.3G	UN0019	II	EXPLOSIVE 1.3G, CORROSIVE, POISON	
	Ammunition, tear-producing with burster, expelling charge or propelling charge	1.4G	UN0301	II	EXPLOSIVE 1.4G, CORROSIVE, POISON	
	Ammunition, toxic, non-explosive, without burster or expelling charge, non-fuzed	6.1	UN2016	II	POISON	
	Ammunition, toxic (water-activated contrivances), with burster, expelling charge or propelling charge, see **Contrivances, water-activated, etc.**					
	Ammunition, toxic with burster, expelling charge, or propelling charge	1.2K	UN0020	II	EXPLOSIVE 1.2K, POISON	
	Ammunition, toxic with burster, expelling charge, or propelling charge	1.3K	UN0021	II	EXPLOSIVE 1.3K, POISON	
	Amyl acetates	3	UN1104	III	FLAMMABLE LIQUID	B1, T1
	Amyl acid phosphate	8	UN2819	III	CORROSIVE	T7
	Amyl alcohols	3	UN1105	II	FLAMMABLE LIQUID	T1
				III	FLAMMABLE LIQUID	B1, B3, T1
	Amyl butyrates	3	UN2620	III	FLAMMABLE LIQUID	B1, T1
	Amyl chlorides	3	UN1107	II	FLAMMABLE LIQUID	T1
	Amyl formates	3	UN1109	III	FLAMMABLE LIQUID	B1, T1
	Amyl mercaptans	3	UN1111	II	FLAMMABLE LIQUID	A3, T8
	Amyl methyl ketone	3	UN1110	III	FLAMMABLE LIQUID	B1, T1

HAZARDOUS MATERIALS TABLE

Sym.	Descriptions and shipping names	Hazard	ID No.	PG	Label(s)	Special provisions
	Amyl nitrate	3	UN1112	III	FLAMMABLE LIQUID	B1, T1
	Amyl nitrites	3	UN1113	III	FLAMMABLE LIQUID	T8
	Amylamines	3	UN1106	II	FLAMMABLE LIQUID, CORROSIVE	T1
	n-Amylene	3	UN1108	III	FLAMMABLE LIQUID, CORROSIVE	B1
	Amyltrichlorosilane	8	UN1728	I	FLAMMABLE LIQUID	T14
				II	CORROSIVE	A7, B2, B6, N34, T8, T26
	Anhydrous ammonia see *Ammonia, anhydrous, liquefied*					
	Anhydrous hydrofluoric acid, see *Hydrogen fluoride, anhydrous*					
	Aniline	6.1	UN1547	II	POISON	T8
	Aniline hydrochloride	6.1	UN1548	III	KEEP AWAY FROM FOOD	
	Aniline oil, see *Aniline*					
	Anisidines	6.1	UN2431	III	KEEP AWAY FROM FOOD	T1
	Anisole	3	UN2222	III	FLAMMABLE LIQUID	B1, T1
	Anisoyl chloride	8	UN1729	II	CORROSIVE	B2, T8
+	**Anti-freeze, liquid, see Flammable liquids, n.o.s.**					
	Antimonous chloride, see *Antimony trichloride*					
	Antimony compounds, inorganic, liquid, n.o.s.	6.1	UN3141	III	KEEP AWAY FROM FOOD	35, T7

HAZARDOUS MATERIALS TABLE

Sym.	Descriptions and shipping names	Hazard	ID No.	PG	Label(s)	Special provisions
	Antimony compounds, Inorganic, solid, n.o.s.	6.1	UN1549	III	KEEP AWAY FROM FOOD	35
	Antimony lactate	6.1	UN1550	III	KEEP AWAY FROM FOOD	
	Antimony pentachloride, liquid	8	UN1730	II	CORROSIVE	B2, T8, T26
	Antimony pentachloride, solutions	8	UN1731	II	CORROSIVE	B2, T8, T27
				III	CORROSIVE	T7, T26
	Antimony pentafluoride	8	UN1732	II	CORROSIVE, POISON	A3, A6, A7, A10, N3, T12, T26
	Antimony potassium tartrate	6.1	UN1551	III	KEEP AWAY FROM FOOD	
	Antimony powder	6.1	UN2871	III	KEEP AWAY FROM FOOD	
	Antimony sulfide and a chlorate, mixtures of	Forbidden				
	Antimony sulfide, solid, see Antimony compounds, Inorganic, n.o.s.					
D	Antimony tribromide, solid	8	NA1549	II	CORROSIVE	B2
D	Antimony tribromide, solution	8	NA1549	II	CORROSIVE	B2
	Antimony trichloride, liquid	8	UN1733	II	CORROSIVE	
	Antimony trichloride, solid	8	UN1733	II	CORROSIVE	B106
D	Antimony trifluoride, solid	8	NA1549	II	CORROSIVE	
D	Antimony trifluoride, solution	8	NA1549	II	CORROSIVE	B2
	Aqua ammonia, see Ammonia solution, etc.					
	Argon, compressed	2.2	UN1006		NONFLAMMABLE GAS	
	Argon, refrigerated liquid (cryogenic liquid)	2.2	UN1951		NONFLAMMABLE GAS	

HAZARDOUS MATERIALS TABLE

Sym.	Descriptions and shipping names	Hazard	ID No.	PG	Label(s)	Special provisions
	Arsenic	6.1	UN1558	II	POISON	
	Arsenic acid, liquid	6.1	UN1553	I	POISON	T18, T27
	Arsenic acid, solid	6.1	UN1554	II	POISON	
	Arsenic bromide	6.1	UN1555	II	POISON	
	Arsenic chloride, see Arsenic trichloride					
	Arsenic compounds, liquid, n.o.s. *including arsenates n.o.s.; arsenites, n.o.s.; arsenic sulfides, n.o.s.; and organic compounds of arsenic, n.o.s.*	6.1	UN1556	I	POISON	
				II	POISON	
				III	KEEP AWAY FROM FOOD	
	Arsenic compounds, solid, n.o.s. *including arsenates, n.o.s.; arsenites, n.o.s.; arsenic sulfides, n.o.s.; and organic compounds of arsenic, n.o.s.*	6.1	UN1557	I	POISON	
				II	POISON	
				III	KEEP AWAY FROM FOOD	
	Arsenic pentoxide	6.1	UN1559	II	POISON	
D	**Arsenic sulfide**	6.1	NA1557	II	POISON	
	Arsenic sulfide and a chlorate, mixtures of	Forbidden				
	Arsenic trichloride	6.1	UN1560	I	POISON	2, B9, B14, B32, B74, T38, T43, T45
	Arsenic trioxide	6.1	UN1561	II	POISON	
D	**Arsenic trisulfide**	6.1	NA1557	II	POISON	

HAZARDOUS MATERIALS TABLE

Sym.	Descriptions and shipping names	Hazard	ID No.	PG	Label(s)	Special provisions
	Arsenic, white, solid, see **Arsenic trioxide**					
	Arsenical dust	6.1	UN1562	II	POISON	
	Arsenical pesticides, liquid, flammable, toxic, *flash point less than 23 degrees C*	3	UN2760	I	FLAMMABLE LIQUID, POISON	
				II	FLAMMABLE LIQUID, POISON	
	Arsenical pesticides, liquid, toxic	6.1	UN2994	I	POISON	T42
				II	POISON	T14
				III	KEEP AWAY FROM FOOD	T14
	Arsenical pesticides, liquid, toxic, flammable *flashpoint not less than 23 degrees C*	6.1	UN2993	I	POISON, FLAMMABLE LIQUID	T42
				II	POISON, FLAMMABLE LIQUID	T14
				III	KEEP AWAY FROM FOOD, FLAMMABLE LIQUID	B1, T14
	Arsenical pesticides, solid, toxic	6.1	UN2759	I	POISON	
				II	POISON	
				III	KEEP AWAY FROM FOOD	
	Arsenious acid, solid, see **Arsenic trioxide**					
	Arsenious and mercuric iodide solution, see **Arsenic compounds, liquid, n.o.s.**					
	Arsine	2.3	UN2188		POISON GAS, FLAMMABLE GAS	1
	Articles, explosive, extremely insensitive or Articles, EEI	1.6N	UN0486	II	EXPLOSIVE 1.6N	101

HAZARDOUS MATERIALS TABLE

Sym.	Descriptions and shipping names	Hazard	ID No.	PG	Label(s)	Special provisions
	Articles, explosive, n.o.s.	1.4S	UN0349	II	EXPLOSIVE 1.4S	101
	Articles, explosive, n.o.s.	1.4B	UN0350	II	EXPLOSIVE 1.4B	101
	Articles, explosive, n.o.s.	1.4C	UN0351	II	EXPLOSIVE 1.4C	101
	Articles, explosive, n.o.s.	1.4D	UN0352	II	EXPLOSIVE 1.4D	101
	Articles, explosive, n.o.s.	1.4G	UN0353	II	EXPLOSIVE 1.4G	101
	Articles, explosive, n.o.s.	1.1L	UN0354	II	EXPLOSIVE 1.1L	101
	Articles, explosive, n.o.s.	1.2L	UN0355	II	EXPLOSIVE 1.2L	101
	Articles, explosive, n.o.s.	1.3L	UN0356	II	EXPLOSIVE 1.3L	101
	Articles, explosive, n.o.s.	1.1C	UN0462	II	EXPLOSIVE 1.1C	101
	Articles, explosive, n.o.s.	1.1D	UN0463	II	EXPLOSIVE 1.1D	101
	Articles, explosive, n.o.s.	1.1E	UN0464	II	EXPLOSIVE 1.1E	101
	Articles, explosive, n.o.s.	1.1F	UN0465	II	EXPLOSIVE 1.1F	101
	Articles, explosive, n.o.s.	1.2C	UN0466	II	EXPLOSIVE 1.2C	101
	Articles, explosive, n.o.s.	1.2D	UN0467	II	EXPLOSIVE 1.2D	101
	Articles, explosive, n.o.s.	1.2E	UN0468	II	EXPLOSIVE 1.2E	101
	Articles, explosive, n.o.s.	1.2F	UN0469	II	EXPLOSIVE 1.2F	101
	Articles, explosive, n.o.s.	1.3C	UN0470	II	EXPLOSIVE 1.3C	101
	Articles, explosive, n.o.s.	1.4E	UN0471	II	EXPLOSIVE 1.4E	101
	Articles, explosive, n.o.s.	1.4F	UN0472	II	EXPLOSIVE 1.4F	101
	Articles, pressurized pneumatic or Hydraulic containing non-flammable gas	2.2	UN3164		NONFLAMMABLE GAS	
	Articles, pyrophoric	1.2L	UN0380	II	EXPLOSIVE 1.2L	

HAZARDOUS MATERIALS TABLE

Sym.	Descriptions and shipping names	Hazard	ID No.	PG	Label(s)	Special provisions
	Articles, pyrotechnic for technical purposes	1.1G	UN0428	II	EXPLOSIVE 1.1G	
	Articles, pyrotechnic for technical purposes	1.2G	UN0429	II	EXPLOSIVE 1.2G	
	Articles, pyrotechnic for technical purposes	1.3G	UN0430	II	EXPLOSIVE 1.3G	
	Articles, pyrotechnic for technical purposes	1.4G	UN0431	II	EXPLOSIVE 1.4G	
	Articles, pyrotechnic for technical purposes	1.4S	UN0432	II	EXPLOSIVE 1.4S	
D	**Asbestos**	9	NA2212	III	CLASS 9	
	Ascaridole (organic peroxide)	Forbidden				
D	**Asphalt**, at or above its flashpoint	3	NA1999	III	FLAMMABLE LIQUID	
D	**Asphalt**, cut back, see **Tars, liquid,** etc.					
	Automobile, motorcycle, tractor, or other self-propelled vehicle, engine, or other mechanical apparatus, see **Engines or Battery** etc.					
	Azaurolic acid (salt of) (dry)	Forbidden				
	Azido guanidine picrate (dry)	Forbidden				
	5-Azido-1-hydroxy tetrazole	Forbidden				
	Azido hydroxy tetrazole (mercury and silver salts)	Forbidden				
	3-Azido-1,2-Propylene glycol dinitrate	Forbidden				
	Azidodithiocarbonic acid	Forbidden				

HAZARDOUS MATERIALS TABLE

Sym.	Descriptions and shipping names	Hazard	ID No.	PG	Label(s)	Special provisions
	Azidoethyl nitrate	Forbidden				
	1-Azindinylphosphine oxide-(tris), see **Tris-(1-aziridinyl) phosphine oxide, solution**					
	Azodicarbonamide	4.1	UN3242	II	FLAMMABLE SOLID	38
	Azotetrazole (dry)	Forbidden				
	Barium	4.3	UN1400	II	DANGEROUS WHEN WET	
	Barium alloys, pyrophoric	4.2	UN1854	I	SPONTANEOUSLY COMBUSTIBLE	
	Barium azide, dry or wetted with less than 50 percent water, by mass	1.1A	UN0224	II	EXPLOSIVE 1.1A, POISON	111, 117
	Barium azide, wetted with not less than 50 percent water, by mass	4.1	UN1571	I	FLAMMABLE SOLID, POISON	A2
	Barium bromate	5.1	UN2719	II	OXIDIZER, POISON	
	Barium chlorate	5.1	UN1445	II	OXIDIZER, POISON	A9, N34, T8
	Barium compounds, n.o.s.	6.1	UN1564	II	POISON	
				III	KEEP AWAY FROM FOOD	
	Barium cyanide	6.1	UN1565	I	POISON	N74, N75
	Barium hypochlorite with more than 22 percent available chlorine	5.1	UN2741	II	OXIDIZER, POISON	A7, A9, N34
	Barium nitrate	5.1	UN1446	II	OXIDIZER, POISON	
	Barium oxide	6.1	UN1884	III	KEEP AWAY FROM FOOD	
	Barium perchlorate	5.1	UN1447	II	OXIDIZER, POISON	T8
	Barium permanganate	5.1	UN1448	II	OXIDIZER, POISON	

HAZARDOUS MATERIALS TABLE

Sym.	Descriptions and shipping names	Hazard	ID No.	PG	Label(s)	Special provisions
D	**Barium peroxide**	5.1	UN1449	II	OXIDIZER, POISON	
	Barium selenate, see **Selenates** or **Selenites**					
	Barium selenite, see **Selenates** or **Selenites**					
	Barium styphnate	1.1A	NA0473	II	EXPLOSIVE 1.1A	111, 117
	Batteries, containing sodium	4.3	UN3292	II	DANGEROUS WHEN WET	
	Batteries, dry, containing potassium hydroxide solid, *electric, storage*	8	UN3028	III	CORROSIVE	
	Batteries, wet, filled with acid, *electric storage*	8	UN2794	III	CORROSIVE	
	Batteries, wet, filled with alkali, *electric storage*	8	UN2795	III	CORROSIVE	
	Batteries, wet, non-spillable, *electric storage*	8	UN2800	III	CORROSIVE	
	Battery, dry, not subject to the requirements of this subchapter					
	Battery fluid, acid	8	UN2796	II	CORROSIVE	A3, A7, B2, B15, N6, N34, T9, T27
	Battery fluid, alkali	8	UN2797	II	CORROSIVE	B2, N6, T8
	Battery lithium type, see **Lithium batteries** etc.					
	Battery-powered vehicle or **Battery-powered equipment wet battery**	9	UN3171		CLASS 9	
	Battery, wet, filled with acid or alkali with automobile (or named self-propelled vehicle or mechanical equipment containing internal combustion engine) see **Vehicles, self-propelled** etc.					
	Battery, wet, with wheelchair, see **Wheelchair, electric**					

HAZARDOUS MATERIALS TABLE

Sym.	Descriptions and shipping names	Hazard	ID No.	PG	Label(s)	Special provisions
+	Benzaldehyde	9	UN1990	III	CLASS 9	B101, T8
	Benzene	3	UN1114	II	FLAMMABLE LIQUID	
	Benzene diazonium chloride (dry)	Forbidden				
	Benzene diazonium nitrate (dry)	Forbidden				
	Benzene phosphorus dichloride, see Phenyl phosphorus dichloride					
	Benzene phosphorus thiodichloride, see Phenyl phosphorus thiodichloride					
	Benzene sulfonyl chloride	8	UN2225	III	CORROSIVE	T8
	Benzene triozonide	Forbidden				
	Benzenethiol, see Phenyl mercaptan					
	Benzidine	6.1	UN1885	II	POISON	
	Benzol derivative pesticides, liquid, flammable, toxic, *flash point less than 23 degrees C*	3	UN2770	I	FLAMMABLE LIQUID, POISON	
				II	FLAMMABLE LIQUID, POISON	
	Benzoic derivative pesticides, liquid, toxic	6.1	UN3004	I	POISON	T42
				II	POISON	T14
				III	KEEP AWAY FROM FOOD	T14

HAZARDOUS MATERIALS TABLE

Sym.	Descriptions and shipping names	Hazard	ID No.	PG	Label(s)	Special provisions
	Benzoic derivative pesticides, liquid, toxic, flammable, *flashpoint not less than 23 degrees C*	6.1	UN3003	I	POISON, FLAMMABLE LIQUID	T42
				II	POISON, FLAMMABLE LIQUID	T14
				III	KEEP AWAY FROM FOOD, FLAMMABLE LIQUID	T14
	Benzoic derivative pesticides, solid, toxic	6.1	UN2769	I	POISON	
				II	POISON	
				III	KEEP AWAY FROM FOOD	
	Benzol, see Benzene					
	Benzonitrile	6.1	UN2224	II	POISON	T14
	Benzoquinone	6.1	UN2587	II	POISON	
	Benzotrichloride	8	UN2226	II	CORROSIVE	B2, B101, T15
	Benzotrifluoride	3	UN2338	II	FLAMMABLE LIQUID	T2
	Benzoxidiazoles (dry)	Forbidden				
	Benzoyl azide	Forbidden				
	Benzoyl chloride	8	UN1736	II	CORROSIVE	B2, T9, T26
	Benzyl bromide	6.1	UN1737	II	POISON, CORROSIVE	A3, A7, N33, N34, T12, T26
	Benzyl chloride	6.1	UN1738	II	POISON, CORROSIVE	A3, A7, B70, N33, N43, T12, T26

HAZARDOUS MATERIALS TABLE

Sym.	Descriptions and shipping names	Hazard	ID No.	PG	Label(s)	Special provisions
	Benzyl chloride *unstabilized*	6.1	UN1738	II	POISON, CORROSIVE	A3, A7, B8, B11, N33, N34, N43, T12, T18, T26
	Benzyl chloroformate	8	UN1739	I	CORROSIVE	A3, A6, B4, N41, T26
	Benzyl iodide	6.1	UN2653	II	POISON	T8
	Benzyldimethylamine	8	UN2619	II	CORROSIVE, FLAMMABLE LIQUID	B2, T1
	Benzylidene chloride	6.1	UN1886	II	POISON	T8
	Beryllium compounds, n.o.s.	6.1	UN1566	II	POISON	
				III	KEEP AWAY FROM FOOD	
	Beryllium nitrate	5.1	UN2464	II	OXIDIZER, POISON	
	Beryllium, powder	6.1	UN1567	II	POISON, FLAMMABLE SOLID	
	Biphenyl triozonide	Forbidden				
	Bipyridilium pesticides, liquid, flammable, toxic, *flash point less than 23 degrees C*	3	UN2782	I	FLAMMABLE LIQUID, POISON	T42
				II	FLAMMABLE LIQUID, POISON	T14
	Bipyridilium pesticides, liquid, toxic	6.1	UN3016	I	POISON	
				II	POISON	T14
				III	KEEP AWAY FROM FOOD	T14

HAZARDOUS MATERIALS TABLE

Sym.	Descriptions and shipping names	Hazard	ID No.	PG	Label(s)	Special provisions
	Bipyridilium pesticides, liquid, toxic, flammable, *flashpoint not less than 23 degrees C*	6.1	UN3015	I	POISON, FLAMMABLE LIQUID	T42
				II	POISON, FLAMMABLE LIQUID	T14
				III	KEEP AWAY FROM FOOD, FLAMMABLE LIQUID	B1, T14
	Bipyridilium pesticides, solid, toxic	6.1	UN2781	I	POISON	
				II	POISON	
				III	KEEP AWAY FROM FOOD	
	Bis (Aminopropyl) piperazine, see *Corrosive liquid, n.o.s.*					
	Bisulfate, aqueous solution	8	UN2837	II	CORROSIVE	A7, B2, N34, T8, T26
				III	CORROSIVE	A7, N34, T7, T26
	Bisulfites, aqueous solutions, n.o.s.	8	UN2693	III	CORROSIVE	T8
	Black powder, compressed or Gunpowder, compressed or Black powder, in pellets or Gunpowder, in pellets	1.1D	UN0028	II	EXPLOSIVE 1.1D	
	Black powder or Gunpowder, *granular or as a meal*	1.1D	UN0027	II	EXPLOSIVE 1.1D	
	Blasting agent, n.o.s., see *Explosives, blasting etc.*					
	Blasting cap assemblies, see *Detonator assemblies, non-electric, for blasting*					
	Blasting caps, electric, see *Detonators, electric for blasting*					

HAZARDOUS MATERIALS TABLE

Sym.	Descriptions and shipping names	Hazard	ID No.	PG	Label(s)	Special provisions
	Blasting caps, non-electric, see **Detonators, non-electric,** for blasting					
	Bleaching powder, see **Calcium hypochlorite mixtures, etc.**					
I	**Blue asbestos** (Crocidolite) or **Brown asbestos** (amosite, mysorite)	9	UN2212	II	CLASS 9	
	Bombs, photo-flash	1.1F	UN0037	II	EXPLOSIVE 1.1F	
	Bombs, photo-flash	1.1D	UN0038	II	EXPLOSIVE 1.1D	
	Bombs, photo-flash	1.2G	UN0039	II	EXPLOSIVE 1.2G	
	Bombs, photo-flash	1.3G	UN0299	II	EXPLOSIVE 1.3G	
	Bombs, smoke, non-explosive, with corrosive liquid, without initiating device	8	UN2028	II	CORROSIVE	
	Bombs, with bursting charge	1.1F	UN0033	II	EXPLOSIVE 1.1F	
	Bombs, with bursting charge	1.1D	UN0034	II	EXPLOSIVE 1.1D	
	Bombs, with bursting charge	1.2D	UN0035	II	EXPLOSIVE 1.2D	
	Bombs, with bursting charge	1.2F	UN0291	II	EXPLOSIVE 1.2F	
	Bombs with flammable liquid, with bursting charge	1.1J	UN0399	II	EXPLOSIVE 1.1J	
	Bombs with flammable liquid, with bursting charge	1.2J	UN0400	II	EXPLOSIVE 1.2J	115
D	**Boosters with detonator**	1.4B	NA0350	II	EXPLOSIVE 1.4B	
	Boosters with detonator	1.1B	UN0225	II	EXPLOSIVE 1.1B	
	Boosters with detonator	1.2B	UN0268	II	EXPLOSIVE 1.2B	
	Boosters, without detonator	1.1D	UN0042	II	EXPLOSIVE 1.1D	

HAZARDOUS MATERIALS TABLE

Sym.	Descriptions and shipping names	Hazard	ID No.	PG	Label(s)	Special provisions
	Boosters, without detonator	1.2D	UN0283		EXPLOSIVE 1.2D	
	Borate and chlorate mixtures, see **Chlorate and borate mixtures**					
	Borneol	4.1	UN1312	III	FLAMMABLE SOLID	A1
+	**Boron tribromide**	8	UN2692	I	CORROSIVE, POISON	2, A3, A7, B9, B14, B32, B74, N34, T38, T43, T45
	Boron trichloride	2.3	UN1741		POISON GAS, CORROSIVE	3, 25, B9, B14
	Boron trifluoride	2.3	UN1008		POISON GAS	2, B9, B14
	Boron trifluoride acetic acid complex	8	UN1742	II	CORROSIVE	B2, B6, T9, T27
	Boron trifluoride diethyl etherate	8	UN2604	I	CORROSIVE, FLAMMABLE LIQUID	A19, T8, T26
	Boron trifluoride dihydrate	8	UN2851	II	CORROSIVE	T9, T27
	Boron trifluoride dimethyl etherate	4.3	UN2965	I	DANGEROUS WHEN WET, CORROSIVE, FLAMMABLE LIQUID	A19, T12, T26
	Boron trifluoride propionic acid complex	8	UN1743	II	CORROSIVE	B2, T9, T27
	Box toe gum, see **Nitrocellulose** *etc.*					
	Bromates, inorganic, aqueous solution, n.o.s.	5.1	UN3213	II	OXIDIZER	T8
	Bromates, inorganic, n.o.s.	5.1	UN1450	II	OXIDIZER	
	Bromine azide	Forbidden				

HAZARDOUS MATERIALS TABLE

Sym.	Descriptions and shipping names	Hazard	ID No.	PG	Label(s)	Special provisions
+	Bromine or Bromine solutions	8	UN1744	I	CORROSIVE, POISON	1, A3, A6, B9, B12, B64, B85, N34, N43, T18, T41
	Bromine chloride	2.3	UN2901		POISON GAS, CORROSIVE, OXIDIZER	2, B9, B12, B14
+	Bromine pentafluoride	5.1	UN1745	I	OXIDIZER, POISON, CORROSIVE	1, B9, B14, B30, B72, T38, T43, T44
+	Bromine trifluoride	5.1	UN1746	I	OXIDIZER, POISON, CORROSIVE	2, B9, B14, B32, B74, T38, T43, T45
	4-Bromo-1,2-dinitrobenzene	Forbidden				
	4-Bromo-1,2-dinitrobenzene (unstable at 59 degrees C.)	Forbidden				
	1-Bromo-3-methylbutane	3	UN2341	III	FLAMMABLE LIQUID	B1, T7, T30
	1-Bromo-3-nitrobenzene (unstable at 56 degrees C)	Forbidden				
–	2-Bromo-2-nitropropane-1,3-diol	6.1	UN3241	III	KEEP AWAY FROM FOOD	46
	Bromoacetic acid, solid	8	UN1938	II	CORROSIVE	A7, N34, T9
	Bromoacetic acid, solution	8	UN1938	II	CORROSIVE	B2, T9
	Bromoacetone	6.1	UN1569	II	POISON	2
	Bromoacetyl bromide	8	UN2513	II	CORROSIVE	B2, T9, T26
	Bromobenzene	3	UN2514	III	FLAMMABLE LIQUID	B1, T1
+	Bromobenzyl cyanides, liquid	6.1	UN1694	I	POISON	T18
	Bromobenzyl cyanides, solid	6.1	UN1694	I	POISON	T18
	2-Bromobutane	3	UN2339	II	FLAMMABLE LIQUID	B1, T1

HAZARDOUS MATERIALS TABLE

Sym.	Descriptions and shipping names	Hazard	ID No.	PG	Label(s)	Special provisions
	Bromochloromethane	6.1	UN1887	III	KEEP AWAY FROM FOOD	T7
	2-Bromoethyl ethyl ether	3	UN2340	II	FLAMMABLE LIQUID	T7
	Bromoform	6.1	UN2515	III	KEEP AWAY FROM FOOD	T7
	Bromomethylpropanes	3	UN2342	II	FLAMMABLE LIQUID	T7, T30
	2-Bromopentane	3	UN2343	II	FLAMMABLE LIQUID	T1
	2-Bromopropanes	3	UN2344	II	FLAMMABLE LIQUID	T7
	3-Bromopropyne	3	UN2345	II	FLAMMABLE LIQUID	T8
	Bromosilane	Forbidden				
	Bromotoluene-alpha, see Benzyl bromide					
	Bromotrifluoroethylene	2.1	UN2419		FLAMMABLE GAS	
	Bromotrifluoromethane, *R13B1*	2.2	UN1009		NONFLAMMABLE GAS	
	Brucine	6.1	UN1570	I	POISON	
	Bursters, *explosive*	1.1D	UN0043	II	EXPLOSIVE 1.1D	
	Butadienes, Inhibited	2.1	UN1010		FLAMMABLE GAS	
	Butane or Butane mixtures *see also* **Petroleum gases, liquefied**	2.1	UN1011		FLAMMABLE GAS	19
	Butane, butane mixtures and mixtures having similar properties in cartridges each not exceeding 500 grams, see **Receptacles,** *etc.*					
	Butanedione	3	UN2346	II	FLAMMABLE LIQUID	T1
	1,2,4-Butanetriol trinitrate	Forbidden				

HAZARDOUS MATERIALS TABLE

Sym.	Descriptions and shipping names	Hazard	ID No.	PG	Label(s)	Special provisions
	Butanols	3	UN1120	II	FLAMMABLE LIQUID	T1
				III	FLAMMABLE LIQUID	B1, T1
	tert-Butoxycarbonyl azide	Forbidden				
	Butoxyl	3	UN2708	III	FLAMMABLE LIQUID	B1, T1
	Butyl acetates	3	UN1123	II	FLAMMABLE LIQUID	T1
				III	FLAMMABLE LIQUID	B1, T1
	Butyl acid phosphate	8	UN1718	III	CORROSIVE	T7
	Butyl alcohols, see Butanols					
	Butyl benzenes	3	UN2709	III	FLAMMABLE LIQUID	B1, T1
	n-Butyl bromide	3	UN1126	II	FLAMMABLE LIQUID	T1
	n-Butyl chloride, see Chlorobutanes					
D	sec-Butyl chloroformate	6.1	NA2742	I	POISON, FLAMMABLE LIQUID, CORROSIVE	2, B9, B14, B32, B74, T38, T43, T45
	n-Butyl chloroformate	6.1	UN2743	I	POISON, CORROSIVE	2, B9, B14, B32, B74, T38, T43, T45
	Butyl ethers, see Dibutyl ethers					
	Butyl ethyl ether, see Ethyl butyl ether					
	n-Butyl formate	3	UN1128	II	FLAMMABLE LIQUID	T1
	tert-Butyl hydroperoxide, with more than 90 percent with water	Forbidden				
	tert-Butyl hypochlorite	4.2	UN3255	I	SPONTANEOUSLY COMBUSTIBLE, CORROSIVE	
	N-n-Butyl Imidazole	6.1	UN2690	II	POISON	T8

HAZARDOUS MATERIALS TABLE

Sym.	Descriptions and shipping names	Hazard	ID No.	PG	Label(s)	Special provisions
	tert-Butyl Isocyanate	6.1	UN2484	I	POISON, FLAMMABLE LIQUID	1, A7, B9, B14, B30, B72, T38, T43, T44
	n-Butyl isocyanate	6.1	UN2485	I	POISON, FLAMMABLE LIQUID	2, A7, B9, B14, B32, B74, B77, T38, T43, T45
	Butyl mercaptans	3	UN2347	II	FLAMMABLE LIQUID	A3, T8
	n-Butyl methacrylate	3	UN2227	III	FLAMMABLE LIQUID	B1, T1
	Butyl methyl ether	3	UN2350	II	FLAMMABLE LIQUID	T8
	Butyl nitrites	3	UN2351	I	FLAMMABLE LIQUID	T8
		3		II	FLAMMABLE LIQUID	T8
		3		III	FLAMMABLE LIQUID	B1, T8
	tert-Butyl peroxyacetate, with more than 76 percent in solution	Forbidden				
	n-Butyl peroxydicarbonate, with more than 52 percent in solution	Forbidden				
	tert-Butyl peroxyisobutyrate, with more than 77 percent in solution	Forbidden				
	Butyl phosphoric acid, see **Butyl acid phosphate**					
	5-tert-Butyl-2,4,6-trinitro-m-xylene or **Musk xylene**	4.1	UN2956	III	FLAMMABLE SOLID	
	Butyl vinyl ether, inhibited	3	UN2352	II	FLAMMABLE LIQUID	B101, T7
	Butylacrylate	3	UN2348	III	FLAMMABLE LIQUID	B1, T8, T31
	n-Butylamine	3	UN1125	II	FLAMMABLE LIQUID, CORROSIVE	B101, T8

HAZARDOUS MATERIALS TABLE

Sym.	Descriptions and shipping names	Hazard	ID No.	PG	Label(s)	Special provisions
	N-Butylaniline	6.1	UN2738	II	POISON	T8
	tert-Butylcyclohexylchloroformate	6.1	UN2747	III	KEEP AWAY FROM FOOD	T8
	Butylene see also Petroleum gases, liquefied	2.1	UN1012		FLAMMABLE GAS	19
	1,2-Butylene oxide, stabilized	3	UN3022	II	FLAMMABLE LIQUID	T8
	Butylpropionate	3	UN1914	III	FLAMMABLE LIQUID	B1, T1
	Butyltoluenes	6.1	UN2667	III	KEEP AWAY FROM FOOD	T2
	Butyltrichlorosilane	8	UN1747	II	CORROSIVE, FLAMMABLE LIQUID	A7, B2, B6, N34, T8, T26
	1,4-Butynediol	6.1	UN2716	III	KEEP AWAY FROM FOOD	A1
	Butyraldehyde	3	UN1129	II	FLAMMABLE LIQUID	T8
	Butyraldoxime	3	UN2840	III	FLAMMABLE LIQUID	B1, T1
	Butyric acid	8	UN2820	III	CORROSIVE	T1
	Butyric anhydride	8	UN2739	III	CORROSIVE	T2
	Butyronitrile	3	UN2411	II	FLAMMABLE LIQUID, POISON	T14
	Butyryl chloride	3	UN2353	II	FLAMMABLE LIQUID, CORROSIVE	B100, T9, T26
	Cacodylic acid	6.1	UN1572	II	POISON	
	Cadmium compounds	6.1	UN2570	I	POISON	
				II	POISON	
				III	KEEP AWAY FROM FOOD	
	Caesium hydroxide	8	UN2682	II	CORROSIVE	
	Caesium hydroxide solution	8	UN2681	II	CORROSIVE	B2, T8

HAZARDOUS MATERIALS TABLE

Sym.	Descriptions and shipping names	Hazard	ID No.	PG	Label(s)	Special provisions
	Calcium	4.3	UN1401	III	CORROSIVE DANGEROUS WHEN WET	T7
	Calcium arsenate	6.1	UN1573	II	POISON	B100
	Calcium arsenate and calcium arsenite, mixtures, solid	6.1	UN1574	II	POISON	
D	Calcium arsenite, solid	6.1	NA1574	II	POISON	
	Calcium bisulfite solution, see Bisulfites, Inorganic, aqueous solutions, n.o.s.					
	Calcium carbide	4.3	UN1402	I	DANGEROUS WHEN WET	A1, A8, B55, B101, B106, N34
				II	DANGEROUS WHEN WET	A1, A8, B55, B101, B106, N34
	Calcium chlorate	5.1	UN1452	II	OXIDIZER	N34
	Calcium chlorate aqueous solution	5.1	UN2429	II	OXIDIZER	A2, N41, T8
	Calcium chlorite	5.1	UN1453	II	OXIDIZER	A9, N34
	Calcium cyanamide with more than 0.1 percent of calcium carbide	4.3	UN1403	III	DANGEROUS WHEN WET	A1, A19, B105
	Calcium cyanide	6.1	UN1575	II	POISON	N79, N80
	Calcium dithionite or Calcium hydrosulfite	4.2	UN1923	II	SPONTANEOUSLY COMBUSTIBLE	A19, A20
	Calcium hydride	4.3	UN1404	I	DANGEROUS WHEN WET	A19, B100, N40
	Calcium hydrosulfite, see Calcium dithionite					

HAZARDOUS MATERIALS TABLE

Sym.	Descriptions and shipping names	Hazard	ID No.	PG	Label(s)	Special provisions
	Calcium hypochlorite, dry or Calcium hypochlorite mixtures dry with more than 39 percent available chlorine (8.8 percent available oxygen)	5.1	UN1748	II	OXIDIZER	A7, A9, N34
	Calcium hypochlorite, hydrated or Calcium hypochlorite, hydrated mixtures, with not less than 5.5 percent but not more than 10 percent water	5.1	UN2880	II	OXIDIZER	
	Calcium hypochlorite mixtures, dry, with more than 10 percent but not more than 39 percent available chlorine	5.1	UN2208	III	OXIDIZER	A1, A29, B103, N34
	Calcium manganese silicon	4.3	UN2844	III	DANGEROUS WHEN WET	A1, A19, B105, B106
	Calcium nitrate	5.1	UN1454	III	OXIDIZER	34
	Calcium oxide	8	UN1910	III	CORROSIVE	
	Calcium perchlorate	5.1	UN1455	II	OXIDIZER	
	Calcium permanganate	5.1	UN1456	II	OXIDIZER	
	Calcium peroxide	5.1	UN1457	II	OXIDIZER	
	Calcium phosphide	4.3	UN1360	I	DANGEROUS WHEN WET, POISON	A8, A19, B100, N40
A	Calcium, pyrophoric or Calcium alloys, pyrophoric	4.2	UN1855	I	SPONTANEOUSLY COMBUSTIBLE	
	Calcium resinate	4.1	UN1313	III	FLAMMABLE SOLID	A1, A19
	Calcium resinate, fused	4.1	UN1314	III	FLAMMABLE SOLID	A1, A19
	Calcium selenate, see Selenates or Selenites					
	Calcium silicide	4.3	UN1405	II	DANGEROUS WHEN WET	A19, B105, B106

HAZARDOUS MATERIALS TABLE

Sym.	Descriptions and shipping names	Hazard	ID No.	PG	Label(s)	Special provisions
	Camphor oil	3	UN1130	III	DANGEROUS WHEN WET	A1, A19, B106, B108
	Camphor, *synthetic*	4.1	UN2717	III	FLAMMABLE LIQUID	B1, T1
	Cannon primers, see **Primers, tubular**				FLAMMABLE SOLID	A1
	Caproic acid	8	UN2829	III	CORROSIVE	T1
	Caps, blasting, see **Detonators,** *etc.*					
	Carbamate pesticides, liquid, flammable, toxic, *flash point less than 23 degrees C*	3	UN2758	I	FLAMMABLE LIQUID, POISON	T42
				II	FLAMMABLE LIQUID, POISON	T14
	Carbamate pesticides, liquid, toxic	6.1	UN2992	I	POISON	T42
				II	POISON	T14
				III	KEEP AWAY FROM FOOD	T14
	Carbamate pesticides, liquid, toxic, flammable, *flash point not less than 23 degrees C*	6.1	UN2991	I	POISON, FLAMMABLE LIQUID	T42
				II	POISON, FLAMMABLE LIQUID	T14
				III	KEEP AWAY FROM FOOD, FLAMMABLE LIQUID	B1, T14
	Carbamate pesticides, solid, toxic	6.1	UN2757	I	POISON	
				II	POISON	
				III	KEEP AWAY FROM FOOD	
	Carbolic acid, see **Phenol, solid** *or* **Phenol, molten**					

HAZARDOUS MATERIALS TABLE

Sym.	Descriptions and shipping names	Hazard	ID No.	PG	Label(s)	Special provisions
	Carbolic acid solutions, see **Phenol solutions**					
I	Carbon, activated	4.2	UN1362	III	SPONTANEOUSLY COMBUSTIBLE	
I	Carbon, *animal or vegetable origin*	4.2	UN1361	II	SPONTANEOUSLY COMBUSTIBLE	
	Carbon bisulfide, see **Carbon disulfide**					
	Carbon dioxide	2.2	UN1013		NONFLAMMABLE GAS	
	Carbon dioxide and nitrous oxide mixtures	2.2	UN1015		NONFLAMMABLE GAS	
	Carbon dioxide and oxygen mixtures	2.2	UN1014		NONFLAMMABLE GAS	
	Carbon dioxide, refrigerated liquid	2.2	UN2187		NONFLAMMABLE GAS	
AW	Carbon dioxide, solid or Dry ice	9	UN1845	III	None	B12
	Carbon disulfide	3	UN1131	I	FLAMMABLE LIQUID, POISON	B16, T18, T26, T29
	Carbon monoxide	2.3	UN1016		POISON GAS, FLAMMABLE GAS	4
	Carbon monoxide and hydrogen mixture	2.3	NA9202		POISON GAS, FLAMMABLE GAS	6
D	Carbon monoxide, refrigerated liquid *(cryogenic liquid)*	2.3			POISON GAS, FLAMMABLE GAS	4
	Carbon tetrabromide	6.1	UN2516	III	KEEP AWAY FROM FOOD	
	Carbon tetrachloride	6.1	UN1846	II	POISON	N36, T8
	Carbonyl chloride, see **Phosgene**					

HAZARDOUS MATERIALS TABLE

Sym.	Descriptions and shipping names	Hazard	ID No.	PG	Label(s)	Special provisions
	Carbonyl fluoride	2.3	UN2417		POISON GAS, CORROSIVE	2
	Carbonyl sulfide	2.3	UN2204		POISON GAS, FLAMMABLE GAS	3, 25, B14
	Cartridge cases, empty primed, see Cases, cartridge, empty, with primer					
	Cartridges, actuating, for aircraft ejector seat catapult, fire extinguisher, canopy removal or apparatus, see Cartridges, power device					
	Cartridges, explosive, see Charges, demolition					
	Cartridges, flash	1.1G	UN0049	II	EXPLOSIVE 1.1G	
	Cartridges, flash	1.3G	UN0050	II	EXPLOSIVE 1.3G	
	Cartridges for weapons, blank	1.1C	UN0326	II	EXPLOSIVE 1.1C	
	Cartridges for weapons, blank	1.2C	UN0413	II	EXPLOSIVE 1.2C	
	Cartridges for weapons, blank or Cartridges, small arms, blank	1.4S	UN0014	II	None	112
	Cartridges for weapons, blank or Cartridges, small arms, blank	1.3C	UN0327	II	EXPLOSIVE 1.3C	
	Cartridges for weapons, blank or Cartridges, small arms, blank	1.4C	UN0338	II	EXPLOSIVE 1.4C	
	Cartridges for weapons, inert projectile	1.2C	UN0328	II	EXPLOSIVE 1.2C	
	Cartridges for weapons, inert projectile or Cartridges, small arms	1.4S	UN0012	II	None	112
	Cartridges for weapons, inert projectile or Cartridges, small arms	1.4C	UN0339	II	EXPLOSIVE 1.4C	

HAZARDOUS MATERIALS TABLE

Sym.	Descriptions and shipping names	Hazard	ID No.	PG	Label(s)	Special provisions
	Cartridges for weapons, inert projectile or Cartridges, small arms	1.3C	UN0417	II	EXPLOSIVE 1.3C	
	Cartridges for weapons, *with bursting charge*	1.1F	UN0005	II	EXPLOSIVE 1.1F	
	Cartridges for weapons, *with bursting charge*	1.1E	UN0006	II	EXPLOSIVE 1.1E	
	Cartridges for weapons, *with bursting charge*	1.2F	UN0007	II	EXPLOSIVE 1.2F	
	Cartridges for weapons, *with bursting charge*	1.2E	UN0321	II	EXPLOSIVE 1.2E	
	Cartridges for weapons, *with bursting charge*	1.4F	UN0348	II	EXPLOSIVE 1.4F	
	Cartridges for weapons, *with bursting charge*	1.4E	UN0412	II	EXPLOSIVE 1.4E	
	Cartridges, oil well	1.3C	UN0277	II	EXPLOSIVE 1.3C	
	Cartridges, oil well	1.4C	UN0278	II	EXPLOSIVE 1.4C	
	Cartridges, power device	1.3C	UN0275	II	EXPLOSIVE 1.3C	110
	Cartridges, power device	1.4C	UN0276	II	EXPLOSIVE 1.4C	110, 112
	Cartridges, power device	1.4S	UN0323	II	EXPLOSIVE 1.4S	
	Cartridges, power device	1.2C	UN0381	II	EXPLOSIVE 1.2C	
	Cartridges, safety, blank, see **Cartridges for weapons, blank** *(UN 0014)*					
	Cartridges, safety, see **Cartridges for weapons**, *other than blank* or **Cartridges, power device** *(UN 0323)*					
	Cartridges, signal	1.3G	UN0054	II	EXPLOSIVE 1.3G	
	Cartridges, signal	1.4G	UN0312	II	EXPLOSIVE 1.4G	
	Cartridges, signal	1.4S	UN0405	II	EXPLOSIVE 1.4S	
D	Cartridges, small arms	ORM-D			None	112

HAZARDOUS MATERIALS TABLE

Sym.	Descriptions and shipping names	Hazard	ID No.	PG	Label(s)	Special provisions
	Cartridges, sporting, see Cartridges for weapons, other than blank					
	Cartridges, starter, jet engine, see Cartridges, power device					
	Cases, cartridge, empty with primer	1.4S	UN0055	II	EXPLOSIVE 1.4S	50
	Cases, cartridges, empty with primer	1.4C	UN0379	II	EXPLOSIVE 1.4C	50
	Cases, combustible, empty, without primer	1.4C	UN0446	II	EXPLOSIVE 1.4C	
	Cases, combustible, empty, without primer	1.3C	UN0447	II	EXPLOSIVE 1.3C	
	Casinghead gasoline see Gasoline					
AW	Castor beans or Castor meal or Castor pomace or Castor flake	9	UN2969	II	None	
	Caustic alkali liquids, n.o.s.	8	UN1719	II	CORROSIVE	B2, T14
				III	CORROSIVE	T7
	Caustic potash, see Potassium hydroxide etc.					
	Caustic soda, (etc.) see Sodium hydroxide etc.					
	Cells, containing sodium	4.3	UN3292	II	DANGEROUS WHEN WET	
	Celluloid, in block, rods, rolls, sheets, tubes, etc., except scrap	4.1	UN2000	III	FLAMMABLE SOLID	
	Celluloid, scrap	4.2	UN2002	III	SPONTANEOUSLY COMBUSTIBLE	
	Cement, see Adhesives containing flammable liquid					
	Cerium, slabs, ingots, or rods	4.1	UN1333	II	FLAMMABLE SOLID	N34
	Cerium, turnings or gritty powder	4.3	UN3078	II	DANGEROUS WHEN WET	A1, B106, B109

HAZARDOUS MATERIALS TABLE

Sym.	Descriptions and shipping names	Hazard	ID No.	PG	Label(s)	Special provisions
	Cesium *or* Caesium	4.3	UN1407	I	DANGEROUS WHEN WET	A19, B100, N34, N40
	Cesium nitrate *or* Caesium nitrate	5.1	UN1451	III	OXIDIZER	A1, A29
D	Charcoal *briquettes, shell, screenings, wood, etc.*	4.2	NA1361	III	SPONTANEOUSLY COMBUSTIBLE	
	Charges, bursting, plastics bonded	1.1D	UN0457	II	EXPLOSIVE 1.1D	
	Charges, bursting, plastics bonded	1.2D	UN0458	II	EXPLOSIVE 1.2D	
	Charges, bursting, plastics bonded	1.4D	UN0459	II	EXPLOSIVE 1.4D	
	Charges, bursting, plastics bonded	1.4S	UN0460	II	EXPLOSIVE 1.4S	
	Charges, demolition	1.1D	UN0048	II	EXPLOSIVE 1.1D	
	Charges, depth	1.1D	UN0056	II	EXPLOSIVE 1.1D	
	Charges, expelling, explosive, for fire extinguishers, see **Cartridges, power device**					
	Charges, explosive, commercial without detonator	1.1D	UN0442	II	EXPLOSIVE 1.1D	
	Charges, explosive, commercial without detonator	1.2D	UN0443	II	EXPLOSIVE 1.2D	
	Charges, explosive, commercial without detonator	1.4D	UN0444	II	EXPLOSIVE 1.4D	
	Charges, explosive, commercial without detonator	1.4S	UN0445	II	EXPLOSIVE 1.4S	
	Charges, propelling	1.1C	UN0271	II	EXPLOSIVE 1.1C	
	Charges, propelling	1.3C	UN0272	II	EXPLOSIVE 1.3C	
	Charges, propelling	1.2C	UN0415	II	EXPLOSIVE 1.2C	
	Charges, propelling	1.4C	UN0491	II	EXPLOSIVE 1.4C	

HAZARDOUS MATERIALS TABLE

Sym.	Descriptions and shipping names	Hazard	ID No.	PG	Label(s)	Special provisions
	Charges, propelling, for cannon	1.3C	UN0242	II	EXPLOSIVE 1.3C	
	Charges, propelling, for cannon	1.1C	UN0279	II	EXPLOSIVE 1.1C	
	Charges, propelling, for cannon	1.2C	UN0414	II	EXPLOSIVE 1.2C	
	Charges, shaped, commercial, *without detonator*	1.1D	UN0059	II	EXPLOSIVE 1.1D	
	Charges, shaped, commercial *without detonator*	1.2D	UN0439	II	EXPLOSIVE 1.2D	
	Charges, shaped, commercial *without detonator*	1.4D	UN0440	II	EXPLOSIVE 1.4D	
	Charges, shaped, commercial *without detonator*	1.4S	UN0441	II	EXPLOSIVE 1.4S	
	Charges, shaped, flexible, linear	1.4D	UN0237	II	EXPLOSIVE 1.4D	
	Charges, shaped, flexible, linear	1.1D	UN0288	II	EXPLOSIVE 1.1D	
	Charges, supplementary explosive	1.1D	UN0060	II	EXPLOSIVE 1.1D	101
	Chemical kit	8	NA1760	II	CORROSIVE	
D	Chemical kits *(must be classified and labelled according to the hazard class of the constituent(s) and must meet the requirements of special provision 15 in 172.102(c)(1))*					
	Chloral, anhydrous, inhibited	6.1	UN2075	II	POISON	B101, T14
	Chlorate and borate mixtures	5.1	UN1458	II	OXIDIZER	A9, N34
				III	OXIDIZER	A9, N34
	Chlorate and magnesium chloride mixtures	5.1	UN1459	II	OXIDIZER	A9, N34, T8
				III	OXIDIZER	A9, N34, T8

HAZARDOUS MATERIALS TABLE

Sym.	Descriptions and shipping names	Hazard	ID No.	PG	Label(s)	Special provisions
	Chlorate of potash, see **Potassium chlorate**					
	Chlorate of soda, see **Sodium chlorate**					
	Chlorates, Inorganic, aqueous solution, n.o.s.	5.1	UN3210	II	OXIDIZER	T8
	Chlorates, Inorganic, n.o.s.	5.1	UN1461	II	OXIDIZER	A9, N34
	Chloric acid aqueous solution, with not more than 10 percent chloric acid	5.1	UN2626	II	OXIDIZER	T25
	Chloride of phosphorus, see **Phosphorus trichloride**					
	Chloride of sulfur, see **Sulfur chloride**					
	Chlorinated lime, see **Calcium hypochlorite mixtures**, etc.					
	Chlorine	2.3	UN1017		POISON GAS, CORROSIVE	2, B9, B14
	Chlorine azide	Forbidden				
D	Chlorine dioxide, hydrate, frozen	5.1	NA9191	II	OXIDIZER, POISON	
	Chlorine dioxide (not hydrate)	Forbidden				
	Chlorine pentafluoride	2.3	UN2548		POISON GAS, OXIDIZER, CORROSIVE	1, B7, B9, B14
	Chlorine trifluoride	2.3	UN1749		POISON GAS, OXIDIZER, CORROSIVE	2, 25, B7, B9, B14
	Chlorite solution with more than 5 percent but less than 16 percent available chlorine	8	UN1908	III	CORROSIVE	A3, A6, A7, B2, N34, T8

HAZARDOUS MATERIALS TABLE

Sym.	Descriptions and shipping names	Hazard	ID No.	PG	Label(s)	Special provisions
	Chlorite solution *with not less than 16 percent available chlorine*	8	UN1908	II	CORROSIVE	A3, A6, A7, B2, N34, T8
	Chlorites, inorganic, n.o.s.	5.1	UN1462	II	OXIDIZER	A7, N34
	1-Chloro-3-bromopropane	6.1	UN2688	III	KEEP AWAY FROM FOOD	T2
	1-*Chloro-1,1-difluoroethane, see* **Chlorodifluoroethanes**					
	1-Chloro-1,1-difluoroethanes, *R142b*	2.1	UN2517		FLAMMABLE GAS	
	3-Chloro-4-methylphenyl isocyanate	6.1	UN2236	II	POISON	
	1-Chloro-1,2,2,2-tetrafluoroethane, *R124*	2.2	UN1021		NONFLAMMABLE GAS	
	4-Chloro-o-toluidine hydrochloride	6.1	UN1579	III	KEEP AWAY FROM FOOD	
	1-Chloro-2,2,2-trifluoroethane, *R133a*	2.2	UN1983		NONFLAMMABLE GAS	
	Chloroacetic acid, molten	6.1	UN3250	II	POISON, CORROSIVE	T9
	Chloroacetic acid, solid	6.1	UN1751	II	POISON, CORROSIVE	A3, A7, N34
	Chloroacetic acid, solution	6.1	UN1750	II	POISON, CORROSIVE	A7, N34, T8, T27
	Chloroacetone, stabilized	6.1	UN1695	II	POISON	2, B9, B14, B32, B74, N12, N32, N34, T38, T43, T45
	Chloroacetone (unstabilized)	Forbidden				
+	Chloroacetonitrile	6.1	UN2668	II	POISON, FLAMMABLE LIQUID	2, B9, B14, B32, B74, T38, T43, T45
+	Chloroacetophenone (CN), *liquid*	6.1	UN1697	II	POISON	A3, N12, N32, N33
	Chloroacetophenone (CN), *solid*	6.1	UN1697	II	POISON	A3, N12, N32, N33, N34

HAZARDOUS MATERIALS TABLE

Sym.	Descriptions and shipping names	Hazard	ID No.	PG	Label(s)	Special provisions
	Chloroacetyl chloride	6.1	UN1752	I	POISON, CORROSIVE	2, A3, A6, A7, B3, B8, B9, B14, B32, B74, B77, N34, N43, T38, T43, T45
	Chloroanilines, liquid	6.1	UN2019	II	POISON	T14
	Chloroanilines, solid	6.1	UN2018	II	POISON	T14, T38
	Chloroanisidines	6.1	UN2233	III	KEEP AWAY FROM FOOD	
	Chlorobenzene	3	UN1134	III	FLAMMABLE LIQUID	B1, T1
	Chlorobenzol, see **Chlorobenzene**					
	Chlorobenzotrifluorides	3	UN2234	III	FLAMMABLE LIQUID	B1, T1
	Chlorobenzyl chlorides	6.1	UN2235	III	KEEP AWAY FROM FOOD	T8
	Chlorobutanes	3	UN1127	II	FLAMMABLE LIQUID	B101, T8
	Chlorocresols, *liquid*	6.1	UN2669	II	POISON	T8
	Chlorocresols, *solid*	6.1	UN2669	II	POISON	
	Chlorodifluorobromomethane, *R12B1*	2.2	UN1974		NONFLAMMABLE GAS	
	Chlorodifluoromethane and chloropenta-fluoroethane mixture with fixed boiling point, with approximately 49 percent chloro-difluoromethane, *R502*	2.2	UN1973		NONFLAMMABLE GAS	
	Chlorodifluoromethane, *R22*	2.2	UN1018		NONFLAMMABLE GAS	T14
	Chlorodinitrobenzenes	6.1	UN1577	II	POISON	2, B9, B14, B32, B74, T38, T43, T45
	2-Chloroethanal	6.1	UN2232	I	POISON	N36, T14
	Chloroform	6.1	UN1888	III	KEEP AWAY FROM FOOD	

HAZARDOUS MATERIALS TABLE

Sym.	Descriptions and shipping names	Hazard	ID No.	PG	Label(s)	Special provisions
	Chloroformates, toxic, corrosive, flammable, n.o.s.	6.1	UN2742	II	POISON, CORROSIVE, FLAMMABLE LIQUID	5
	Chloroformates, toxic, corrosive, n.o.s.	6.1	UN3277	II	POISON, CORROSIVE	T12, T26
	Chloromethyl chloroformate	6.1	UN2745	II	POISON, CORROSIVE	T18
	Chloromethyl ethyl ether	3	UN2354	II	FLAMMABLE LIQUID, POISON	T8
	Chloronitroanilines	6.1	UN2237	III	KEEP AWAY FROM FOOD	
+	Chloronitrobenzene, ortho, liquid	6.1	UN1578	II	POISON	T14
+	Chloronitrobenzenes meta or para, solid	6.1	UN1578	II	POISON	T14
	Chloronitrotoluenes liquid	6.1	UN2433	III	KEEP AWAY FROM FOOD	
	Chloronitrotoluenes, solid	6.1	UN2433	III	KEEP AWAY FROM FOOD	
	Chloropentafluoroethane, R115	2.2	UN1020		NONFLAMMABLE GAS	
	Chlorophenolates, liquid or Phenolates, liquid	8	UN2904	III	CORROSIVE	
	Chlorophenolates, solid or Phenolates, solid	8	UN2905	III	CORROSIVE	T7
	Chlorophenols, liquid	6.1	UN2021	III	KEEP AWAY FROM FOOD	T7
	Chlorophenols, solid	6.1	UN2020	III	KEEP AWAY FROM FOOD	
	Chlorophenyltrichlorosilane	8	UN1753	II	CORROSIVE	A7, B2, B6, N34, T8, T26
+	Chloropicrin	6.1	UN1580	I	POISON	2, B7, B9, B14, B32, B46, B74, T38, T43, T45
	Chloropicrin and methyl bromide mixtures	2.3	UN1581		POISON GAS	2, B9, B14
	Chloropicrin and methyl chloride mixtures	2.3	UN1582		POISON GAS	2

HAZARDOUS MATERIALS TABLE

Sym.	Descriptions and shipping names	Hazard	ID No.	PG	Label(s)	Special provisions
	Chloropicrin mixture, flammable (pressure not exceeding 14.7 psia at 115 degrees F flash point below 100 degrees F) see Toxic liquids, flammable, etc.					
	Chloropicrin mixtures, n.o.s.	6.1	UN1583	I	POISON	
				II	POISON	5
D	Chloropivaloyl chloride	6.1	NA9263	III	KEEP AWAY FROM FOOD	
				I	POISON, CORROSIVE	2, B9, B14, B32, B74, T38, T43, T45
	Chloroplatinic acid, solid	8	UN2507	III	CORROSIVE	B57, T15
	Chloroprene, Inhibited	3	UN1991	I	FLAMMABLE LIQUID, POISON	
	Chloroprene, uninhibited	Forbidden				
	2-Chloropropane	3	UN2356	I	FLAMMABLE LIQUID	N36, T14
	3-Chloropropanol-1	6.1	UN2849	III	KEEP AWAY FROM FOOD	T8
	2-Chloropropene	3	UN2456	I	FLAMMABLE LIQUID	A3, N36, T20
	2-Chloropropionic acid	8	UN2511	III	CORROSIVE	T8
	2-Chloropyridine	6.1	UN2822	II	POISON	T14
	Chlorosilanes, corrosive, flammable, n.o.s.	8	UN2986	II	CORROSIVE, FLAMMABLE LIQUID	B100
	Chlorosilanes, corrosive, n.o.s.	8	UN2987	II	CORROSIVE	B2
	Chlorosilanes, flammable, corrosive, n.o.s.	3	UN2985	II	FLAMMABLE LIQUID, CORROSIVE	B100, T18, T26

HAZARDOUS MATERIALS TABLE

Sym.	Descriptions and shipping names	Hazard	ID No.	PG	Label(s)	Special provisions
	Chlorosilanes, water-reactive, flammable, corrosive, n.o.s.	4.3	UN2988	I	DANGEROUS WHEN WET, FLAMMABLE LIQUID, CORROSIVE	A2
+	Chlorosulfonic acid (with or without sulfur trioxide)	8	UN1754	I	CORROSIVE, POISON	2, A3, A6, A10, B9, B10, B14, B32, B74, T38, T43, T45
	Chlorotoluenes	3	UN2238	III	FLAMMABLE LIQUID	B1, T1
	Chlorotoluidines *liquid*	6.1	UN2239	III	KEEP AWAY FROM FOOD	
	Chlorotoluidines *solid*	6.1	UN2239	III	KEEP AWAY FROM FOOD	T7
	Chlorotrifluoromethane and trifluoromethane azeotropic mixture with approximately 60 percent chlorotrifluoromethane, R503	2.2	UN2599		NONFLAMMABLE GAS	
	Chlorotrifluoromethane, R13	2.2	UN1022		NONFLAMMABLE GAS	
D	Chromic acid, solid	5.1	NA1463	II	OXIDIZER, CORROSIVE	
	Chromic acid solution	8	UN1755	II	CORROSIVE	B2, T9, T27
				III	CORROSIVE	T8, T26
	Chromic anhydride, see **Chromium trioxide, anhydrous**					
	Chromic fluoride, solid	8	UN1756	II	CORROSIVE	B2, T8
	Chromic fluoride, solution	8	UN1757	II	CORROSIVE	T7
				III	CORROSIVE	
	Chromium nitrate	5.1	UN2720	III	OXIDIZER	A1, A29
	Chromium oxychloride	8	UN1758	I	CORROSIVE	A3, A6, A7, B10, N34, T12, T26
	Chromium trioxide, anhydrous	5.1	UN1463	II	OXIDIZER, CORROSIVE	B106

HAZARDOUS MATERIALS TABLE

Sym.	Descriptions and shipping names	Hazard	ID No.	PG	Label(s)	Special provisions
	Chromosulfuric acid	8	UN2240	I	CORROSIVE	A3, A6, A7, B4, B6, N34, T12, T27
	Chromyl chloride, see **Chromium oxychloride**					
	Cigar and cigarette lighters, charged with fuel, see **Lighters for cigars, cigarettes**, etc.					
	Coal briquettes, hot	Forbidden				
	Coal gas	2.3	UN1023		POISON GAS, FLAMMABLE GAS	3
	Coal tar distillates, flammable	3	UN1136	II	FLAMMABLE LIQUID	T8, T31
				III	FLAMMABLE LIQUID	B1, T7, T30
	Coal tar dye, corrosive, liquid, n.o.s, see **Dyes, liquid or solid, n.o.s.** or **Dye Intermediates, liquid or solid, n.o.s.**, *corrosive*					
	Coating solution	3	UN1139	II	FLAMMABLE LIQUID	T7, T30
				III	FLAMMABLE LIQUID	B1, T7, T30
	Cobalt naphthenates, powder	4.1	UN2001	III	FLAMMABLE SOLID	A19
	Cobalt resinate, precipitated	4.1	UN1318	III	FLAMMABLE SOLID	A1, A19
	Coke, hot	Forbidden				
	Collodion, see **Nitrocellulose** *etc.*					
D	**Combustible liquid, n.o.s.**	Combustible liquid	NA1993	III	None	T1
	Components, explosive train, n.o.s.	1.2B	UN0382	II	EXPLOSIVE 1.2B	101

HAZARDOUS MATERIALS TABLE

Sym.	Descriptions and shipping names	Hazard	ID No.	PG	Label(s)	Special provisions
	Components, explosive train, n.o.s.	1.4B	UN0383	II	EXPLOSIVE 1.4B	101
	Components, explosive train, n.o.s.	1.4S	UN0384	II	EXPLOSIVE 1.4S	101
	Components, explosive train, n.o.s.	1.1B	UN0461	II	EXPLOSIVE 1.1B	101
	Composition B, see Hexolite, etc.					
D	**Compounds, cleaning liquid**	8	NA1760	I	CORROSIVE	A7, B10, T42
				II	CORROSIVE	B2, N37, T14
				III	CORROSIVE	N37, T7
D	**Compounds, cleaning liquid**	3	NA1993	I	FLAMMABLE LIQUID	T42
				II	FLAMMABLE LIQUID	T8, T31
				III	FLAMMABLE LIQUID	B1, B52, T7, T30
D	**Compounds, tree killing, liquid** *or* **Compounds, weed killing, liquid**	8	NA1760	I	CORROSIVE	A7, B10, T42
				II	CORROSIVE	B2, N37, T14
				III	CORROSIVE	N37, T7
D	**Compounds, tree killing, liquid** *or* **Compounds, weed killing, liquid**	3	NA1993	I	FLAMMABLE LIQUID	T42
				II	FLAMMABLE LIQUID	T8, T31
				III	FLAMMABLE LIQUID	B1, B52, T7, T30
D	**Compounds, tree killing, liquid** *or* **Compounds, weed killing, liquid**	6.1	NA2810	I	POISON	
				II	POISON	
				III	KEEP AWAY FROM FOOD	
	Compressed gas, oxidizing, n.o.s.	2.2	UN3156		NONFLAMMABLE GAS, OXIDIZER	
	Compressed gases, flammable, n.o.s.	2.1	UN1954		FLAMMABLE GAS	

HAZARDOUS MATERIALS TABLE

Sym.	Descriptions and shipping names	Hazard	ID No.	PG	Label(s)	Special provisions
	Compressed gases, n.o.s.	2.2	UN1956		NONFLAMMABLE GAS	
	Compressed gases, toxic, flammable, n.o.s. *Inhalation hazard Zone A*	2.3	UN1953		POISON GAS, FLAMMABLE GAS	1
	Compressed gases, toxic, flammable, n.o.s. *Inhalation hazard Zone B*	2.3	UN1953		POISON GAS, FLAMMABLE GAS	2, B9, B14
	Compressed gases, toxic, flammable, n.o.s. *Inhalation Hazard Zone C*	2.3	UN1953		POISON GAS, FLAMMABLE GAS	3, B14
	Compressed gases, toxic, flammable, n.o.s. *Inhalation Hazard Zone D*	2.3	UN1953		POISON GAS, FLAMMABLE GAS	4
	Compressed gases, toxic, n.o.s. *Inhalation Hazard Zone A*	2.3	UN1955		POISON GAS	1
	Compressed gases, toxic, n.o.s. *Inhalation Hazard Zone B*	2.3	UN1955		POISON GAS	2, B9, B14
	Compressed gases, toxic, n.o.s. *Inhalation Hazard Zone C*	2.3	UN1955		POISON GAS	3, B14
	Compressed gases, toxic, n.o.s. *Inhalation Hazard Zone D*	2.3	UN1955		POISON GAS	4
D	**Consumer commodity**	ORM-D			None	
	Contrivances, water-activated, *with burster, expelling charge or propelling charge*	1.2L	UN0248	II	EXPLOSIVE 1.2L	101
	Contrivances, water-activated, *with burster, expelling charge or propelling charge*	1.3L	UN0249	II	EXPLOSIVE 1.3L	101
	Copper acetoarsenite	6.1	UN1585	II	POISON	
	Copper acetylide	Forbidden				
	Copper amine azide	Forbidden				

HAZARDOUS MATERIALS TABLE

Sym.	Descriptions and shipping names	Hazard	ID No.	PG	Label(s)	Special provisions
	Copper arsenite	6.1	UN1586	II	POISON	
	Copper based pesticides, liquid, flammable, toxic, *flash point less than 23 degrees C*	3	UN2776	I	FLAMMABLE LIQUID, POISON	
				II	FLAMMABLE LIQUID, POISON	
	Copper based pesticides, liquid, toxic	6.1	UN3010	I	POISON	T42
				II	POISON	T14
				III	KEEP AWAY FROM FOOD	T14
	Copper based pesticides, liquid, toxic, flammable *flashpoint not less than 23 degrees C*	6.1	UN3009	I	POISON, FLAMMABLE LIQUID	T42
				II	POISON, FLAMMABLE LIQUID	T14
				III	KEEP AWAY FROM FOOD, FLAMMABLE LIQUID	B1, T14
	Copper based pesticides, solid, toxic	6.1	UN2775	I	POISON	
				II	POISON	
				III	KEEP AWAY FROM FOOD	
	Copper chlorate	5.1	UN2721	II	OXIDIZER	A1
	Copper chloride	8	UN2802	III	CORROSIVE	
	Copper cyanide	6.1	UN1587	II	POISON	
	Copper selenate, see **Selenates** or **Selenites**					
	Copper selenite, see **Selenates** or **Selenites**					
	Copper tetramine nitrate	Forbidden				

HAZARDOUS MATERIALS TABLE

Sym.	Descriptions and shipping names	Hazard	ID No.	PG	Label(s)	Special provisions
AW	**Copra**	4.2	UN1363	III	SPONTANEOUSLY COMBUSTIBLE	
	Cord, detonating, *flexible*	1.1D	UN0065	II	EXPLOSIVE 1.1D	102
	Cord, detonating, *flexible*	1.4D	UN0289	II	EXPLOSIVE 1.4D	
	Cord detonating or Fuse detonating *metal clad*	1.2D	UN0102	II	EXPLOSIVE 1.2D	
	Cord, detonating or Fuse, detonating *metal clad*	1.1D	UN0290	II	EXPLOSIVE 1.1D	
	Cord, detonating, mild effect or Fuse, detonating, mild effect *metal clad*	1.4D	UN0104	II	EXPLOSIVE 1.4D	
	Cord, igniter	1.4G	UN0066	II	EXPLOSIVE 1.4G	
	Cordeau detonant fuse, see **Cord, detonating, etc.;** **Cord, detonating,** *flexible*					
	Cordite, see **Powder, smokeless**					
	Corrosive liquid, acidic, Inorganic, n.o.s.	8	UN3264	I	CORROSIVE	B10
				II	CORROSIVE	B2, T14
				III	CORROSIVE	T7
	Corrosive liquid, acidic, organic, n.o.s.	8	UN3265	I	CORROSIVE	B10
				II	CORROSIVE	B2, T14
				III	CORROSIVE	T7
	Corrosive liquid, basic, Inorganic, n.o.s.	8	UN3266	I	CORROSIVE	B10
				II	CORROSIVE	B2, T14
				III	CORROSIVE	T7

HAZARDOUS MATERIALS TABLE

Sym.	Descriptions and shipping names	Hazard	ID No.	PG	Label(s)	Special provisions
	Corrosive liquid, basic, organic, n.o.s.	8	UN3267	I	CORROSIVE	B10
				II	CORROSIVE	B2, T14
				III	CORROSIVE	T7
	Corrosive liquid, self-heating, n.o.s.	8	UN3301	I	CORROSIVE, SPONTANEOUSLY COMBUSTIBLE	B10
				II	CORROSIVE, SPONTANEOUSLY COMBUSTIBLE	B2
	Corrosive liquids, flammable, n.o.s.	8	UN2920	I	CORROSIVE, FLAMMABLE LIQUID	B10, T42
				II	CORROSIVE, FLAMMABLE LIQUID	B2, T15, T26
	Corrosive liquids, n.o.s.	8	UN1760	I	CORROSIVE	A7, B10, T42
				II	CORROSIVE	B2, T14
				III	CORROSIVE	T7
	Corrosive liquids, oxidizing, n.o.s.	8	UN3093	I	CORROSIVE, OXIDIZER	A7, B10
				II	CORROSIVE, OXIDIZER	B3
	Corrosive liquids, toxic, n.o.s.	8	UN2922	I	CORROSIVE, POISON	
				II	CORROSIVE, POISON	
				III	CORROSIVE, KEEP AWAY FROM FOOD	
	Corrosive liquids, water-reactive, n.o.s.	8	UN3094	I	CORROSIVE, DANGEROUS WHEN WET	
				II	CORROSIVE, DANGEROUS WHEN WET	

Sym.	Descriptions and shipping names	Hazard	ID No.	PG	Label(s)	Special provisions
	Corrosive solid, acidic, inorganic, n.o.s.	8	UN3260	I	CORROSIVE	
				II	CORROSIVE	
				III	CORROSIVE	
	Corrosive solid, acidic, organic, n.o.s.	8	UN3261	I	CORROSIVE	
				II	CORROSIVE	
				III	CORROSIVE	
	Corrosive solid, basic, inorganic, n.o.s.	8	UN3262	I	CORROSIVE	
				II	CORROSIVE	
				III	CORROSIVE	
	Corrosive solid, basic, organic, n.o.s.	8	UN3263	I	CORROSIVE	
				II	CORROSIVE	
				III	CORROSIVE	
	Corrosive solids, flammable, n.o.s.	8	UN2921	I	CORROSIVE, FLAMMABLE SOLID	B106
				II	CORROSIVE, FLAMMABLE SOLID	
	Corrosive solids, n.o.s.	8	UN1759	I	CORROSIVE	
				II	CORROSIVE	
				III	CORROSIVE	
	Corrosive solids, oxidizing, n.o.s.	8	UN3084	I	CORROSIVE, OXIDIZER	B100
				II	CORROSIVE, OXIDIZER	B100

HAZARDOUS MATERIALS TABLE

Sym.	Descriptions and shipping names	Hazard	ID No.	PG	Label(s)	Special provisions
	Corrosive solids, self-heating, n.o.s.	8	UN3095	II	CORROSIVE, SPONTANEOUSLY COMBUSTIBLE	B100
				II	CORROSIVE, SPONTANEOUSLY COMBUSTIBLE	
	Corrosive solids, toxic, n.o.s.	8	UN2923	I	CORROSIVE, POISON	
				II	CORROSIVE, POISON	
				III	CORROSIVE, KEEP AWAY FROM FOOD	
	Corrosive solids, water-reactive, n.o.s.	8	UN3096	I	CORROSIVE, DANGEROUS WHEN WET	B105
				II	CORROSIVE, DANGEROUS WHEN WET	B105
DW	Cotton	9	NA1365	III	CLASS 9	W41
AIW	Cotton waste, oily	4.2	UN1364	III	SPONTANEOUSLY COMBUSTIBLE	N9
AIW	Cotton, wet	4.2	UN1365	III	SPONTANEOUSLY COMBUSTIBLE	
	Coumarin derivative pesticides, liquid, flammable, toxic, *flashpoint less than 23 degrees C*	3	UN3024	I	FLAMMABLE LIQUID, POISON	
				II	FLAMMABLE LIQUID, POISON	
	Coumarin derivative pesticides, liquid, toxic	6.1	UN3026	I	POISON	
				II	POISON	
				III	KEEP AWAY FROM FOOD	

HAZARDOUS MATERIALS TABLE

Sym.	Descriptions and shipping names	Hazard	ID No.	PG	Label(s)	Special provisions
	Coumarin derivative pesticides, liquid, toxic, flammable *flashpoint not less than 23 degrees C*	6.1	UN3025	I	POISON, FLAMMABLE LIQUID	
				II	POISON, FLAMMABLE LIQUID	
				III	KEEP AWAY FROM FOOD, FLAMMABLE LIQUID	B1
	Coumarin derivative pesticides, solid, toxic	6.1	UN3027	I	POISON	
				II	POISON	
				III	KEEP AWAY FROM FOOD	
	Cresols	6.1	UN2076	II	POISON, CORROSIVE	B110, T8
	Cresylic acid	6.1	UN2022	II	POISON, CORROSIVE	B110, T8
	Crotonaldehyde, stabilized	6.1	UN1143	I	POISON, FLAMMABLE LIQUID	2, B9, B14, B32, B74, B77, T38, T43, T45
	Crotonic acid *liquid*	8	UN2823	III	CORROSIVE	
	Crotonic acid, *solid*	8	UN2823	III	CORROSIVE	
	Crotonylene	3	UN1144	I	FLAMMABLE LIQUID	T20
	Cupriethylenediamine solution	8	UN1761	II	CORROSIVE, POISON	T8, T26
				III	CORROSIVE, KEEP AWAY FROM FOOD	T7
	Cutters, cable, explosive	1.4S	UN0070	II	EXPLOSIVE 1.4S	
	Cyanide or cyanide mixtures, dry, see Cyanides, inorganic, solid, n.o.s.					

HAZARDOUS MATERIALS TABLE

Sym.	Descriptions and shipping names	Hazard	ID No.	PG	Label(s)	Special provisions
	Cyanide solutions, n.o.s.	6.1	UN1935	I	POISON	B37, T18, T26
				II	POISON	T18, T26
				III	KEEP AWAY FROM FOOD	T18, T26
	Cyanides, inorganic, solid, n.o.s.	6.1	UN1588	I	POISON	N74, N75
				II	POISON	N74, N75
				III	KEEP AWAY FROM FOOD	N74, N75
	Cyanogen bromide	6.1	UN1889	I	POISON, CORROSIVE	A6, A8
	Cyanogen chloride, inhibited	2.3	UN1589		POISON GAS, CORROSIVE	1
	Cyanogen, liquefied	2.3	UN1026		POISON GAS, FLAMMABLE GAS	2
	Cyanuric chloride	8	UN2670	II	CORROSIVE	
	Cyanuric triazide	Forbidden				
	Cyclobutane	2.1	UN2601		FLAMMABLE GAS	
	Cyclobutyl chloroformate	6.1	UN2744	II	POISON, CORROSIVE	T18
	1,5,9-Cyclododecatriene	6.1	UN2518	III	KEEP AWAY FROM FOOD	T7
	Cycloheptane	3	UN2241	II	FLAMMABLE LIQUID	T1
	Cycloheptatriene	3	UN2603	II	FLAMMABLE LIQUID, POISON	T14
	Cycloheptene	3	UN2242	II	FLAMMABLE LIQUID	B1, T7
	Cyclohexane	3	UN1145	II	FLAMMABLE LIQUID	B101, T8
	Cyclohexanone	3	UN1915	III	FLAMMABLE LIQUID	B1, T1

HAZARDOUS MATERIALS TABLE

Sym.	Descriptions and shipping names	Hazard	ID No.	PG	Label(s)	Special provisions
	Cyclohexene	3	UN2256	II	FLAMMABLE LIQUID	B101, T7
	Cyclohexenyltrichlorosilane	8	UN1762	II	CORROSIVE	A7, B2, N34, T8, T26
	Cyclohexyl acetate	3	UN2243	III	FLAMMABLE LIQUID	B1, T1
+	Cyclohexyl isocyanate	6.1	UN2488	II	POISON	2, B9, B14, B32, B74, B77, T38, T43, T45
	Cyclohexyl mercaptan	3	UN3054	III	FLAMMABLE LIQUID	B1, T1
	Cyclohexylamine	8	UN2357	II	CORROSIVE, FLAMMABLE LIQUID	B101, T8, T26
	Cyclohexyltrichlorosilane	8	UN1763	II	CORROSIVE	A7, B2, N34, T8, T26
	Cyclonite and cyclotetramethylenetetranitramine mixtures, wetted or desensitized see RDX and HMX mixtures, wetted or desensitized etc.					
	Cyclonite and HMX mixtures, wetted or desensitized see RDX and HMX mixtures, wetted or desensitized etc.					
	Cyclonite and octogen mixtures, wetted or desensitized see RDX and HMX mixtures, wetted or desensitized etc.					
	Cyclonite, see Cyclotrimethylenetrinitramine, etc.					
	Cyclooctadiene phosphines, see 9-Phosphabicyclononanes					
	Cyclooctadienes	3	UN2520	III	FLAMMABLE LIQUID	B1, T1
	Cyclooctatetraene	3	UN2358	II	FLAMMABLE LIQUID	T8
	Cyclopentane	3	UN1146	II	FLAMMABLE LIQUID	B101, T14

HAZARDOUS MATERIALS TABLE

Sym.	Descriptions and shipping names	Hazard	ID No.	PG	Label(s)	Special provisions
	Cyclopentane, methyl, see Methylcyclopentane					
	Cyclopentanol	3	UN2244	III	FLAMMABLE LIQUID	B1, T1
	Cyclopentanone	3	UN2245	III	FLAMMABLE LIQUID	B1, T1
	Cyclopentene	3	UN2246	II	FLAMMABLE LIQUID	B101, T13
	Cyclopropane, liquefied	2.1	UN1027		FLAMMABLE GAS	
	Cyclotetramethylene tetranitramine (dry or unphlegmatized) (HMX)	Forbidden				
	Cyclotetramethylenetetranitramine, desensitized or Octogen, desensitized or HMX, desensitized	1.1D	UN0484	II	EXPLOSIVE 1.1D	
	Cyclotetramethylenetetranitramine, wetted or HMX, wetted or Octogen, wetted with *not less than 15 percent water, by mass*	1.1D	UN0226	II	EXPLOSIVE 1.1D	
	Cyclotrimethylenenitramine and octogen, mixtures, wetted *or* desensitized *see* RDX and HMX mixtures, wetted *or* desensitized *etc.*					
	Cyclotrimethylenetrinitramine and cyclotetramethylenetetramine mixtures, wetted *or* desensitized *see* RDX and HMX mixtures, wetted *or* desensitized *etc.*					
	Cyclotrimethylenetrinitramine and HMX mixtures, wetted *or* desensitized *see* RDX and HMX mixtures, wetted *or* desensitized *etc.*					

HAZARDOUS MATERIALS TABLE

Sym.	Descriptions and shipping names	Hazard	ID No.	PG	Label(s)	Special provisions
	Cyclotrimethylenetrinitramine, desensitized or Cyclonite, desensitized or Hexogen, desensitized or RDX, desensitized	1.1D	UN0483	II	EXPLOSIVE 1.1D	
	Cyclotrimethylenetrinitramine, wetted or Cyclonite, wetted or Hexogen, wetted or RDX, wetted with not less than 15 percent water by mass	1.1D	UN0072	II	EXPLOSIVE 1.1D	
	Cymenes	3	UN2046	III	FLAMMABLE LIQUID	B1, T1
	Decaborane	4.1	UN1868	II	FLAMMABLE SOLID, POISON	A19, A20
	Decahydronaphthalene	3	UN1147	III	FLAMMABLE LIQUID	B1, T1
	n-Decane	3	UN2247	III	FLAMMABLE LIQUID	B1, T1
	Deflagrating metal salts of aromatic nitroderivatives, n.o.s.	1.3C	UN0132	II	EXPLOSIVE 1.3C	
	Delay electric igniter, see Igniters					
D	Denatured alcohol	3	NA1986	I	FLAMMABLE LIQUID, POISON	T8, T31
				II	FLAMMABLE LIQUID, POISON	T8, T31
				III	FLAMMABLE LIQUID, KEEP AWAY FROM FOOD	B1, T8, T31
D	Denatured alcohol	3	NA1987	II	FLAMMABLE LIQUID	T8, T31
				III	FLAMMABLE LIQUID	B1, T7, T30
	Depth charges, see Charges, depth					
	Detonating relays, see Detonators, etc.					

HAZARDOUS MATERIALS TABLE

Sym.	Descriptions and shipping names	Hazard	ID No.	PG	Label(s)	Special provisions
	Detonator assemblies, non-electric for blasting	1.1B	UN0360	II	EXPLOSIVE 1.1B	
	Detonator assemblies, non-electric, for blasting	1.4B	UN0361	II	EXPLOSIVE 1.4B	103
	Detonators, electric, for blasting	1.1B	UN0030	II	EXPLOSIVE 1.1B	
	Detonators, electric, for blasting	1.4B	UN0255	II	EXPLOSIVE 1.4B	103
	Detonators, electric, for blasting	1.4S	UN0456	II	EXPLOSIVE 1.4S	104
	Detonators for ammunition	1.1B	UN0073	II	EXPLOSIVE 1.1B	
	Detonators for ammunition	1.2B	UN0364	II	EXPLOSIVE 1.2B	
	Detonators for ammunition	1.4B	UN0365	II	EXPLOSIVE 1.4B	103
	Detonators for ammunition	1.4S	UN0366	II	EXPLOSIVE 1.4S	104
	Detonators, non-electric, for blasting	1.1B	UN0029	II	EXPLOSIVE 1.1B	
	Detonators, non-electric, for blasting	1.4B	UN0267	II	EXPLOSIVE 1.4B	103
	Detonators, non-electric, for blasting	1.4S	UN0455	II	EXPLOSIVE 1.4S	104
	Deuterium	2.1	UN1957		FLAMMABLE GAS	
	Devices, small, hydrocarbon gas powered or Hydrocarbon gas refills for small devices with release device	2.1	UN3150		FLAMMABLE GAS	
	Di-n-amylamine	3	UN2841	III	FLAMMABLE LIQUID, KEEP AWAY FROM FOOD	B1, T8
	Di-n-butyl peroxydicarbonate, with more than 52 percent in solution	Forbidden				
	Di-n-butylamine	8	UN2248	II	CORROSIVE, FLAMMABLE LIQUID	T8

HAZARDOUS MATERIALS TABLE

Sym.	Descriptions and shipping names	Hazard	ID No.	PG	Label(s)	Special provisions
	2,2-Di-(tert-butylperoxy) butane, with more than 55 percent in solution	Forbidden				
	Di-(tert-butylperoxy) phthalate, with more than 55 percent in solution	Forbidden				
	2,2-Di-(4,4-di-tert-butylperoxycyclohexyl) propane, with more than 42 percent with inert solid	Forbidden				
	Di-2,4-dichlorobenzoyl peroxide, with more than 75 percent with water	Forbidden				
	1,2-Di-(dimethylamino)ethane	3	UN2372	II	FLAMMABLE LIQUID	T8
	Di-2-ethylhexyl phosphoric acid, see **Diisooctyl acid phosphate**					
	Di-(1-hydroxytetrazole) (dry)	Forbidden				
	Di-(1-naphthoyl) peroxide	Forbidden				
	a,a'-Di-(nitroxy) methylether	Forbidden				
	Di-(beta-nitroxyethyl) ammonium nitrate	Forbidden				
	Diacetone alcohol	3	UN1148	II	FLAMMABLE LIQUID	T1
				III	FLAMMABLE LIQUID	B1, T1
	Diacetone alcohol peroxides, with more than 57 percent in solution with more than 9 percent hydrogen peroxide, less than 26 percent diacetone alcohol and less than 9 percent water; total active oxygen content more than 9 percent by mass	Forbidden				

HAZARDOUS MATERIALS TABLE

Sym.	Descriptions and shipping names	Hazard	ID No.	PG	Label(s)	Special provisions
	Diacetyl, see **Butanedione**					
	Diacetyl peroxide, solid, or with more than 25 percent in solution	Forbidden				
	Diallylamine	3	UN2359	II	FLAMMABLE LIQUID, POISON, CORROSIVE	T8
	Diallylether	3	UN2360	II	FLAMMABLE LIQUID, POISON	N12, T8
	4,4'-Diaminodiphenyl methane	6.1	UN2651	III	KEEP AWAY FROM FOOD	
	p-Diazidobenzene	Forbidden				
	1,2-Diazidoethane	Forbidden				
	1,1'-Diazoaminonaphthalene	Forbidden				
	Diazoaminotetrazole (dry)	Forbidden				
	Diazodinitrophenol (dry)	Forbidden				
	Diazodinitrophenol, wetted with not less than 40 percent water or mixture of alcohol and water, by mass	1.1A	UN0074	II	EXPLOSIVE 1.1A	111, 117
	Diazodiphenylmethane	Forbidden				
	Diazonium nitrates (dry)	Forbidden				
	Diazonium perchlorates (dry)	Forbidden				

HAZARDOUS MATERIALS TABLE

Sym.	Descriptions and shipping names	Hazard	ID No.	PG	Label(s)	Special provisions
	1,3-Diazopropane	Forbidden				
	Dibenzyl peroxydicarbonate, with more than 87 percent with water	Forbidden				
	Dibenzyldichlorosilane	8	UN2434	II	CORROSIVE	B2, T8, T26
	Diborane	2.3	UN1911		POISON GAS, FLAMMABLE GAS	1
D	Diborane mixtures	2.1	NA1911		FLAMMABLE GAS	5
	Dibromoacetylene	Forbidden				
	Dibromobenzene	3	UN2711	III	FLAMMABLE LIQUID	B1, T1
	1,2-Dibromobutan-3-one	6.1	UN2648	II	POISON	
	Dibromochloropropane	6.1	UN2872	III	KEEP AWAY FROM FOOD	T7
	Dibromodifluoromethane, R12B2	9	UN1941	III	None	T22
A	1,2-Dibromoethane, see Ethylene dibromide					
	Dibromomethane	6.1	UN2664	III	KEEP AWAY FROM FOOD	T7
	Dibutyl ethers	3	UN1149	III	FLAMMABLE LIQUID	B1, T1
	Dibutylaminoethanol	6.1	UN2873	III	KEEP AWAY FROM FOOD	T1
	N,N-Dichlorazodicarbonamidine (salts of) (dry)	Forbidden				
	1,1-Dichloro-1-nitroethane	6.1	UN2650	II	POISON	T8
D	3,5-Dichloro-2,4,6-trifluoropyridine	6.1	NA9264	I	POISON	2, B9, B14, B32, B74, T38, T43, T45
	Dichloroacetic acid	8	UN1764	II	CORROSIVE	A3, A6, A7, B2, N34, T9, T27

HAZARDOUS MATERIALS TABLE

Sym.	Descriptions and shipping names	Hazard	ID No.	PG	Label(s)	Special provisions
	1,3-Dichloroacetone	6.1	UN2649	II	POISON	
	Dichloroacetyl chloride	8	UN1765	II	CORROSIVE	A3, A6, A7, B2, B6, N34, T8, T26
	Dichloroacetylene	Forbidden				
	Dichloroanilines, liquid	6.1	UN1590	II	POISON	T14
	Dichloroanilines, solid	6.1	UN1590	II	POISON	T14
	o-Dichlorobenzene	6.1	UN1591	III	KEEP AWAY FROM FOOD	T7
D	Dichlorobutene	8	NA2920	I	CORROSIVE, FLAMMABLE LIQUID	
	2,2'-Dichlorodiethyl ether	6.1	UN1916	II	POISON, FLAMMABLE LIQUID	N33, N34, T8
	Dichlorodifluoromethane and difluoroethane azeotropic mixture *with approximately 74 percent dichlorodifluoromethane, R500*	2.2	UN2602		NONFLAMMABLE GAS	
	Dichlorodifluoromethane, *R12*	2.2	UN1028		NONFLAMMABLE GAS	T25
	Dichlorodimethyl ether, symmetrical	6.1	UN2249	I	POISON	B101, T7
	1,1-Dichloroethane	3	UN2362	II	FLAMMABLE LIQUID	
	1,2-Dichloroethane, see Ethylene dichloride					
	Dichloroethyl sulfide	Forbidden				
	Dichloroethylene	3	UN1150	II	FLAMMABLE LIQUID	T14
	Dichlorofluoromethane, *R21*	2.2	UN1029		NONFLAMMABLE GAS	
	Dichloroisocyanuric acid, dry or Dichloroisocyanuric acid salts	5.1	UN2465	II	OXIDIZER	28

HAZARDOUS MATERIALS TABLE

Sym.	Descriptions and shipping names	Hazard	ID No.	PG	Label(s)	Special provisions
	Dichloroisopropyl ether	6.1	UN2490	II	POISON	T8
	Dichloromethane	6.1	UN1593	III	KEEP AWAY FROM FOOD	N36, T13
	Dichloropentanes	3	UN1152	III	FLAMMABLE LIQUID	B1, T1
	Dichlorophenyl Isocyanates	6.1	UN2250	II	POISON	
	Dichlorophenyltrichlorosilane	8	UN1766	II	CORROSIVE	A7, B2, B6, N34, T8, T26
	Dichloropropane, see Propylene dichloride					
	1,3-Dichloropropanol-2	6.1	UN2750	II	POISON	T8
	Dichloropropene and propylene dichloride mixture, see Propylene dichloride					
	Dichloropropenes	3	UN2047	II	FLAMMABLE LIQUID	T8
				III	FLAMMABLE LIQUID	B1, T8
	Dichlorosilane	2.3	UN2189		POISON GAS, FLAMMABLE GAS, CORROSIVE	2, B9, B14
	Dichlorotetrafluoroethane, R114	2.2	UN1958		NONFLAMMABLE GAS	
	Dichlorovinylchloroarsine	Forbidden				
	Dicycloheptadiene, see 2,5-Norbornadiene					
	Dicyclohexylamine	8	UN2565	III	CORROSIVE	T8
	Dicyclohexylammonium nitrite	4.1	UN2687	III	FLAMMABLE SOLID	
	Dicyclopentadiene	3	UN2048	III	FLAMMABLE LIQUID	B1, T1
	Didymium nitrate	5.1	UN1465	III	OXIDIZER	A1
D	Dieldrin	6.1	NA2761	II	POISON	

HAZARDOUS MATERIALS TABLE

Sym.	Descriptions and shipping names	Hazard	ID No.	PG	Label(s)	Special provisions
D	**Diesel fuel**	3	NA1993	III	None	B1
	Diethanol nitrosamine dinitrate (dry)	Forbidden				
	Diethoxymethane	3	UN2373	II	FLAMMABLE LIQUID	T8
	3,3-Diethoxypropene	3	UN2374	II	FLAMMABLE LIQUID	T1
	Diethyl carbonate	3	UN2366	III	FLAMMABLE LIQUID	B1, T1
	Diethyl cellosolve, see Ethylene glycol diethyl ether					
	Diethyl ether or **Ethyl ether**	3	UN1155	I	FLAMMABLE LIQUID	T21
	Diethyl ketone	3	UN1156	II	FLAMMABLE LIQUID	T1
	Diethyl peroxydicarbonate, with more than 27 percent in solution	Forbidden				
	Diethyl sulfate	6.1	UN1594	II	POISON	B101, T14
	Diethyl sulfide	3	UN2375	II	FLAMMABLE LIQUID	B101, T14
	Diethylamine	3	UN1154	II	FLAMMABLE LIQUID, CORROSIVE	B101, N34, T8
	Diethylaminoethanol	3	UN2686	III	FLAMMABLE LIQUID	B1, T1
	Diethylaminopropylamine	3	UN2684	III	FLAMMABLE LIQUID, CORROSIVE	B1, T8
	N,N-Diethylaniline	6.1	UN2432	III	KEEP AWAY FROM FOOD	T2
	Diethylbenzene	3	UN2049	III	FLAMMABLE LIQUID	B1, T1
	Diethyldichlorosilane	8	UN1767	II	CORROSIVE, FLAMMABLE LIQUID	A7, B6, B100, N34, T8, T26
	Diethylene glycol dinitrate	Forbidden				

HAZARDOUS MATERIALS TABLE

Sym.	Descriptions and shipping names	Hazard	ID No.	PG	Label(s)	Special provisions
	Diethyleneglycol dinitrate, desensitized with not less than 25 percent non-volatile water-insoluble phlegmatizer, by mass	1.1D	UN0075	II	EXPLOSIVE 1.1D	
	Diethylenetriamine	8	UN2079	II	CORROSIVE	B2, T8
	N,N-Diethylethylenediamine	8	UN2685	II	CORROSIVE, FLAMMABLE LIQUID	T8
	Diethyl/gold bromide	Forbidden				
	Diethylthiophosphoryl chloride	8	UN2751	II	CORROSIVE	B2, T8
	Diethylzinc	4.2	UN1366	I	SPONTANEOUSLY COMBUSTIBLE	B11, T28, T40
	Difluorochloroethanes, see 1-Chloro-1,1-difluoroethanes					
	1,1-Difluoroethane, R152a	2.1	UN1030		FLAMMABLE GAS	
	1,1-Difluoroethylene, R1132a	2.1	UN1959		FLAMMABLE GAS	
	Difluoromethane	2.1	UN3252		FLAMMABLE GAS	
	Difluorophosphoric acid, anhydrous	8	UN1768	II	CORROSIVE	A6, A7, B2, N5, N34, T9, T27
	2,3-Dihydropyran	3	UN2376	II	FLAMMABLE LIQUID	T7
	1,8-Dihydroxy-2,4,5,7-tetranitroanthraquinone (chrysamminic acid)	Forbidden				
	Diiodoacetylene	Forbidden				
	Diisobutyl ketone	3	UN1157	III	FLAMMABLE LIQUID	B1, T1
	Diisobutylamine	3	UN2361	III	FLAMMABLE LIQUID, CORROSIVE	B1, T1

HAZARDOUS MATERIALS TABLE

Sym.	Descriptions and shipping names	Hazard	ID No.	PG	Label(s)	Special provisions
	Diisobutylene, isomeric compounds	3	UN2050	II	FLAMMABLE LIQUID	T1
	Diisooctyl acid phosphate	8	UN1902	III	CORROSIVE	T7
	Diisopropyl ether	3	UN1159	II	FLAMMABLE LIQUID	B101, T8
	Diisopropylamine	3	UN1158	II	FLAMMABLE LIQUID, CORROSIVE	B101, T8
	Diisopropylbenzene hydroperoxide, with more than 72 percent in solution	Forbidden				
	Diketene, inhibited	6.1	UN2521	I	POISON, FLAMMABLE LIQUID	2, B9, B14, B32, B74, T38, T43, T45
	1,2-Dimethoxyethane	3	UN2252	II	FLAMMABLE LIQUID	T1
	1,1-Dimethoxyethane	3	UN2377	II	FLAMMABLE LIQUID	T13
	Dimethyl carbonate	3	UN1161	II	FLAMMABLE LIQUID	T8
	Dimethyl chlorothiophosphate, see Dimethyl thiophosphoryl chloride					
	2,5-Dimethyl-2,5-dihydroperoxy hexane, with more than 82 percent with water	Forbidden				
	Dimethyl disulfide	3	UN2381	II	FLAMMABLE LIQUID	T8
	Dimethyl ether	2.1	UN1033		FLAMMABLE GAS	
	Dimethyl-N-propylamine	3	UN2266	II	FLAMMABLE LIQUID, CORROSIVE	T14, T26
	Dimethyl sulfate	6.1	UN1595	I	POISON, CORROSIVE	2, B9, B14, B32, B74, B77, T38, T43, T45
	Dimethyl sulfide	3	UN1164	II	FLAMMABLE LIQUID	B100, T14
	Dimethyl thiophosphoryl chloride	6.1	UN2267	II	POISON, CORROSIVE	T7
	Dimethylamine, anhydrous	2.1	UN1032		FLAMMABLE GAS	

HAZARDOUS MATERIALS TABLE

Sym.	Descriptions and shipping names	Hazard	ID No.	PG	Label(s)	Special provisions
	Dimethylamine solution	3	UN1160	II	FLAMMABLE LIQUID, CORROSIVE	T8, T34
	2-Dimethylaminoacetonitrile	3	UN2378	II	FLAMMABLE LIQUID, POISON	T8
	2-Dimethylaminoethanol	8	UN2051	II	CORROSIVE, FLAMMABLE LIQUID	B2, T8
	Dimethylaminoethyl methacrylate	6.1	UN2522	II	POISON	T8
	N,N-Dimethylaniline	6.1	UN2253	II	POISON	T8
	2,3-Dimethylbutane	3	UN2457	II	FLAMMABLE LIQUID	T13
	1,3-Dimethylbutylamine	3	UN2379	II	FLAMMABLE LIQUID, CORROSIVE	T8
	Dimethylcarbamoyl chloride	8	UN2262	II	CORROSIVE	B2, T8
	Dimethylcyclohexanes	3	UN2263	II	FLAMMABLE LIQUID	T1
	Dimethylcyclohexylamine	8	UN2264	II	CORROSIVE, FLAMMABLE LIQUID	B2, T8
	Dimethyldichlorosilane	3	UN1162	II	FLAMMABLE LIQUID, CORROSIVE	B77, T15, T26
	Dimethyldiethoxysilane	3	UN2380	II	FLAMMABLE LIQUID	T8
	Dimethyldioxanes	3	UN2707	II	FLAMMABLE LIQUID	T8, T31
	N,N-Dimethylformamide	3	UN2265	III	FLAMMABLE LIQUID	B1, T7, T30
	Dimethylhexane dihydroperoxide (dry)	Forbidden				B1, T1

HAZARDOUS MATERIALS TABLE

Sym.	Descriptions and shipping names	Hazard	ID No.	PG	Label(s)	Special provisions
	Dimethylhydrazine, symmetrical	6.1	UN2382	I	POISON, FLAMMABLE LIQUID	2, A7, B9, B14, B32, B74, B77, T38, T43, T45
	Dimethylhydrazine, unsymmetrical	6.1	UN1163	I	POISON, FLAMMABLE LIQUID, CORROSIVE	2, B9, B14, B32, B74, B79, T38, T43, T45
	2,2-Dimethylpropane	2.1	UN2044		FLAMMABLE GAS	
	Dimethylzinc	4.2	UN1370		SPONTANEOUSLY COMBUSTIBLE	B11, B16, T28, T29, T40
	Dinitro-o-cresol, *solid*	6.1	UN1598	II	POISON	T14
	Dinitro-o-cresol, *solution*	6.1	UN1598	II	POISON	T14
	1,3-Dinitro-5,5-dimethyl hydantoin	Forbidden				
	Dinitro-7,8-dimethylglycoluril (dry)	Forbidden				
	1,3-Dinitro-4,5-dinitrosobenzene	Forbidden				
	1,4-Dinitro-1,1,4,4-tetramethylolbutanetetranitrate (dry)	Forbidden				
	2,4-Dinitro-1,3,5-trimethylbenzene	Forbidden				
	Dinitroanilines	6.1	UN1596	II	POISON	T14
	Dinitrobenzenes, *liquid*	6.1	UN1597	II	POISON	11, T14
	Dinitrobenzenes, *solid*	6.1	UN1597	II	POISON	11
	Dinitrochlorobenzene, see Chlorodinitrobenzene					

HAZARDOUS MATERIALS TABLE

Sym.	Descriptions and shipping names	Hazard	ID No.	PG	Label(s)	Special provisions
	1,2-Dinitroethane	Forbidden				
	1,1-Dinitroethane (dry)	Forbidden				
	Dinitrogen tetroxide, liquefied	2.3	UN1067		POISON GAS, OXIDIZER, CORROSIVE	1, B7, B12, B14, B45, B46, B61, B66, B67, B77
	Dinitroglycoluril or Dingu	1.1D	UN0489	II	EXPLOSIVE 1.1D	
	Dinitromethane	Forbidden				
	Dinitrophenol, *dry or wetted with less than 15 percent water, by mass*	1.1D	UN0076	II	EXPLOSIVE 1.1D, POISON	
	Dinitrophenol solutions	6.1	UN1599	II	POISON	T8
				III	KEEP AWAY FROM FOOD	T7
	Dinitrophenol, wetted with not less than 15 percent water, by mass	4.1	UN1320	I	FLAMMABLE SOLID, POISON	23, A8, A19, A20, N41
	Dinitrophenolates alkali metals, *dry or wetted with less than 15 percent water, by mass*	1.3C	UN0077	II	EXPLOSIVE 1.3C, POISON	
	Dinitrophenolates, wetted with not less than 15 percent water, by mass	4.1	UN1321	I	FLAMMABLE SOLID, POISON	23, A8, A19, A20, N41
	Dinitropropylene glycol	Forbidden				
	Dinitroresorcinol, *dry or wetted with less than 15 percent water, by mass*	1.1D	UN0078	II	EXPLOSIVE 1.1D	
	2,4-Dinitroresorcinol (heavy metal salts of) (dry)	Forbidden				

HAZARDOUS MATERIALS TABLE

Sym.	Descriptions and shipping names	Hazard	ID No.	PG	Label(s)	Special provisions
	4,6-Dinitroresorcinol (heavy metal salts of) (dry)	Forbidden				
	Dinitroresorcinol, wetted with not less than 15 percent water, by mass	4.1	UN1322		FLAMMABLE SOLID	23, A8, A19, A20, N41
	3,5-Dinitrosalicylic acid (lead salt) (dry)	Forbidden				
	Dinitrosobenzene	1.3C	UN0406	II	EXPLOSIVE 1.3C	
	Dinitrosobenzylamidine and salts of (any)	Forbidden				
	2,2-Dinitrostilbene	Forbidden				
	Dinitrotoluenes, *liquid*	6.1	UN2038	II	POISON	T8
	Dinitrotoluenes, *molten*	6.1	UN1600	II	POISON	B100, T14
	Dinitrotoluenes, *solid*	6.1	UN2038	II	POISON	T8
	1,9-Dinitroxy pentamethylene-2,4, 6,8-tetramine (dry)	Forbidden				
	Dioxane	3	UN1165	II	FLAMMABLE LIQUID	T8
	Dioxolane	3	UN1166	II	FLAMMABLE LIQUID	T8
	Dipentene	3	UN2052	III	FLAMMABLE LIQUID	B1, T1
	Diphenylamine chloroarsine	6.1	UN1698	I	POISON	A8, B14, B32, N33, N34
	Diphenylchloroarsine, *liquid*	6.1	UN1699	I	POISON	A8, B14, B32, N33, N34
	Diphenylchloroarsine, *solid*	6.1	UN1699	I	POISON	
	Diphenyldichlorosilane	8	UN1769	II	CORROSIVE	A7, B2, N34, T8, T26

HAZARDOUS MATERIALS TABLE

Sym.	Descriptions and shipping names	Hazard	ID No.	PG	Label(s)	Special provisions
I	Diphenylmethane-4,4'diisocyanate	6.1	UN2489	III	KEEP AWAY FROM FOOD	T8
	Diphenylmethyl bromide	8	UN1770	II	CORROSIVE	
	Dipicryl sulfide, *dry or wetted with less than 10 percent water, by mass*	1.1D	UN0401		EXPLOSIVE 1.1D	
	Dipicryl sulfide, *wetted with not less than 10 percent water, by mass*	4.1	UN2852	I	FLAMMABLE SOLID	A2, N41
	Dipicrylamine, *see* Hexanitrodiphenylamine					
	Dipropionyl peroxide, *with more than 28 percent in solution*	Forbidden				
	Dipropyl ether	3	UN2384	II	FLAMMABLE LIQUID	T1
	Dipropyl ketone	3	UN2710	III	FLAMMABLE LIQUID	B1, T1
	Dipropylamine	3	UN2383	II	FLAMMABLE LIQUID, CORROSIVE	T8
	Disinfectants, liquid, corrosive n.o.s.	8	UN1903	II	CORROSIVE	B2
				III	CORROSIVE	
	Disinfectants, liquid, toxic, n.o.s.	6.1	UN3142	I	POISON	A4, T42
				II	POISON	T14
				III	KEEP AWAY FROM FOOD	T7
	Disinfectants, solid, toxic, n.o.s.	6.1	UN1601	II	POISON	
				III	KEEP AWAY FROM FOOD	
	Disodium trioxosilicate, pentahydrate	8	UN3253	III	CORROSIVE	
	Dispersant gases, n.o.s. *see* Refrigerant gases, n.o.s.					

HAZARDOUS MATERIALS TABLE

Sym.	Descriptions and shipping names	Hazard	ID No.	PG	Label(s)	Special provisions
	Dithiocarbamate pesticides, liquid, flammable, toxic, *flash point less than 23 degrees C*	3	UN2772	I	FLAMMABLE LIQUID, POISON	
				II	FLAMMABLE LIQUID, POISON	
	Dithiocarbamate pesticides, liquid, toxic	6.1	UN3006	I	POISON	T42
				II	POISON	T14
				III	KEEP AWAY FROM FOOD	T14
	Dithiocarbamate pesticides, liquid, toxic, flammable, *flash point not less than 23 degrees C*	6.1	UN3005	I	POISON, FLAMMABLE LIQUID	T42
				II	POISON, FLAMMABLE LIQUID	T14
				III	KEEP AWAY FROM FOOD, FLAMMABLE LIQUID	
	Dithiocarbamate pesticides, solid, toxic	6.1	UN2771	I	POISON	T14
				II	POISON	
				III	KEEP AWAY FROM FOOD	
	Divinyl ether, inhibited	3	UN1167	I	FLAMMABLE LIQUID	T14
D	Dodecylbenzenesulfonic acid	8	NA2584	II	CORROSIVE	B2
	Dodecyltrichlorosilane	8	UN1771	II	CORROSIVE	A7, B2, B6, N34, T8, T26
	Dry Ice, *see* Carbon dioxide, solid					

HAZARDOUS MATERIALS TABLE

Sym.	Descriptions and shipping names	Hazard	ID No.	PG	Label(s)	Special provisions
	Dyes, liquid, corrosive n.o.s *or* Dye intermediates, liquid, corrosive, n.o.s.	8	UN2801	II	CORROSIVE	11, B2, T14
	Dyes, liquid, toxic, n.o.s *or* Dye Intermediates, liquid, toxic, n.o.s.	6.1	UN1602	II	POISON	11, T7
				III	KEEP AWAY FROM FOOD	
	Dyes, solid, corrosive, n.o.s. *or* Dye Intermediates, solid, corrosive, n.o.s.	8	UN3147	II	CORROSIVE	
				III	CORROSIVE	
	Dyes, solid, toxic, n.o.s. *or* Dye Intermediates, solid, toxic, n.o.s.	6.1	UN3143	I	POISON	A5
				II	POISON	
				III	KEEP AWAY FROM FOOD	
	Dynamite, *see* **Explosive, blasting, type A**					
	Electrolyte (acid or alkali) for batteries, see **Battery fluid, acid** *or* **Battery fluid, alkali**					
	Elevated temperature liquid, flammable, **n.o.s.**, *with flash point above 37.8 C, at or above its flash point*	3	UN3256	III	FLAMMABLE LIQUID	T1
	Elevated temperature liquid, **n.o.s.**, *at or above 100 C and below its flash point*	9	UN3257	III	CLASS 9	T1
	Elevated temperature solid, **n.o.s.**, *at or above 240 C, see section 173.247(h)(4)*	9	UN3258	III	CLASS 9	
	Engine starting fluid, *with flammable gas*	2.1	UN1960		FLAMMABLE GAS	
	Engines, internal combustion, *including when fitted in machinery or vehicles*	9	UN3166		CLASS 9	

HAZARDOUS MATERIALS TABLE

Sym.	Descriptions and shipping names	Hazard	ID No.	PG	Label(s)	Special provisions
	Environmentally hazardous substances, liquid, n.o.s.	9	UN3082	III	CLASS 9	8, N50, T1
	Environmentally hazardous substances, solid, n.o.s.	9	UN3077	III	CLASS 9	8, N50, B54
	Epibromohydrin	6.1	UN2558	I	POISON	T18, T26
	Epichlorohydrin	6.1	UN2023	II	POISON	T14
	1,2-Epoxy-3-ethoxypropane	3	UN2752	III	FLAMMABLE LIQUID	B1, T1
	Esters, n.o.s.	3	UN3272	II	FLAMMABLE LIQUID	T8
				III	FLAMMABLE LIQUID	B1, T7
	Etching acid, liquid, n.o.s., see **Hydrofluoric acid, solution etc.**					
	Ethane, compressed	2.1	UN1035		FLAMMABLE GAS	
D	Ethane-Propane mixture, refrigerated liquid	2.1	NA1961		FLAMMABLE GAS	
	Ethane, refrigerated liquid	2.1	UN1961		FLAMMABLE GAS	
	Ethanol amine dinitrate	Forbidden				
	Ethanol or Ethyl alcohol or Ethanol solutions or Ethyl alcohol solutions	3	UN1170	II	FLAMMABLE LIQUID	T1
				III	FLAMMABLE LIQUID	B1, T1
	Ethanolamine or Ethanolamine solutions	8	UN2491	III	CORROSIVE	T7
	Ether, see **Diethyl ether**					
	Ethers, n.o.s.	3	UN3271	II	FLAMMABLE LIQUID	T8
				III	FLAMMABLE LIQUID	B1, T7
	Ethyl acetate	3	UN1173	II	FLAMMABLE LIQUID	T2

HAZARDOUS MATERIALS TABLE

Sym.	Descriptions and shipping names	Hazard	ID No.	PG	Label(s)	Special provisions
	Ethyl acrylate, Inhibited	3	UN1917	II	FLAMMABLE LIQUID	T8
	Ethyl alcohol, see Ethanol					
	Ethyl aldehyde, see Acetaldehyde					
	Ethyl amyl ketone	3	UN2271	III	FLAMMABLE LIQUID	B1, T1
	N-Ethyl-N-benzylaniline	6.1	UN2274	III	KEEP AWAY FROM FOOD	T2
	Ethyl borate	3	UN1176	II	FLAMMABLE LIQUID	T8
	Ethyl bromide	6.1	UN1891	II	POISON	B100, T17
	Ethyl bromoacetate	6.1	UN1603	II	POISON	T14
	Ethyl butyl ether	3	UN1179	II	FLAMMABLE LIQUID	B1, B101, T1
	Ethyl butyrate	3	UN1180	III	FLAMMABLE LIQUID	B1, T1
	Ethyl chloride	2.1	UN1037		FLAMMABLE GAS	B77
	Ethyl chloroacetate	6.1	UN1181	II	POISON	T14
	Ethyl chloroformate	6.1	UN1182	I	POISON, FLAMMABLE LIQUID, CORROSIVE	2, A3, A6, A7, B9, B14, B32, B74, N34, T38, T43, T45
	Ethyl 2-chloropropionate	3	UN2935	III	FLAMMABLE LIQUID	B1, T1
	Ethyl chlorothioformate	8	UN2826	II	CORROSIVE, POISON, FLAMMABLE LIQUID	2, B9, B14, B32, B74, T38, T43, T45
	Ethyl crotonate	3	UN1862	II	FLAMMABLE LIQUID	T1
	Ethyl cyanoacetate	6.1	UN2666	III	KEEP AWAY FROM FOOD	
	Ethyl ether, see Diethyl ether					
	Ethyl fluoride	2.1	UN2453		FLAMMABLE GAS	
+	Ethyl formate	3	UN1190	II	FLAMMABLE LIQUID	T8

HAZARDOUS MATERIALS TABLE

Sym.	Descriptions and shipping names	Hazard	ID No.	PG	Label(s)	Special provisions
	Ethyl hydroperoxide	Forbidden				
	Ethyl isobutyrate	3	UN2385	II	FLAMMABLE LIQUID	T1
+	Ethyl isocyanate	3	UN2481	I	FLAMMABLE LIQUID, POISON	1, A7, B9, B14, B30, B72, T38, T43, T44
	Ethyl lactate	3	UN1192	III	FLAMMABLE LIQUID	B1, T1
	Ethyl mercaptan	3	UN2363	I	FLAMMABLE LIQUID	T21
	Ethyl methacrylate	3	UN2277	II	FLAMMABLE LIQUID	T1
	Ethyl methyl ether	2.1	UN1039		FLAMMABLE GAS	
	Ethyl methyl ketone or Methyl ethyl ketone	3	UN1193	II	FLAMMABLE LIQUID	T8
	Ethyl nitrite solutions	3	UN1194	I	FLAMMABLE LIQUID, POISON	
	Ethyl orthoformate	3	UN2524	III	FLAMMABLE LIQUID	B1, T7
	Ethyl oxalate	6.1	UN2525	III	KEEP AWAY FROM FOOD	T1
	Ethyl perchlorate	Forbidden				
D	Ethyl phosphonothioic dichloride, anhydrous	6.1	NA2927	I	POISON, CORROSIVE	2, B9, B14, B32, B74, T38, T43, T45
D	Ethyl phosphonous dichloride, anhydrous *pyrophoric liquid*	6.1	NA2845	I	POISON, SPONTANEOUSLY COMBUSTIBLE	2, B9, B14, B32, B74, T38, T43, T45
D	Ethyl phosphorodichloridate	6.1	NA2927	I	POISON, CORROSIVE	2, B9, B14, B32, B74, T38, T43, T45
	Ethyl propionate	3	UN1195	II	FLAMMABLE LIQUID	T1
	Ethyl propyl ether	3	UN2615	II	FLAMMABLE LIQUID	B101, T8
	Ethyl silicate, see Tetraethyl silicate					

HAZARDOUS MATERIALS TABLE

Sym.	Descriptions and shipping names	Hazard	ID No.	PG	Label(s)	Special provisions
	Ethylacetylene, Inhibited	2.1	UN2452		FLAMMABLE GAS	
	Ethylamine	2.1	UN1036		FLAMMABLE GAS	B77
	Ethylamine, aqueous solution with not less than 50 percent but not more than 70 percent ethylamine	3	UN2270	II	FLAMMABLE LIQUID, CORROSIVE	T14
	N-Ethylaniline	6.1	UN2272	III	KEEP AWAY FROM FOOD	T2
	2-Ethylaniline	6.1	UN2273	III	KEEP AWAY FROM FOOD	T2
	Ethylbenzene	3	UN1175	II	FLAMMABLE LIQUID	T1
	N-Ethylbenzytoluidines liquid	6.1	UN2753	III	KEEP AWAY FROM FOOD	T14
	N-Ethylbenzytoluidines solid	6.1	UN2753	III	KEEP AWAY FROM FOOD	
	2-Ethylbutanol	3	UN2275	III	FLAMMABLE LIQUID	B1, T1
	2-Ethylbutyl acetate	3	UN1177	III	FLAMMABLE LIQUID	B1, T1
	2-Ethylbutyraldehyde	3	UN1178	II	FLAMMABLE LIQUID	B1, T1
	Ethyldichloroarsine	6.1	UN1892	I	POISON	2, B9, B14, B32, B74, T38, T43, T45
	Ethyldichlorosilane	4.3	UN3138	I	DANGEROUS WHEN WET, CORROSIVE, FLAMMABLE LIQUID	A2, A3, A7, N34, T18, T26
	Ethylene, acetylene and propylene in mixtures, refrigerated liquid with at least 71.5 percent ethylene with not more than 22.5 percent acetylene and not more than 6 percent propylene	2.1	UN1135		FLAMMABLE GAS	
	Ethylene chlorohydrin	6.1	UN1135	I	POISON, FLAMMABLE LIQUID	2, B9, B14, B32, B74, T38, T43, T45
	Ethylene, compressed	2.1	UN1962		FLAMMABLE GAS	

HAZARDOUS MATERIALS TABLE

Sym.	Descriptions and shipping names	Hazard	ID No.	PG	Label(s)	Special provisions
	Ethylene diamine diperchlorate	Forbidden				
	Ethylene dibromide	6.1	UN1605	I	POISON	2, B9, B14, B32, B74, B77, T38, T43, T45
	Ethylene dibromide and methyl bromide liquid mixtures, see Methyl bromide and ethylene dibromide, liquid mixtures					
	Ethylene dichloride	3	UN1184	II	FLAMMABLE LIQUID, POISON	T14
	Ethylene glycol diethyl ether	3	UN1153	III	FLAMMABLE LIQUID	B1, T1
	Ethylene glycol dinitrate	Forbidden				
	Ethylene glycol monobutyl ether	6.1	UN2369	III	KEEP AWAY FROM FOOD	T1
	Ethylene glycol monoethyl ether	3	UN1171	III	FLAMMABLE LIQUID	B1, T1
	Ethylene glycol monoethyl ether acetate	3	UN1172	III	FLAMMABLE LIQUID	B1, T1
	Ethylene glycol monomethyl ether	3	UN1188	III	FLAMMABLE LIQUID	B1, T1
	Ethylene glycol monomethyl ether acetate	3	UN1189	III	FLAMMABLE LIQUID	B1, T1
	Ethylene oxide and carbon dioxide mixture with more than 9 percent but not more than 87 percent ethylene oxide	2.3	UN3300		POISON GAS, FLAMMABLE GAS	4
	Ethylene oxide and carbon dioxide mixtures with not more than 9 percent ethylene oxide	2.1	UN1041		FLAMMABLE GAS	
	Ethylene oxide and carbon dioxide mixtures with not more than 8.8 percent ethylene oxide	2.2	UN1952		NONFLAMMABLE GAS	
	Ethylene oxide and chlorotetrafluoroethane mixture with not more than 8.8 percent ethylene oxide	2.2	UN3297		NONFLAMMABLE GAS	

HAZARDOUS MATERIALS TABLE

Sym.	Descriptions and shipping names	Hazard	ID No.	PG	Label(s)	Special provisions
	Ethylene oxide and dichlorodifluoromethane mixture, *with not more than 12.5 percent ethylene oxide*	2.2	UN3070		NONFLAMMABLE GAS	
	Ethylene oxide and pentafluoroethane mixture *with not more than 7.9 percent ethylene oxide*	2.2	UN3298		NONFLAMMABLE GAS	
	Ethylene oxide and propylene oxide mixtures, *with not more than 30 percent ethylene oxide*	3	UN2983	I	FLAMMABLE LIQUID, POISON	5, A11, N4, N34, T24, T29
	Ethylene oxide and tetrafluoroethane mixture *with not more than 5.6 percent ethylene oxide*	2.2	UN3299		NONFLAMMABLE GAS	
	Ethylene oxide or Ethylene oxide with nitrogen up to a total pressure of 1MPa (10 bar) at 50 degrees C	2.3	UN1040		POISON GAS, FLAMMABLE GAS	4, 25
	Ethylene, refrigerated liquid *(cryogenic liquid)*	2.1	UN1038		FLAMMABLE GAS	
	Ethylenediamine	8	UN1604	II	CORROSIVE, FLAMMABLE LIQUID	T14
	Ethyleneimine, inhibited	6.1	UN1185	I	POISON, FLAMMABLE LIQUID	1, B9, B14, B30, B72, B77, N25, N32, T38, T43, T44
	Ethylhexaldehyde, see Octyl aldehydes etc.					
	2-Ethylhexyl chloroformate	6.1	UN2748	II	POISON, CORROSIVE	T12
	2-Ethylhexylamine	3	UN2276	III	FLAMMABLE LIQUID, CORROSIVE	B1, T2
	Ethylphenyldichlorosilane	8	UN2435	II	CORROSIVE	A7, B2, N34, T8, T26
	1-Ethylpiperidine	3	UN2386	II	FLAMMABLE LIQUID, CORROSIVE	T8

HAZARDOUS MATERIALS TABLE

Sym.	Descriptions and shipping names	Hazard	ID No.	PG	Label(s)	Special provisions
	N-Ethyltoluidines	6.1	UN2754	II	POISON	T14
	Ethyltrichlorosilane	3	UN1196	II	FLAMMABLE LIQUID, CORROSIVE	A7, B100, N34, T15, T26
	Etiologic agent, see Infectious substances, etc.)					
	Explosive articles, see Articles, explosive, n.o.s. etc.					
	Explosive, blasting, type A	1.1D	UN0081	II	EXPLOSIVE 1.1D	
	Explosive, blasting, type B	1.1D	UN0082	II	EXPLOSIVE 1.1D	
	Explosive, blasting, type B or Agent blasting, Type B	1.5D	UN0331	II	EXPLOSIVE 1.5D	105, 106
	Explosive, blasting, type C	1.1D	UN0083	II	EXPLOSIVE 1.1D	
	Explosive, blasting, type D	1.1D	UN0084	II	EXPLOSIVE 1.1D	
	Explosive, blasting, type E	1.1D	UN0241	II	EXPLOSIVE 1.1D	
	Explosive, blasting, type E or Agent blasting, Type E	1.5D	UN0332	II	EXPLOSIVE 1.5D	105, 106
	Explosive, forbidden. See Sec. 173.54	Forbidden				
D	**Explosive pest control devices**	1.1E	NA0006	II	EXPLOSIVE 1.1E	
D	**Explosive pest control devices**	1.4E	NA0412	II	EXPLOSIVE 1.4E	
	Explosive substances, see Substances, explosive, n.o.s. etc.					
	Explosives, slurry, see Explosive, blasting, type E					

HAZARDOUS MATERIALS TABLE

Sym.	Descriptions and shipping names	Hazard	ID No.	PG	Label(s)	Special provisions
	Explosives, water gels, see **Explosive, blasting, type E**					
	Extracts, aromatic, liquid	3	UN1169	II	FLAMMABLE LIQUID	T7, T30
				III	FLAMMABLE LIQUID	B1, T7, T30
	Extracts, flavoring, liquid	3	UN1197	II	FLAMMABLE LIQUID	T7, T30
				III	FLAMMABLE LIQUID	B1, T7, T30
	Fabric with animal or vegetable oil, see **Fibers or fabrics, etc.**					
	Ferric arsenate	6.1	UN1606	II	POISON	
	Ferric arsenite	6.1	UN1607	II	POISON	
	Ferric chloride, anhydrous	8	UN1773	III	CORROSIVE	
	Ferric chloride, solution	8	UN2582	III	CORROSIVE	B15, T8
	Ferric nitrate	5.1	UN1466	III	OXIDIZER	A1, A29
	Ferrocerium	4.1	UN1323	II	FLAMMABLE SOLID	A19
	Ferrosilicon, with 30 percent or more but less than 90 percent silicon	4.3	UN1408	III	DANGEROUS WHEN WET, KEEP AWAY FROM FOOD	A1, A19
	Ferrous arsenate	6.1	UN1608	II	POISON	
D	Ferrous chloride, solid	8	NA1759	II	CORROSIVE	
D	Ferrous chloride, solution	8	NA1760	II	CORROSIVE	B3
	Ferrous metal borings or Ferrous metal shavings or Ferrous metal turnings or Ferrous metal cuttings *in a form liable to self-heating*	4.2	UN2793	III	SPONTANEOUSLY COMBUSTIBLE	A1, A19, B101
	Fertilizer ammoniating solution *with free ammonia*	2.2	UN1043		NONFLAMMABLE GAS	

HAZARDOUS MATERIALS TABLE

Sym.	Descriptions and shipping names	Hazard	ID No.	PG	Label(s)	Special provisions
A|W	Fibers or Fabrics, animal or vegetable or Synthetic, n.o.s. with animal or vegetable oil	4.2	UN1373	III	SPONTANEOUSLY COMBUSTIBLE	
	Fibers or Fabrics impregnated with weakly nitrated nitrocellulose, n.o.s.	4.1	UN1353	III	FLAMMABLE SOLID	A1
	Films, nitrocellulose base, from which gelatine has been removed; film scrap, see Celluloid scrap					
	Films, nitrocellulose base, gelatine coated (except scrap)	4.1	UN1324	III	FLAMMABLE SOLID	
	Fire extinguisher charges, corrosive liquid	8	UN1774	II	CORROSIVE	N41
	Fire extinguisher charges, expelling, explosive, see Cartridges, power device					
	Fire extinguishers containing compressed or liquefied gas	2.2	UN1044		NONFLAMMABLE GAS	
	Firelighters, solid with flammable liquid	4.1	UN2623	III	FLAMMABLE SOLID	A19
	Fireworks			III	FLAMMABLE SOLID	A1, A19
	Fireworks	1.1G	UN0333	II	EXPLOSIVE 1.1G	108
	Fireworks	1.2G	UN0334	II	EXPLOSIVE 1.2G	108
	Fireworks	1.3G	UN0335	II	EXPLOSIVE 1.3G	108
	Fireworks	1.4G	UN0336	II	EXPLOSIVE 1.4G	108
	Fireworks	1.4S	UN0337	II	EXPLOSIVE 1.4S	108
	Fish meal, stabilized or Fish scrap, stabilized	9	UN2216	III	None	
W	Fish meal, unstabilized or Fish scrap, unstabilized	4.2	UN1374	II	SPONTANEOUSLY COMBUSTIBLE	A1, A19

HAZARDOUS MATERIALS TABLE

Sym.	Descriptions and shipping names	Hazard	ID No.	PG	Label(s)	Special provisions
	Fissile radioactive materials, see **Radioactive material, fissile, n.o.s.**					
	Flammable compressed gas, see **Compressed or Liquefied gas, flammable,** *etc.*					
	Flammable compressed gas (small receptacles not fitted with a dispersion device, not refillable), see **Receptacles,** *etc.*					
	Flammable gas in lighters, see **Lighters or lighter refills, containing flammable gas**					
	Flammable liquid, toxic, corrosive, n.o.s.	3	UN3286	I	FLAMMABLE LIQUID, POISON, CORROSIVE	T14
				II	FLAMMABLE LIQUID, POISON, CORROSIVE	T42
	Flammable liquids, corrosive, n.o.s.	3	UN2924	I	FLAMMABLE LIQUID, CORROSIVE	T15, T26
				II	FLAMMABLE LIQUID, CORROSIVE	B1, T15, T26
				III	FLAMMABLE LIQUID, CORROSIVE	
	Flammable liquids, n.o.s.	3	UN1993	I	FLAMMABLE LIQUID	T42
				II	FLAMMABLE LIQUID	T8, T31
				III	FLAMMABLE LIQUID	B1, B52, T7, T30

HAZARDOUS MATERIALS TABLE

Sym.	Descriptions and shipping names	Hazard	ID No.	PG	Label(s)	Special provisions
	Flammable liquids, toxic, n.o.s.	3	UN1992	I	FLAMMABLE LIQUID, POISON	T42
				II	FLAMMABLE LIQUID, POISON	T18
				III	FLAMMABLE LIQUID, KEEP AWAY FROM FOOD	B1, T18
	Flammable solid, corrosive, inorganic, n.o.s.	4.1	UN3180	II	FLAMMABLE SOLID, CORROSIVE	A1, B106
				III	FLAMMABLE SOLID, CORROSIVE	A1, B106
	Flammable solid, inorganic, n.o.s.	4.1	UN3178	II	FLAMMABLE SOLID	A1
				III	FLAMMABLE SOLID	A1
	Flammable solid, organic, molten, n.o.s.	4.1	UN3176	II	FLAMMABLE SOLID	T9
				III	FLAMMABLE SOLID	T9
	Flammable solid, toxic, inorganic, n.o.s.	4.1	UN3179	II	FLAMMABLE SOLID, POISON	A1, B106
				III	FLAMMABLE SOLID, KEEP AWAY FROM FOOD	A1, B106
	Flammable solids, corrosive, organic, n.o.s.	4.1	UN2925	II	FLAMMABLE SOLID, CORROSIVE	A1, B106
				III	FLAMMABLE SOLID, CORROSIVE	A1, B106
	Flammable solids, organic, n.o.s.	4.1	UN1325	II	FLAMMABLE SOLID	A1
				III	FLAMMABLE SOLID	A1

HAZARDOUS MATERIALS TABLE

Sym.	Descriptions and shipping names	Hazard	ID No.	PG	Label(s)	Special provisions
	Flammable solids, toxic, organic, n.o.s.	4.1	UN2926	II	FLAMMABLE SOLID, POISON	A1, B106
				III	FLAMMABLE SOLID, KEEP AWAY FROM FOOD	A1, B106
	Flares, aerial	1.3G	UN0093	II	EXPLOSIVE 1.3G	
	Flares, aerial	1.4G	UN0403	II	EXPLOSIVE 1.4G	
	Flares, aerial	1.4S	UN0404	II	EXPLOSIVE 1.4S	
	Flares, aerial	1.1G	UN0420	II	EXPLOSIVE 1.1G	
	Flares, aerial	1.2G	UN0421	II	EXPLOSIVE 1.2G	
	Flares, airplane, see Flares, aerial					
	Flares, signal, see Cartridges, signal					
	Flares, surface	1.3G	UN0092	II	EXPLOSIVE 1.3G	
	Flares, surface	1.1G	UN0418	II	EXPLOSIVE 1.1G	
	Flares, surface	1.2G	UN0419	II	EXPLOSIVE 1.2G	
	Flares, water-activated, see Contrivances, water-activated, etc.					
	Flash powder	1.1G	UN0094	II	EXPLOSIVE 1.1G	
	Flash powder	1.3G	UN0305	II	EXPLOSIVE 1.3G	
	Flue dusts, poisonous, see Arsenical dust					
	Fluoric acid, see Hydrofluoric acid, solution, etc.					
	Fluorine, compressed	2.3	UN1045		POISON GAS, OXIDIZER, CORROSIVE	1
	Fluoroacetic acid	6.1	UN2642	I	POISON	B100

HAZARDOUS MATERIALS TABLE

Sym.	Descriptions and shipping names	Hazard	ID No.	PG	Label(s)	Special provisions
	Fluoroanilines	6.1	UN2941	III	KEEP AWAY FROM FOOD	T8
	Fluorobenzene	3	UN2387	II	FLAMMABLE LIQUID	B101, T8
	Fluoroboric acid	8	UN1775	II	CORROSIVE	A6, A7, B2, B15, N3, N34, T15, T27
	Fluorophosphoric acid anhydrous	8	UN1776	II	CORROSIVE	A6, A7, B2, N3, N34, T9, T27
	Fluorosilicates, n.o.s.	6.1	UN2856	III	KEEP AWAY FROM FOOD	T8
	Fluorosilicic acid	8	UN1778	II	CORROSIVE	A6, A7, B2, B15, N3, N34, T12, T27
	Fluorosulfonic acid	8	UN1777	I	CORROSIVE	A3, A6, A7, A10, B6, B10, N3, T9, T27
	Fluorotoluenes	3	UN2388	II	FLAMMABLE LIQUID	T8
	Forbidden materials. See 173.21	Forbidden				
	Formaldehyde, solutions, flammable	3	UN1198	III	FLAMMABLE LIQUID, CORROSIVE	B1, T8
	Formaldehyde, solutions, *with not less than 25 percent formaldehyde*	8	UN2209	III	CORROSIVE	T1
	Formalin, see **Formaldehyde, solutions**					
	Formic acid	8	UN1779	II	CORROSIVE	B2, B12, B28, T8
	Fracturing devices, explosive, *without detonators for oil wells*	1.1D	UN0099	II	EXPLOSIVE 1.1D	

HAZARDOUS MATERIALS TABLE

Sym.	Descriptions and shipping names	Hazard	ID No.	PG	Label(s)	Special provisions
	Fuel, aviation, turbine engine	3	UN1863	I	FLAMMABLE LIQUID	T7
		3		II	FLAMMABLE LIQUID	T1
		3		III	FLAMMABLE LIQUID	B1, T1
D	**Fuel oil** (No. 1, 2, 4, 5, or 6)	3	NA1993	III	FLAMMABLE LIQUID	B1
	Fulminate of mercury (dry)	Forbidden				
	Fulminate of mercury, wet, see **Mercury fulminate, etc.**					
	Fulminating gold	Forbidden				
	Fulminating mercury	Forbidden				
	Fulminating platinum	Forbidden				
	Fulminating silver	Forbidden				
	Fulminic acid	Forbidden				
	Fumaryl chloride	8	UN1780	II	CORROSIVE	B2, T8, T26
	Furan	3	UN2389	I	FLAMMABLE LIQUID	T18
	Furfural	3	UN1199	III	FLAMMABLE LIQUID	B1, T1
	Furfuryl alcohol	6.1	UN2874	III	KEEP AWAY FROM FOOD	T2
	Furfurylamine	3	UN2526	III	FLAMMABLE LIQUID, CORROSIVE	B1, T1

HAZARDOUS MATERIALS TABLE

Sym.	Descriptions and shipping names	Hazard	ID No.	PG	Label(s)	Special provisions
	Fuse, detonating, metal clad, see Cord, detonating, metal clad					
	Fuse, detonating, mild effect, metal clad, see Cord, detonating, mild effect, metal clad					
	Fuse, igniter tubular metal clad	1.4G	UN0103	II	EXPLOSIVE 1.4G	
	Fuse, Instantaneous, non-detonating or Quickmatch	1.3G	UN0101	II	EXPLOSIVE 1.3G	
	Fuse, safety	1.4S	UN0105	II	EXPLOSIVE 1.4S	
D	Fusee (railway or highway)	4.1	NA1325	II	FLAMMABLE SOLID	T1
	Fusel oil	3	UN1201	II	FLAMMABLE LIQUID	B1, T1
				III	FLAMMABLE LIQUID	
	Fuses, tracer, see Tracers for ammunition					
	Fuzes, combination, percussion and time, see Fuzes, detonating (UN 0257, UN 0367); Fuzes, Igniting (UN 0317, UN 0368)					
	Fuzes, detonating	1.1B	UN0106	II	EXPLOSIVE 1.1B	
	Fuzes, detonating	1.2B	UN0107	II	EXPLOSIVE 1.2B	
	Fuzes, detonating	1.4B	UN0257	II	EXPLOSIVE 1.4B	116
	Fuzes, detonating	1.4S	UN0367	II	EXPLOSIVE 1.4S	116
	Fuzes, detonating, with protective features	1.1D	UN0408	II	EXPLOSIVE 1.1D	
	Fuzes, detonating, with protective features	1.2D	UN0409	II	EXPLOSIVE 1.2D	
	Fuzes, detonating, with protective features	1.4D	UN0410	II	EXPLOSIVE 1.4D	116
	Fuzes, Igniting	1.3G	UN0316	II	EXPLOSIVE 1.3G	
	Fuzes, Igniting	1.4G	UN0317	II	EXPLOSIVE 1.4G	

HAZARDOUS MATERIALS TABLE

Sym.	Descriptions and shipping names	Hazard	ID No.	PG	Label(s)	Special provisions
	Fuzes, igniting	1.4S	UN0368	II	EXPLOSIVE 1.4S	
	Galactsan trinitrate	Forbidden				
	Gallium	8	UN2803	III	CORROSIVE	
	Gas generator assemblies (aircraft), containing a non-flammable non-toxic gas and a propellant cartridge	2.2			NONFLAMMABLE GAS	
D	Gas identification set	2.3	NA9035		POISON GAS	6
	Gas oil or Diesel fuel or Heating oil, light	3	UN1202	III	FLAMMABLE LIQUID	B1, T7, T30
	Gas, refrigerated liquid, n.o.s.	2.2	UN3158		NONFLAMMABLE GAS	
	Gas sample, non-pressurized, flammable, n.o.s., *not refrigerated liquid*	2.1	UN3167		FLAMMABLE GAS	
	Gas sample, non-pressurized, toxic, flammable, n.o.s., *not refrigerated liquid*	2.3	UN3168		POISON GAS, FLAMMABLE GAS	
	Gas sample, non-pressurized, toxic, n.o.s., *not refrigerated liquid*	2.3	UN3169		POISON GAS	
D	Gasohol gasoline mixed with ethyl alcohol, with not more than 20 percent alcohol	3	NA1203	II	FLAMMABLE LIQUID	
	Gasoline	3	UN1203	II	FLAMMABLE LIQUID	B33, B101, T8
	Gasoline, casinghead, see Gasoline					
	Gelatine, blasting, see Explosive, blasting, type A					
	Gelatine dynamites, see Explosive, blasting, type A					
	Germane	2.3	UN2192		POISON GAS, FLAMMABLE GAS	2, 25

HAZARDOUS MATERIALS TABLE

Sym.	Descriptions and shipping names	Hazard	ID No.	PG	Label(s)	Special provisions
	Glycerol-1,3-dinitrate	Forbidden				
	Glycerol gluconate trinitrate	Forbidden				
	Glycerol lactate trinitrate	Forbidden				
	Glycerol alpha-monochlorohydrin	6.1	UN2689	III	KEEP AWAY FROM FOOD	T2
	Glyceryl trinitrate, see **Nitroglycerin**, etc.					
	Glycidaldehyde	3	UN2622	II	FLAMMABLE LIQUID, POISON	T8
D	Grenades, empty primed	1.4S	NA0349	II	None	
	Grenades, hand or rifle, with bursting charge	1.1D	UN0284	II	EXPLOSIVE 1.1D	
	Grenades, hand or rifle, with bursting charge	1.2D	UN0285	II	EXPLOSIVE 1.2D	
	Grenades, hand or rifle, with bursting charge	1.1F	UN0292	II	EXPLOSIVE 1.1F	
	Grenades, hand or rifle, with bursting charge	1.2F	UN0293	II	EXPLOSIVE 1.2F	
	Grenades, illuminating, see **Ammunition, Illuminating**, etc.					
	Grenades, practice, hand or rifle	1.4S	UN0110	II	EXPLOSIVE 1.4S	
	Grenades, practice, hand or rifle	1.3G	UN0318	II	EXPLOSIVE 1.3G	
	Grenades, practice, hand or rifle	1.2G	UN0372	II	EXPLOSIVE 1.2G	
	Grenades practice Hand or rifle	1.4G	UN0452	II	EXPLOSIVE 1.4G	
	Grenades, smoke, see **Ammunition, smoke**, etc.					
	Guanidine nitrate	5.1	UN1467	III	OXIDIZER	A1

HAZARDOUS MATERIALS TABLE

Sym.	Descriptions and shipping names	Hazard	ID No.	PG	Label(s)	Special provisions
	Guanyl nitrosaminoguanylidene hydrazine (dry)	Forbidden				
	Guanyl nitrosaminoguanylidene hydrazine, wetted with not less than 30 percent water, by mass	1.1A	UN0113	II	EXPLOSIVE 1.1A	111, 117
	Guanyl nitrosaminoguanyltetrazene (dry)	Forbidden				
	Guanyl nitrosaminoguanyltetrazene, wetted or Tetrazene, wetted with not less than 30 percent water or mixture of alcohol and water, by mass	1.1A	UN0114	II	EXPLOSIVE 1.1A	111, 117
	Gunpowder, compressed or Gunpowder in pellets, see Black powder (UN 0028)					
	Gunpowder, granular or as a meal, see Black powder (UN 0027)					
	Hafnium powder, dry	4.2	UN2545	I	SPONTANEOUSLY COMBUSTIBLE	B100
				II	SPONTANEOUSLY COMBUSTIBLE	A19, A20, B101, B106, N34
				III	SPONTANEOUSLY COMBUSTIBLE	B105, B106
	Hafnium powder, wetted with not less than 25 percent water (a visible excess of water must be present) (a) mechanically produced, particle size less than 53 microns; (b) chemically produced, particle size less than 840 microns	4.1	UN1326	II	FLAMMABLE SOLID	A6, A19, A20, N34

HAZARDOUS MATERIALS TABLE

Sym.	Descriptions and shipping names	Hazard	ID No.	PG	Label(s)	Special provisions
	Halogenated Irritating liquids, n.o.s.	6.1	UN1610	I	POISON	T42
				II	POISON	T14
				III	KEEP AWAY FROM FOOD	T14
	Hand signal device, see Signal devices, hand					
	Hazardous substances, liquid or solid, n.o.s., see Environmentally hazardous substances, etc.					
D	Hazardous waste, liquid, n.o.s.	9	NA3082	III	CLASS 9	
D	Hazardous waste, solid, n.o.s.	9	NA3077	III	CLASS 9	B54
	Helium, compressed	2.2	UN1046		NONFLAMMABLE GAS	
	Helium-oxygen mixture, see Rare gases and oxygen mixtures					
	Helium, refrigerated liquid (cryogenic liquid)	2.2	UN1963		NONFLAMMABLE GAS	
	Heptafluoropropane	2.2	UN3296		NONFLAMMABLE GAS	
	n-Heptaldehyde	3	UN3056	III	FLAMMABLE LIQUID	B1, T1
	Heptanes	3	UN1206	II	FLAMMABLE LIQUID	T2
	n-Heptene	3	UN2278	II	FLAMMABLE LIQUID	B101, T8
	Hexachloroacetone	6.1	UN2661	III	KEEP AWAY FROM FOOD	T8
	Hexachlorobenzene	6.1	UN2729	III	KEEP AWAY FROM FOOD	
	Hexachlorobutadiene	6.1	UN2279	III	KEEP AWAY FROM FOOD	T7
	Hexachlorocyclopentadiene	6.1	UN2646	I	POISON	2, B9, B14, B32, B74, B77, T38, T43, T45
	Hexachlorophene	6.1	UN2875	III	KEEP AWAY FROM FOOD	

HAZARDOUS MATERIALS TABLE

Sym.	Descriptions and shipping names	Hazard	ID No.	PG	Label(s)	Special provisions
	Hexadecyltrichlorosilane	8	UN1781	II	CORROSIVE	A7, B2, B6, N34, T8
	Hexadienes	3	UN2458	II	FLAMMABLE LIQUID	B101, T7
	Hexaethyl tetraphosphate and compressed gas mixtures	2.3	UN1612		POISON GAS	3
	Hexaethyl tetraphosphate *liquid*	6.1	UN1611	I	POISON	A4
				II	POISON	N76
				III	KEEP AWAY FROM FOOD	N77
	Hexaethyl tetraphosphate, *solid*	6.1	UN1611	I	POISON	
				II	POISON	N76
				III	KEEP AWAY FROM FOOD	N77
	Hexafluoroacetone	2.3	UN2420		POISON GAS, CORROSIVE	2, B9, B14
	Hexafluoroacetone hydrate	6.1	UN2552	II	POISON	T14
	Hexafluoroethane, *R116*	2.2	UN2193		NONFLAMMABLE GAS	
	Hexafluorophosphoric acid	8	UN1782	II	CORROSIVE	A6, A7, B2, N3, N34, T9, T27
	Hexafluoropropylene oxide	2.2	NA1956		NONFLAMMABLE GAS	
	Hexafluoropropylene, *R1216*	2.2	UN1858		NONFLAMMABLE GAS	
	Hexaldehyde	3	UN1207	III	FLAMMABLE LIQUID	B1, T1
	Hexamethylene diisocyanate	6.1	UN2281	II	POISON	B101, T14
D	Hexamethylene triperoxide diamine (dry)	Forbidden				
	Hexamethylenediamine, solid	8	UN2280	III	CORROSIVE	

HAZARDOUS MATERIALS TABLE

Sym.	Descriptions and shipping names	Hazard	ID No.	PG	Label(s)	Special provisions
	Hexamethylenediamine solution	8	UN1783	II	CORROSIVE	T8
				III	CORROSIVE	T7
	Hexamethyleneimine	3	UN2493	II	FLAMMABLE LIQUID, CORROSIVE	B101, T8
	Hexamethylenetetramine	4.1	UN1328	III	FLAMMABLE SOLID	A1
	Hexamethylol benzene hexanitrate	Forbidden				
	Hexanes	3	UN1208	II	FLAMMABLE LIQUID	B101, T8
	2,2',4,4',6,6'-Hexanitro-3,3'-dihydroxyazobenzene (dry)	Forbidden				
	Hexanitroazoxy benzene	Forbidden				
	N,N'-(hexanitrodiphenyl) ethylene dinitramine (dry)	Forbidden				
	Hexanitrodiphenyl urea	Forbidden				
	2,2',3',4,4',6-Hexanitrodiphenylamine	Forbidden				
	Hexanitrodiphenylamine or Dipicrylamine or Hexyl	1.1D	UN0079	II	EXPLOSIVE 1.1D	
	2,3',4,4',6,6'-Hexanitrodiphenylether	Forbidden				
	Hexanitroethane	Forbidden				
	Hexanitrooxanilide	Forbidden				

HAZARDOUS MATERIALS TABLE

Sym.	Descriptions and shipping names	Hazard	ID No.	PG	Label(s)	Special provisions
	Hexanitrostilbene	1.1D	UN0392	II	EXPLOSIVE 1.1D	
	Hexanoic acid, see Corrosive liquids, n.o.s.					
	Hexanols	3	UN2282	III	FLAMMABLE LIQUID	B1, T1
	1-Hexene	3	UN2370	II	FLAMMABLE LIQUID	B101, T8
	Hexogen and cyclotetramethylenetetranitramine mixtures, wetted or desensitized *see RDX and HMX mixtures, wetted or desensitized etc.*					
	Hexogen and HMX mixtures, wetted or desensitized *see RDX and HMX mixtures, wetted or desensitized etc.*					
	Hexogen and octogen mixtures, wetted or desensitized *see RDX and HMX mixtures, wetted or desensitized etc.*					
	Hexogen, *see Cyclotrimethylenetrinitramine, etc.*					
	Hexolite, *or* Hexotol *dry or wetted with less than 15 percent water, by mass*	1.1D	UN0118	II	EXPLOSIVE 1.1D	
	Hexotonal	1.1D	UN0393	II	EXPLOSIVE 1.1D	
	Hexyl, *see Hexanitrodiphenylamine*					
	Hexytrichlorosilane	8	UN1784	II	CORROSIVE	A7, B2, B6, N34, T8, T26
	High explosives, see individual explosives' entries					
	HMX, *see Cyclotetramethylenetetranitramine, etc.*					

HAZARDOUS MATERIALS TABLE

Sym.	Descriptions and shipping names	Hazard	ID No.	PG	Label(s)	Special provisions
	Hydrazine, anhydrous or **Hydrazine aqueous solutions with more than 64 percent hydrazine, by mass**	8	UN2029	I	CORROSIVE, FLAMMABLE LIQUID, POISON	A3, A6, A7, A10, B16, B53, T25
	Hydrazine, aqueous solution with not more than 37 percent hydrazine, by mass	6.1	UN3293	III	KEEP AWAY FROM FOOD	T7
	Hydrazine azide	Forbidden				
	Hydrazine chlorate	Forbidden				
	Hydrazine dicarbonic acid diazide	Forbidden				
	Hydrazine hydrate or **Hydrazine aqueous solutions, with not less than 37 percent but not more than 64 percent hydrazine, by mass**	8	UN2030	II	CORROSIVE, POISON	B16, B53, B110, T15
	Hydrazine perchlorate	Forbidden				
	Hydrazine selenate	Forbidden				
	Hydriodic acid, anhydrous, see Hydrogen iodide, anhydrous					
	Hydriodic acid, solution	8	UN1787	II	CORROSIVE	A3, A6, B2, N41, T9, T27
				III	CORROSIVE	T8, T26
	Hydrobromic acid, anhydrous, see Hydrogen bromide, anhydrous					

HAZARDOUS MATERIALS TABLE

Sym.	Descriptions and shipping names	Hazard	ID No.	PG	Label(s)	Special provisions
	Hydrobromic acid solution, with more than 49 percent hydrobromic acid	8	UN1788	II	CORROSIVE	B2, B15, N41, T9, T27
				III	CORROSIVE	T8, T26
	Hydrobromic acid solution, with not more than 49 percent hydrobromic acid	8	UN1788	II	CORROSIVE	A3, A6, B2, B15, N41, T9, T27
				III	CORROSIVE	T8, T26
	Hydrocarbon gases, compressed, n.o.s. or Hydrocarbon gases mixtures, compressed, n.o.s.	2.1	UN1964		FLAMMABLE GAS	
	Hydrocarbon gases, liquefied, n.o.s. or Hydrocarbon gases mixtures, liquefied, n.o.s.	2.1	UN1965		FLAMMABLE GAS	
	Hydrocarbons, liquid, n.o.s.	3	UN3295	I	FLAMMABLE LIQUID	T8, T31
				II	FLAMMABLE LIQUID	T8, T31
				III	FLAMMABLE LIQUID	B1, T7, T30
	Hydrochloric acid, anhydrous, see Hydrogen chloride, anhydrous					
	Hydrochloric acid, solution	8	UN1789	II	CORROSIVE	A3, A6, B3, B15, N41, T9, T27
				III	CORROSIVE	T8, T26
	Hydrocyanic acid, anhydrous, see Hydrogen cyanide etc.					
	Hydrocyanic acid, aqueous solutions or Hydrogen cyanide, aqueous solutions with not more than 20 percent hydrogen cyanide	6.1	UN1613	I	POISON	2, B12, B61, B65, B77, B82

HAZARDOUS MATERIALS TABLE

Sym.	Descriptions and shipping names	Hazard	ID No.	PG	Label(s)	Special provisions
D	**Hydrocyanic acid, aqueous solutions with less than 5 percent hydrogen cyanide**	6.1	NA1613	II	POISON	B12, T18, T26
	Hydrocyanic acid, liquefied, see **Hydrogen cyanide, etc.**	Forbidden				
	Hydrocyanic acid (prussic), unstabilized	Forbidden				
	Hydrofluoric acid and Sulfuric acid mixtures	8	UN1786	I	CORROSIVE, POISON	A6, A7, B15, B23, N5, N34, T18, T27
	Hydrofluoric acid, anhydrous, see **Hydrogen fluoride, anhydrous**					
	Hydrofluoric acid, solution, with more than 60 percent strength	8	UN1790	I	CORROSIVE, POISON	A6, A7, B4, B12, B15, B23, N5, N34, T18, T27
	Hydrofluoric acid, solution, with not more than 60 percent strength	8	UN1790	II	CORROSIVE, POISON	A6, A7, B12, B15, B110, N5, N34, T18, T27
	Hydrofluoroboric acid, see **Fluoroboric acid**					
	Hydrofluorosilicic acid, see **Fluorosilicic acid**					
	Hydrogen and Methane mixtures, compressed	2.1	UN2034		FLAMMABLE GAS	
	Hydrogen bromide, anhydrous	2.3	UN1048		POISON GAS, CORROSIVE	3, B14
	Hydrogen chloride, anhydrous	2.3	UN1050		POISON GAS, CORROSIVE	3
	Hydrogen chloride, refrigerated liquid	2.3	UN2186		POISON GAS, CORROSIVE	3, B6
	Hydrogen, compressed	2.1	UN1049		FLAMMABLE GAS	

HAZARDOUS MATERIALS TABLE

Sym.	Descriptions and shipping names	Hazard	ID No.	PG	Label(s)	Special provisions
	Hydrogen cyanide, solution in alcohol with not more than 45 percent hydrogen cyanide	6.1	UN3294	I	POISON, FLAMMABLE LIQUID	T18, T26
	Hydrogen cyanide, stabilized with less than 3 percent water	6.1	UN1051	I	POISON, FLAMMABLE LIQUID	1, B12, B35, B61, B65, B77, B82
	Hydrogen cyanide, stabilized, with less than 3 percent water and absorbed in a porous inert material	6.1	UN1614	I	POISON	5
	Hydrogen fluoride, anhydrous	8	UN1052	I	CORROSIVE, POISON	3, B7, B12, B46, B71, B77, T24, T27
	Hydrogen iodide, anhydrous	2.3	UN2197		POISON GAS	3, 25, B14
	Hydrogen iodide solution, see **Hydriodic acid, solution**					
	Hydrogen peroxide and peroxyacetic acid mixtures, stabilized with acids, water and not more than 5 percent peroxyacetic acid	5.1	UN3149	II	OXIDIZER, CORROSIVE	A2, A3, A6, B12, B53, B104, B110, T14
	Hydrogen peroxide, aqueous solutions with more than 40 percent but not more than 60 percent hydrogen peroxide (stabilized as necessary)	5.1	UN2014	II	OXIDIZER, CORROSIVE	12, A3, A6, B12, B53, B80, B81, B85, B104, B110, T14, T37
	Hydrogen peroxide, aqueous solutions with not less than 8 percent but less than 20 percent hydrogen peroxide (stabilized as necessary)	5.1	UN2984	III	OXIDIZER	17, A1, B104, T8, T37
	Hydrogen peroxide, aqueous solutions with not less than 20 percent but not more than 40 percent hydrogen peroxide (stabilized as necessary)	5.1	UN2014	II	OXIDIZER, CORROSIVE	A2, A3, A6, B12, B53, B104, B110, T14, T37

HAZARDOUS MATERIALS TABLE

Sym.	Descriptions and shipping names	Hazard	ID No.	PG	Label(s)	Special provisions
	Hydrogen peroxide, stabilized or **Hydrogen peroxide aqueous solutions, stabilized** with more than 60 percent hydrogen peroxide	5.1	UN2015		OXIDIZER, CORROSIVE	12, A3, A6, B12, B53, B80, B81, B85, T15, T37
	Hydrogen, refrigerated liquid (cryogenic liquid)	2.1	UN1966		FLAMMABLE GAS	1
	Hydrogen selenide, anhydrous	2.3	UN2202		POISON GAS, FLAMMABLE GAS	
	Hydrogen sulfate, see Sulfuric acid					
	Hydrogen sulfide, liquefied	2.3	UN1053		POISON GAS, FLAMMABLE GAS	2, B9, B14
	Hydrogendifluorides, n.o.s. solid	8	UN1740	II	CORROSIVE	N3, N34
		8	UN1740	III	CORROSIVE	N3, N34
	Hydrogendifluorides, n.o.s. solutions	8	UN1740	II	CORROSIVE	N3, N34
		8	UN1740	III	CORROSIVE	N3, N34
	Hydroquinone	6.1	UN2662	III	KEEP AWAY FROM FOOD	
	Hydrosilicofluoric acid, see Fluorosilicic acid					
	Hydroxyl amine iodide		Forbidden			
	Hydroxylamine sulfate	8	UN2865	III	CORROSIVE	
	Hypochlorite solutions with more than 5 percent but less than 16 percent available chlorine	8	UN1791	III	CORROSIVE	B104, N34, T7
	Hypochlorite solutions with 16 percent or more available chlorine	8	UN1791	II	CORROSIVE	A7, B2, B15, N34, T7
	Hypochlorites, inorganic, n.o.s.	5.1	UN3212	II	OXIDIZER	

HAZARDOUS MATERIALS TABLE

Sym.	Descriptions and shipping names	Hazard	ID No.	PG	Label(s)	Special provisions
	Hyponitrous acid	Forbidden				
	Igniter fuse, metal clad, see Fuse, Igniter, tubular, metal clad					
	Igniters	1.1G	UN0121	II	EXPLOSIVE 1.1G	
	Igniters	1.2G	UN0314	II	EXPLOSIVE 1.2G	
	Igniters	1.3G	UN0315	II	EXPLOSIVE 1.3G	
	Igniters	1.4G	UN0325	II	EXPLOSIVE 1.4G	
	Igniters	1.4S	UN0454	II	EXPLOSIVE 1.4S	
	3,3'-Iminodipropylamine	8	UN2269	III	CORROSIVE	T8
	Infectious substances, affecting animals only	6.2	UN2900		INFECTIOUS SUBSTANCE	
	Infectious substances, affecting humans	6.2	UN2814		INFECTIOUS SUBSTANCE	
	Inflammable, see Flammable					
	Initiating explosives (dry)	Forbidden				
	Inositol hexanitrate (dry)	Forbidden				
D	Insecticide gases *flammable* n.o.s.	2.1	NA1954		FLAMMABLE GAS	
	Insecticide gases, n.o.s.	2.2	UN1968		NONFLAMMABLE GAS	
	Insecticide gases, toxic, n.o.s.	2.3	UN1967		POISON GAS	3
	Inulin trinitrate (dry)	Forbidden				
	Iodine azide (dry)	Forbidden				

HAZARDOUS MATERIALS TABLE

Sym.	Descriptions and shipping names	Hazard	ID No.	PG	Label(s)	Special provisions
	Iodine monochloride	8	UN1792	II	CORROSIVE	B6, N41, T8, T26
	Iodine pentafluoride	5.1	UN2495	I	OXIDIZER, POISON, CORROSIVE	
	2-Iodobutane	3	UN2390	II	FLAMMABLE LIQUID	T8
	Iodomethylpropanes	3	UN2391	II	FLAMMABLE LIQUID	T8
	Iodopropanes	3	UN2392	III	FLAMMABLE LIQUID	B1, T8
	Iodoxy compounds (dry)	Forbidden				
	Iridium nitratopentamine iridium nitrate	Forbidden				
	Iron chloride, see Ferric chloride					
	Iron oxide, spent, or Iron sponge, spent obtained from coal gas purification	4.2	UN1376	III	SPONTANEOUSLY COMBUSTIBLE	B18
	Iron pentacarbonyl	6.1	UN1994	I	POISON, FLAMMABLE LIQUID	1, B9, B14, B30, B72, B77, T38, T43, T44
	Iron sesquichloride, see Ferric chloride					
	Irritating material, see Tear gas substances, etc.					
	Isobutane or Isobutane mixtures see also Petroleum gases, liquefied	2.1	UN1969		FLAMMABLE GAS	19
	Isobutanol or Isobutyl alcohol	3	UN1212	III	FLAMMABLE LIQUID	B1, T1
	Isobutyl acetate	3	UN1213	II	FLAMMABLE LIQUID	T1
	Isobutyl acrylate	3	UN2527	III	FLAMMABLE LIQUID	B1, T1
	Isobutyl alcohol, see Isobutanol					
	Isobutyl aldehyde, see Isobutyraldehyde					

HAZARDOUS MATERIALS TABLE

Sym.	Descriptions and shipping names	Hazard	ID No.	PG	Label(s)	Special provisions
D	Isobutyl chloroformate	6.1	NA2742	I	POISON, FLAMMABLE LIQUID, CORROSIVE	2, B9, B14, B32, B74, T38, T43, T45
	Isobutyl formate	3	UN2393	II	FLAMMABLE LIQUID	T1
	Isobutyl isobutyrate	3	UN2528	III	FLAMMABLE LIQUID	B1, T1
+	Isobutyl isocyanate	3	UN2486	II	FLAMMABLE LIQUID, POISON	1, B9, B14, B30, B72, T38, T43, T44
	Isobutyl methacrylate	3	UN2283	III	FLAMMABLE LIQUID	B1, T1
	Isobutyl propionate	3	UN2394	III	FLAMMABLE LIQUID	B1, T1
	Isobutylamine	3	UN1214	II	FLAMMABLE LIQUID, CORROSIVE	B101, T8
	Isobutylene see also Petroleum gases, liquefied	2.1	UN1055		FLAMMABLE GAS	19
	Isobutyraldehyde or Isobutyl aldehyde	3	UN2045	II	FLAMMABLE LIQUID	T8
	Isobutyric acid	3	UN2529	III	FLAMMABLE LIQUID, CORROSIVE	B1, T1
	Isobutyric anhydride	3	UN2530	III	FLAMMABLE LIQUID, CORROSIVE	B1, T1
	Isobutyronitrile	3	UN2284	II	FLAMMABLE LIQUID, POISON	T17
	Isobutyryl chloride	3	UN2395	II	FLAMMABLE LIQUID, CORROSIVE	B100, T9, T26
	Isocyanates, flammable, toxic, n.o.s. or Isocyanate solutions, flammable, toxic, n.o.s. *flashpoint less than 23 degrees C*	3	UN2478	II	FLAMMABLE LIQUID, POISON	5, A3, A7, T15

HAZARDOUS MATERIALS TABLE

Sym.	Descriptions and shipping names	Hazard	ID No.	PG	Label(s)	Special provisions
	Isocyanates, toxic, flammable, n.o.s. or **Isocyanate solutions, toxic, flammable, n.o.s.**, flash point not less than 23 degrees C but not more than 61 degrees C and boiling point less than 300 degrees C	6.1	UN3080	II	POISON, FLAMMABLE LIQUID	T15
	Isocyanates, toxic, n.o.s. or **Isocyanate, solutions, toxic, n.o.s.**, flash point more than 61 degrees C and boiling point less than 300 degrees C	6.1	UN2206	II	POISON	T15
				III	KEEP AWAY FROM FOOD	T8
	Isocyanatobenzotrifluorides	6.1	UN2285	II	POISON	5, B101, T14
	Isoheptenes	3	UN2287	II	FLAMMABLE LIQUID	T7
	Isohexenes	3	UN2288	II	FLAMMABLE LIQUID	T7
	Isooctane, see Octanes					
	Isooctenes	3	UN1216	II	FLAMMABLE LIQUID	T8
	Isopentane, see Pentane					
	Isopentanoic acid, see Corrosive liquids, n.o.s.					
	Isopentenes	3	UN2371	I	FLAMMABLE LIQUID	T20
	Isophorone diisocyanate	6.1	UN2290	III	KEEP AWAY FROM FOOD	T7
	Isophoronediamine	8	UN2289	III	CORROSIVE	T8
	Isoprene, inhibited	3	UN1218	I	FLAMMABLE LIQUID	T20
	Isopropanol or Isopropyl alcohol	3	UN1219	II	FLAMMABLE LIQUID	T1
	Isopropenyl acetate	3	UN2403	II	FLAMMABLE LIQUID	T1
	Isopropenylbenzene	3	UN2303	III	FLAMMABLE LIQUID	B1, T1

HAZARDOUS MATERIALS TABLE

Sym.	Descriptions and shipping names	Hazard	ID No.	PG	Label(s)	Special provisions
	Isopropyl acetate	3	UN1220	II	FLAMMABLE LIQUID	T1
	Isopropyl acid phosphate	8	UN1793	III	CORROSIVE	T7
	Isopropyl alcohol, see **Isopropanol**					
	Isopropyl butyrate	3	UN2405	III	FLAMMABLE LIQUID	B1, T1
	Isopropyl chloroacetate	3	UN2947	III	FLAMMABLE LIQUID	B1, T1
	Isopropyl chloroformate	6.1	UN2407	I	POISON, FLAMMABLE LIQUID, CORROSIVE	2, B9, B14, B32, B74, B77, T38, T43, T45
	Isopropyl 2-chloropropionate	3	UN2934	III	FLAMMABLE LIQUID	B1, T1
	Isopropyl isobutyrate	3	UN2406	III	FLAMMABLE LIQUID	T1
+	**Isopropyl isocyanate**	3	UN2483	I	FLAMMABLE LIQUID, POISON	1, B9, B14, B30, B72, T38, T43, T44
	Isopropyl mercaptan, see **Propanethiols**					
	Isopropyl nitrate	3	UN1222	II	FLAMMABLE LIQUID	T25
	Isopropyl phosphoric acid, see **Isopropyl acid phosphate**					
	Isopropyl propionate	3	UN2409	II	FLAMMABLE LIQUID	T1
	Isopropylamine	3	UN1221	I	FLAMMABLE LIQUID, CORROSIVE	T20
	Isopropylbenzene	3	UN1918	III	FLAMMABLE LIQUID	B1, T1
	Isopropylcumyl hydroperoxide, with more than 72 percent in solution	Forbidden				
	Isosorbide dinitrate mixture with not less than 60 percent lactose, mannose, starch or calcium hydrogen phosphate	4.1	UN2907	II	FLAMMABLE SOLID	
	Isosorbide-5-mononitrate	4.1	UN3251	III	FLAMMABLE SOLID	

HAZARDOUS MATERIALS TABLE

Sym.	Descriptions and shipping names	Hazard	ID No.	PG	Label(s)	Special provisions
	Isothiocyanic acid	Forbidden				
	Jet fuel, see **Fuel aviation, turbine engine**					
D	**Jet perforating guns, charged oil well, with detonator**	1.1D	NA0124	II	EXPLOSIVE 1.1D	D55, 56
D	**Jet perforating guns, charged oil well, with detonator**	1.4D	NA0494	II	EXPLOSIVE 1.4D	55, 56
	Jet perforating guns, charged *oil well, without detonator*	1.1D	UN0124	II	EXPLOSIVE 1.1D	55
	Jet perforating guns, charged, *oil well, without detonator*	1.4D	UN0494	II	EXPLOSIVE 1.4D	55, 114
	Jet perforators, see **Charges, shaped, commercial etc.**					
	Jet tappers, without detonator, see **Charges, shaped commercial, etc.**					
	Jet thrust igniters, for rocket motors or Jato, see **Igniters**					
	Jet thrust unit (Jato), see **Rocket motors**					
	Kerosene	3	UN1223	III	FLAMMABLE LIQUID	B1, T1
	Ketones, liquid, n.o.s.	3	UN1224	I	FLAMMABLE LIQUID	T8, T31
				II	FLAMMABLE LIQUID	T8, T31
				III	FLAMMABLE LIQUID	B1, T7, T30
	Krypton, compressed	2.2	UN1056		NONFLAMMABLE GAS	
	Krypton, refrigerated liquid *(cryogenic liquid)*	2.2	UN1970		NONFLAMMABLE GAS	

HAZARDOUS MATERIALS TABLE

Sym.	Descriptions and shipping names	Hazard	ID No.	PG	Label(s)	Special provisions
	Lacquer base or lacquer chips, nitrocellulose, dry, see **Nitrocellulose**, etc. (UN 2557)					
	Lacquer base or lacquer chips, plastic, wet with alcohol or solvent, see **Nitrocellulose** (UN 2059, UN 2060, UN 2555, UN2556) or **Paint** etc. (UN1263)					
	Lead acetate	6.1	UN1616	III	KEEP AWAY FROM FOOD	
	Lead arsenates	6.1	UN1617	II	POISON	
	Lead arsenites	6.1	UN1618	II	POISON	
	Lead azide (dry)	Forbidden				
	Lead azide, **wetted** with not less than 20 percent water or mixture of alcohol and water, by mass	1.1A	UN0129	II	EXPLOSIVE 1.1A	111, 117
	Lead compounds, soluble, n.o.s.	6.1	UN2291	III	KEEP AWAY FROM FOOD	
	Lead cyanide	6.1	UN1620	II	POISON	
	Lead dioxide	5.1	UN1872	III	OXIDIZER	A1
	Lead dross, see **Lead sulfate**, with more than 3 percent free acid					
D	Lead mononitroresorcinate	1.1A	NA0473	II	EXPLOSIVE 1.1A	111, 117
	Lead nitrate	5.1	UN1469	II	OXIDIZER, POISON	
	Lead nitroresorcinate (dry)	Forbidden				
	Lead perchlorate, solid	5.1	UN1470	II	OXIDIZER, POISON	T8
	Lead perchlorate, solution	5.1	UN1470	II	OXIDIZER, POISON	T8

HAZARDOUS MATERIALS TABLE

Sym.	Descriptions and shipping names	Hazard	ID No.	PG	Label(s)	Special provisions
	Lead peroxide, see **Lead dioxide**					
	Lead phosphite, dibasic	4.1	UN2989	II	FLAMMABLE SOLID	
				III	FLAMMABLE SOLID	
	Lead picrate (dry)	Forbidden				
	Lead styphnate (dry)	Forbidden				
	Lead styphnate, wetted or **Lead trinitroresorcinate, wetted** with not less than 20 percent water or mixture of alcohol and water, by mass	1.1A	UN0130	II	EXPLOSIVE 1.1A	111, 117
	Lead sulfate with more than 3 percent free acid	8	UN1794	II	CORROSIVE	
	Lead trinitroresorcinate, see **Lead styphnate, etc.**					
	Life-saving appliances, not self inflating containing dangerous goods as equipment	9	UN3072		None	
	Life-saving appliances, self inflating	9	UN2990		None	
	Lighter replacement cartridges containing liquefied petroleum gases (and similar devices, each not exceeding 65 grams), see **Lighters or lighter refills etc.** containing flammable gas					
D	**Lighters for cigars, cigarettes, etc., with lighter fluids**	3	NA1226	II	FLAMMABLE LIQUID	N10
	Lighters, fuse	1.4S	UN0131	II	EXPLOSIVE 1.4S	

HAZARDOUS MATERIALS TABLE

Sym.	Descriptions and shipping names	Hazard	ID No.	PG	Label(s)	Special provisions
	Lighters or Lighter refills cigarettes, containing flammable gas	2.1	UN1057		FLAMMABLE GAS	N10
	Lime, unslaked, see **Calcium oxide**					
	Liquefied gas, flammable, n.o.s.	2.1	UN3161		FLAMMABLE GAS	
	Liquefied gas, n.o.s.	2.2	UN3163		NONFLAMMABLE GAS	
	Liquefied gas, oxidizing, n.o.s.	2.2	UN3157		NONFLAMMABLE GAS, OXIDIZER	
	Liquefied gas, toxic, flammable, n.o.s. *Inhalation Hazard Zone A*	2.3	UN3160		POISON GAS, FLAMMABLE GAS	1
	Liquefied gas, toxic, flammable, n.o.s. *Inhalation Hazard Zone B*	2.3	UN3160		POISON GAS, FLAMMABLE GAS	2, B9, B14
	Liquefied gas, toxic, flammable, n.o.s. *Inhalation Hazard Zone C*	2.3	UN3160		POISON GAS, FLAMMABLE GAS	3, B14
	Liquefied gas, toxic, flammable, n.o.s. *Inhalation Hazard Zone D*	2.3	UN3160		POISON GAS, FLAMMABLE GAS	4
	Liquefied gas, toxic, n.o.s. *Inhalation Hazard Zone A*	2.3	UN3162		POISON GAS	1
	Liquefied gas, toxic, n.o.s. *Inhalation Hazard Zone B*	2.3	UN3162		POISON GAS	2, B9, B14
	Liquefied gas, toxic, n.o.s. *Inhalation Hazard Zone C*	2.3	UN3162		POISON GAS	3, B14
	Liquefied gas, toxic, n.o.s. *Inhalation Hazard Zone D*	2.3	UN3162		POISON GAS	4
	Liquefied gases, non-flammable charged with nitrogen, carbon dioxide or air	2.2	UN1058		NONFLAMMABLE GAS	
	Liquefied hydrocarbon gas, see **Hydrocarbon gases, liquefied, n.o.s.**, etc.					

HAZARDOUS MATERIALS TABLE

Sym.	Descriptions and shipping names	Hazard	ID No.	PG	Label(s)	Special provisions
	Liquefied natural gas, see Methane, etc. (UN 1972)					
	Liquefied petroleum gas see **Petroleum gases, liquefied**					
	Lithium	4.3	UN1415	I	DANGEROUS WHEN WET	A7, A19, B100, N45
	Lithium acetylide ethylenediamine complex, see Water reactive solid etc.					
	Lithium alkyls	4.2	UN2445	I	SPONTANEOUSLY COMBUSTIBLE, DANGEROUS WHEN WET	B11, T28, T40
	Lithium aluminum hydride	4.3	UN1410	I	DANGEROUS WHEN WET	A19, B100,
	Lithium aluminum hydride, ethereal	4.3	UN1411	I	DANGEROUS WHEN WET, FLAMMABLE LIQUID	A2, A3, A11, N34
	Lithium batteries, contained in equipment	9	UN3091	II	CLASS 9	29
	Lithium battery	9	UN3090	II	CLASS 9	29
	Lithium borohydride	4.3	UN1413	I	DANGEROUS WHEN WET	A19, B100, N40
	Lithium ferrosilicon	4.3	UN2830	II	DANGEROUS WHEN WET	A19, B105, B106
	Lithium hydride	4.3	UN1414	I	DANGEROUS WHEN WET	A19, B100, N40
	Lithium hydride, fused solid	4.3	UN2805	II	DANGEROUS WHEN WET	A8, A19, A20, B101, B106
	Lithium hydroxide, monohydrate or Lithium hydroxide, solid	8	UN2680	II	CORROSIVE	B2, T8
	Lithium hydroxide, solution	8	UN2679	III	CORROSIVE	T8

Sym.	Descriptions and shipping names	Hazard	ID No.	PG	Label(s)	Special provisions
	Lithium hypochlorite, dry or Lithium hypochlorite mixtures, dry	5.1	UN1471	II	OXIDIZER	A9, N34
	Lithium in cartridges, see Lithium					
	Lithium nitrate	5.1	UN2722	III	OXIDIZER	A1
	Lithium nitride	4.3	UN2806	I	DANGEROUS WHEN WET	A19, B101, B106, N40
	Lithium peroxide	5.1	UN1472	II	OXIDIZER	A9, N34
	Lithium silicon	4.3	UN1417	II	DANGEROUS WHEN WET	A19, A20, B105, B106
	LNG, see Methane etc. (UN 1972)					
	London purple	6.1	UN1621	II	POISON	
	LPG, see Petroleum gases, liquefied					
	Lye, see Sodium hydroxide, solutions					
	Magnesium alkyls	4.2	UN3053	I	SPONTANEOUSLY COMBUSTIBLE	B11, T28, T29, T40
	Magnesium aluminum phosphide	4.3	UN1419	I	DANGEROUS WHEN WET, POISON	A19, B100, N34, N40
	Magnesium arsenate	6.1	UN1622	II	POISON	
	Magnesium bisulfite solution, see Bisulfites, aqueous solutions, n.o.s.					
	Magnesium bromate	5.1	UN1473	II	OXIDIZER	A1
	Magnesium chlorate	5.1	UN2723	II	OXIDIZER	
+	**Magnesium diamide**	4.2	UN2004	II	SPONTANEOUSLY COMBUSTIBLE	A8, A19, A20

HAZARDOUS MATERIALS TABLE

Sym.	Descriptions and shipping names	Hazard	ID No.	PG	Label(s)	Special provisions
	Magnesium diphenyl	4.2	UN2005	I	SPONTANEOUSLY COMBUSTIBLE	
	Magnesium dross, wet or hot	Forbidden				
	Magnesium fluorosilicate	6.1	UN2853	III	KEEP AWAY FROM FOOD	
	Magnesium granules, coated particle size not less than 149 microns	4.3	UN2950	III	DANGEROUS WHEN WET	A1, A19, B108
	Magnesium hydride	4.3	UN2010	I	DANGEROUS WHEN WET	A19, B100, N40
	Magnesium or Magnesium alloys with more than 50 percent magnesium in pellets, turnings or ribbons	4.1	UN1869	III	FLAMMABLE SOLID	A1
	Magnesium nitrate	5.1	UN1474	III	OXIDIZER	A1
	Magnesium perchlorate	5.1	UN1475	II	OXIDIZER	
	Magnesium peroxide	5.1	UN1476	II	OXIDIZER	
	Magnesium phosphide	4.3	UN2011	I	DANGEROUS WHEN WET, POISON	A19, N40
	Magnesium, powder or Magnesium alloys, powder	4.3	UN1418	I	DANGEROUS WHEN WET, SPONTANEOUSLY COMBUSTIBLE	A19, B56
				II	DANGEROUS WHEN WET, SPONTANEOUSLY COMBUSTIBLE	A19, B56, B101, B106
				III	DANGEROUS WHEN WET, SPONTANEOUSLY COMBUSTIBLE	A19, B56, B106, B108
	Magnesium scrap, see Magnesium, etc. (UN 1869)					

HAZARDOUS MATERIALS TABLE

Sym.	Descriptions and shipping names	Hazard	ID No.	PG	Label(s)	Special provisions
	Magnesium silicide	4.3	UN2624	II	DANGEROUS WHEN WET	A19, A20, B105, B106
	Magnetized material, see section 173.21					
D	**Maleic acid**	8	NA2215	III	CORROSIVE	
	Maleic anhydride	8	UN2215	III	CORROSIVE	T7
	Malononitrile	6.1	UN2647	II	POISON	
	Mancozeb (manganese ethylenebisdithiocarbamate complex with zinc) see Maneb					
	Maneb or Maneb preparations with not less than 60 percent maneb	4.2	UN2210	III	SPONTANEOUSLY COMBUSTIBLE, DANGEROUS WHEN WET	A1, A19, B105
	Maneb stabilized or Maneb preparations, stabilized against self-heating	4.3	UN2968	III	DANGEROUS WHEN WET	54, A1, A19, B108
	Manganese nitrate	5.1	UN2724	III	OXIDIZER	A1
	Manganese resinate	4.1	UN1330	III	FLAMMABLE SOLID	A1
	Mannitan tetranitrate	Forbidden				
	Mannitol hexanitrate (dry)	Forbidden				
D	**Mannitol hexanitrate, wetted or Nitromannite, wetted with not less than 40 percent water, by mass or mixture of alcohol and water**	1.1A	NA0133	II	EXPLOSIVE 1.1A	111
	Marine pollutants, liquid or solid, n.o.s., see Environmentally hazardous substances, liquid or solid, n.o.s.					

HAZARDOUS MATERIALS TABLE

Sym.	Descriptions and shipping names	Hazard	ID No.	PG	Label(s)	Special provisions
	Matches, block, see Matches, 'strike anywhere'					
	Matches, fusee	4.1	UN2254	III	FLAMMABLE SOLID	
	Matches, safety (book, card or strike on box)	4.1	UN1944	III	FLAMMABLE SOLID	
	Matches, strike anywhere	4.1	UN1331	III	FLAMMABLE SOLID	
	Matches, wax, Vesta	4.1	UN1945	III	FLAMMABLE SOLID	
	Matting acid, see Sulfuric acid					
	Medicine, liquid, flammable, toxic, n.o.s.	3	UN3248	II	FLAMMABLE LIQUID, POISON	36
				III	FLAMMABLE LIQUID, KEEP AWAY FROM FOOD	36
	Medicine, liquid, toxic, n.o.s.	6.1	UN1851	II	POISON	
				III	KEEP AWAY FROM FOOD	36
	Medicine, solid, toxic, n.o.s.	6.1	UN3249	II	POISON	36
				III	KEEP AWAY FROM FOOD	
D	Medicines, corrosive, liquid, n.o.s.	8	NA1760	II	CORROSIVE	B3
				III	CORROSIVE	
D	Medicines, corrosive, solid, n.o.s.	8	NA1759	II	CORROSIVE	
				III	CORROSIVE	
D	Medicines, flammable, liquid, n.o.s.	3	NA1993	I	FLAMMABLE LIQUID	
				II	FLAMMABLE LIQUID	
				III	FLAMMABLE LIQUID	B1
D	Medicines, flammable, solid, n.o.s.	4.1	NA1325	II	FLAMMABLE SOLID	

HAZARDOUS MATERIALS TABLE

Sym.	Descriptions and shipping names	Hazard	ID No.	PG	Label(s)	Special provisions
D	Medicines, oxidizing substance, solid, n.o.s.	5.1	NA1479	II	OXIDIZER	
	Memtetrahydrophthalic anhydride, see Corrosive liquids, n.o.s.					
	Mercaptans, liquid, flammable, toxic, n.o.s. or Mercaptan mixtures, liquid, flammable, toxic, n.o.s.	3	UN1228	II	FLAMMABLE LIQUID, POISON	T13
				III	FLAMMABLE LIQUID, KEEP AWAY FROM FOOD	B1, T8
	Mercaptans, liquid, toxic, flammable, n.o.s. or Mercaptan mixtures, liquid, toxic, flammable, n.o.s., *flash point not less than 23 degrees C*	6.1	UN3071	II	POISON, FLAMMABLE LIQUID	T14
	5-Mercaptotetrazol-1-acetic acid	1.4C	UN0448	II	EXPLOSIVE 1.4C	
	Mercuric arsenate	6.1	UN1623	II	POISON	
	Mercuric chloride	6.1	UN1624	II	POISON	
	Mercuric compounds, see Mercury compounds, etc.					
	Mercuric nitrate	6.1	UN1625	II	POISON	N73
	Mercuric potassium cyanide	6.1	UN1626	I	POISON	N74, N75
	Mercuric sulfocyanate, see Mercury thiocyanate					
	Mercurol, see Mercury nucleate					
+	Mercurous azide	Forbidden				
	Mercurous compounds, see Mercury compounds, etc.					
	Mercurous nitrate	6.1	UN1627	II	POISON	

HAZARDOUS MATERIALS TABLE

Sym.	Descriptions and shipping names	Hazard	ID No.	PG	Label(s)	Special provisions
A, W	**Mercury**	8	UN2809	III	CORROSIVE	
	Mercury acetate	6.1	UN1629	II	POISON	
	Mercury acetylide	Forbidden				
	Mercury ammonium chloride	6.1	UN1630	II	POISON	
	Mercury based pesticides, liquid, flammable, toxic, *flash point less than 23 degrees C*	3	UN2778	I	FLAMMABLE LIQUID, POISON	
				II	FLAMMABLE LIQUID, POISON	
	Mercury based pesticides, liquid, toxic	6.1	UN3012	I	POISON	T42
				II	POISON	T14
				III	KEEP AWAY FROM FOOD	T14
	Mercury based pesticides, liquid, toxic, flammable, *flashpoint not less than 23 degrees C*	6.1	UN3011	I	POISON, FLAMMABLE LIQUID	T42
				II	POISON, FLAMMABLE LIQUID	T14
				III	KEEP AWAY FROM FOOD, FLAMMABLE LIQUID	T14
	Mercury based pesticides, solid, toxic	6.1	UN2777	I	POISON	
				II	POISON	
				III	KEEP AWAY FROM FOOD	
	Mercury benzoate	6.1	UN1631	II	POISON	

HAZARDOUS MATERIALS TABLE

Sym.	Descriptions and shipping names	Hazard	ID No.	PG	Label(s)	Special provisions
	Mercury bromides	6.1	UN1634	II	POISON	
	Mercury compounds, liquid, n.o.s.	6.1	UN2024	I	POISON	
				II	POISON	
				III	KEEP AWAY FROM FOOD	
	Mercury compounds, solid, n.o.s.	6.1	UN2025	I	POISON	
				II	POISON	
				III	KEEP AWAY FROM FOOD	
A	Mercury contained in manufactured articles	8	UN2809	I	CORROSIVE	
	Mercury cyanide	6.1	UN1636	II	POISON	N74, N75
	Mercury fulminate, wetted with not less than 20 percent water, or mixture of alcohol and water, by mass	1.1A	UN0135	II	EXPLOSIVE 1.1A	111, 117
	Mercury gluconate	6.1	UN1637	II	POISON	
	Mercury iodide	6.1	UN1638	II	POISON	
	Mercury iodide aquabasic ammonobasic (Iodide of Millon's base)	Forbidden				
	Mercury iodide, solution	6.1	UN1638	II	POISON	
	Mercury nitride	Forbidden				
	Mercury nucleate	6.1	UN1639	II	POISON	
	Mercury oleate	6.1	UN1640	II	POISON	
	Mercury oxide	6.1	UN1641	II	POISON	
	Mercury oxycyanide	Forbidden				

HAZARDOUS MATERIALS TABLE

Sym.	Descriptions and shipping names	Hazard	ID No.	PG	Label(s)	Special provisions
	Mercury oxycyanide, desensitized	6.1	UN1642	II	POISON	
	Mercury potassium iodide	6.1	UN1643	II	POISON	
	Mercury salicylate	6.1	UN1644	II	POISON	
	Mercury sulfates	6.1	UN1645	II	POISON	
	Mercury thiocyanate	6.1	UN1646	II	POISON	
	Mesityl oxide	3	UN1229	III	FLAMMABLE LIQUID	B1, T1
	Metal alkyl halides, n.o.s. or Metal aryl halides, n.o.s.	4.2	UN3049	I	SPONTANEOUSLY COMBUSTIBLE	B9, B11, T28, T29, T40
	Metal alkyl hydrides, n.o.s. or Metal aryl hydrides, n.o.s.	4.2	UN3050	I	SPONTANEOUSLY COMBUSTIBLE	B9, B11, T28, T29, T40
D	Metal alkyl, solution, n.o.s.	3	NA9195	II	FLAMMABLE LIQUID	
	Metal alkyls, n.o.s. or Metal aryls, n.o.s.	4.2	UN2003	I	SPONTANEOUSLY COMBUSTIBLE	B11, T42
	Metal carbonyls, n.o.s.	6.1	UN2281	I	POISON	T14
				II	POISON	T7
				III	KEEP AWAY FROM FOOD	
	Metal catalyst, dry	4.2	UN2881	I	SPONTANEOUSLY COMBUSTIBLE	N34
				II	SPONTANEOUSLY COMBUSTIBLE	N34
				III	SPONTANEOUSLY COMBUSTIBLE	N34
	Metal catalyst, wetted with a visible excess of liquid	4.2	UN1378	II	SPONTANEOUSLY COMBUSTIBLE	A2, A8, N34

HAZARDOUS MATERIALS TABLE

Sym.	Descriptions and shipping names	Hazard	ID No.	PG	Label(s)	Special provisions
	Metal hydrides, flammable, n.o.s.	4.1	UN3182	II	FLAMMABLE SOLID	A1
				III	FLAMMABLE SOLID	A1
	Metal hydrides, water reactive, n.o.s.	4.3	UN1409	I	DANGEROUS WHEN WET	A19, B100, N34, N40
				II	DANGEROUS WHEN WET	A19, B101, B106, N34, N40
	Metal powder, self-heating, n.o.s.	4.2	UN3189	II	SPONTANEOUSLY COMBUSTIBLE	
				III	SPONTANEOUSLY COMBUSTIBLE	
	Metal powders, flammable, n.o.s.	4.1	UN3089	II	FLAMMABLE SOLID	
				III	FLAMMABLE SOLID	
	Metal salts of methyl nitramine (dry)	Forbidden				
	Metal salts of organic compounds, flammable, n.o.s.	4.1	UN3181	II	FLAMMABLE SOLID	A1
				III	FLAMMABLE SOLID	A1
	Metaldehyde	4.1	UN1332	III	FLAMMABLE SOLID	A1
	Metallic substance, water-reactive, n.o.s.	4.3	UN3208	I	DANGEROUS WHEN WET	B101, B106
				II	DANGEROUS WHEN WET	B101, B106
				III	DANGEROUS WHEN WET	B105, B108
	Metallic substance, water-reactive, self-heating, n.o.s.	4.3	UN3209	I	DANGEROUS WHEN WET, SPONTANEOUSLY COMBUSTIBLE	B100

HAZARDOUS MATERIALS TABLE

Sym.	Descriptions and shipping names	Hazard	ID No.	PG	Label(s)	Special provisions
	Methacrylaldehyde	3	UN2396	II	DANGEROUS WHEN WET, SPONTANEOUSLY COMBUSTIBLE	B101, B106
	Methacrylic acid, Inhibited	8	UN2531	III	CORROSIVE	45, T8
	Methacrylonitrile, Inhibited	3	UN3079	I	DANGEROUS WHEN WET, SPONTANEOUSLY COMBUSTIBLE	B101, B106
					FLAMMABLE LIQUID, POISON	T8
					FLAMMABLE LIQUID, POISON	2, B9, B14, B32, B74, T38, T43, T45
+	Methallyl alcohol	3	UN2614	III	FLAMMABLE LIQUID	B1, T1
	Methane and hydrogen, mixtures, see Hydrogen and methane, mixtures, etc.					
	Methane, compressed or Natural gas, compressed (with high methane content)	2.1	UN1971		FLAMMABLE GAS	
	Methane, refrigerated liquid (cryogenic liquid) or Natural gas, refrigerated liquid (cryogenic liquid), with high methane content	2.1	UN1972		FLAMMABLE GAS	
	Methanesulfonyl chloride	6.1	UN3246	I	POISON, CORROSIVE	T24, T26
I	Methanol, or Methyl alcohol	3	UN1230	II	FLAMMABLE LIQUID, POISON	T8
D	Methanol, or Methyl alcohol	3	UN1230	II	FLAMMABLE LIQUID	T8
	Methazoic acid	Forbidden				
	4-Methoxy-4-methylpentan-2-one	3	UN2293	III	FLAMMABLE LIQUID	B1, T1
	1-Methoxy-2-propanol	3	UN3092	III	FLAMMABLE LIQUID	B1, T1

HAZARDOUS MATERIALS TABLE

Sym.	Descriptions and shipping names	Hazard	ID No.	PG	Label(s)	Special provisions
+	**Methoxymethyl isocyanate**	3	UN2605	I	FLAMMABLE LIQUID, POISON	1, B9, B14, B30, B72, T38, T43, T44
	Methyl acetate	3	UN1231	II	FLAMMABLE LIQUID	B101, T8
	Methyl acetylene and propadiene mixtures, stabilized	2.1	UN1060		FLAMMABLE GAS	
	Methyl acrylate, inhibited	3	UN1919	II	FLAMMABLE LIQUID	T8
	Methyl alcohol, see Methanol					
	Methyl allyl chloride	3	UN2554	II	FLAMMABLE LIQUID	B101, T8
	Methyl amyl ketone, see Amyl methyl ketone					
	Methyl benzoate	6.1	UN2938	III	KEEP AWAY FROM FOOD	T1
	Methyl bromide	2.3	UN1062		POISON GAS	3, B14
	Methyl bromide and chloropicrin mixtures with more than 2 percent chloropicrin, see Chloropicrin and methyl bromide mixtures					
	Methyl bromide and chloropicrin mixtures with not more than 2 percent chloropicrin, see Methyl bromide					
	Methyl bromide and ethylene dibromide mixtures, liquid	6.1	UN1647	I	POISON	2, B9, B14, B32, B74, N65, T38, T43, T45
	Methyl bromoacetate	6.1	UN2643	II	POISON	B100, T8
	2-Methyl-1-butene	3	UN2459	I	FLAMMABLE LIQUID	T14
	2-Methyl-2-butene	3	UN2460	II	FLAMMABLE LIQUID	T14
	3-Methyl-1-butene	3	UN2561	I	FLAMMABLE LIQUID	T20
	Methyl tert-butyl ether	3	UN2398	II	FLAMMABLE LIQUID	B101, T14
	Methyl butyrate	3	UN1237	II	FLAMMABLE LIQUID	T1

HAZARDOUS MATERIALS TABLE

Sym.	Descriptions and shipping names	Hazard	ID No.	PG	Label(s)	Special provisions
	Methyl chloride	2.1	UN1063		FLAMMABLE GAS	
	Methyl chloride and chloropicrin mixtures, see **Chloropicrin and methyl chloride mixtures**					
	Methyl chloride and methylene chloride mixtures	2.1	UN1912		FLAMMABLE GAS	
	Methyl chloroacetate	6.1	UN2295	II	POISON	T11
	Methyl chlorocarbonate, see **Methyl chloroformate**					
	Methyl chloroform, see **1,1,1-Trichloroethane**					
	Methyl chloroformate	6.1	UN1238	I	POISON, FLAMMABLE LIQUID, CORROSIVE	1, B9, B14, B30, B72, N34, T38, T43, T44
	Methyl chloromethyl ether	6.1	UN1239	I	POISON, FLAMMABLE LIQUID	1, B9, B14, B30, B72, T38, T43, T44
	Methyl 2-chloropropionate	3	UN2933	III	FLAMMABLE LIQUID	B1, T7
	Methyl dichloroacetate	6.1	UN2299	III	KEEP AWAY FROM FOOD	T1
	Methyl ethyl ether, see **Ethyl methyl ether**					
	Methyl ethyl ketone, see **Ethyl methyl ketone**					
	Methyl ethyl ketone peroxide, in solution with more than 9 percent by mass active oxygen	Forbidden				
	2-Methyl-5-ethylpyridine	6.1	UN2300	III	KEEP AWAY FROM FOOD	T7
	Methyl fluoride	2.1	UN2454		FLAMMABLE GAS	
	Methyl formate	3	UN1243	I	FLAMMABLE LIQUID	T20
	Methyl iodide	6.1	UN2644		POISON	2, B9, B14, B32, B74, T38, T43, T45

HAZARDOUS MATERIALS TABLE

Sym.	Descriptions and shipping names	Hazard	ID No.	PG	Label(s)	Special provisions
	Methyl isobutyl carbinol	3	UN2053	III	FLAMMABLE LIQUID	B1, T1
	Methyl isobutyl ketone	3	UN1245	II	FLAMMABLE LIQUID	T1
	Methyl isobutyl ketone peroxide, in solution with more than 9 percent by mass active oxygen	Forbidden				
	Methyl isocyanate	6.1	UN2480	I	POISON, FLAMMABLE LIQUID	1, A7, B9, B14, B30, B72, T38, T43, T44
	Methyl isopropenyl ketone, Inhibited	3	UN1246	II	FLAMMABLE LIQUID	T7
	Methyl isothiocyanate	3	UN2477	II	FLAMMABLE LIQUID, POISON	2, B9, B14, B32, B74, T38, T43, T45
	Methyl isovalerate	3	UN2400	II	FLAMMABLE LIQUID	T1
	Methyl magnesium bromide, in ethyl ether	4.3	UN1928	I	DANGEROUS WHEN WET, FLAMMABLE LIQUID	
	Methyl mercaptan	2.3	UN1064		POISON GAS, FLAMMABLE GAS	3, 25, B7, B9, B14
	Methyl mercaptopropionaldehyde, see Thia-4-pentanal					
+	Methyl methacrylate monomer, Inhibited	3	UN1247	II	FLAMMABLE LIQUID	T8
	Methyl nitramine (dry)	Forbidden				
	Methyl nitrate	Forbidden				
	Methyl nitrite	Forbidden				
	Methyl norbornene dicarboxylic anhydride, see Corrosive liquids, n.o.s.					

HAZARDOUS MATERIALS TABLE

Sym.	Descriptions and shipping names	Hazard	ID No.	PG	Label(s)	Special provisions
	Methyl orthosilicate	6.1	UN2606	I	POISON, FLAMMABLE LIQUID	2, B9, B14, B32, T38, T43, T45
D	Methyl parathion *liquid*	6.1	NA3018	II	POISON	N76, T14
D	Methyl parathion *solid*	6.1	NA2783	II	POISON	N77
D	Methyl phosphonic dichloride	6.1	NA9206	I	POISON, CORROSIVE	2, A3, B9, B14, B32, B74, N34, N43, T38, T43, T45
	Methyl phosphonothioic dichloride, anhydrous, see *Corrosive liquid, n.o.s.*					
D	Methyl phosphonous dichloride, pyrophoric *liquid*	6.1	NA2845	I	POISON, SPONTANEOUSLY COMBUSTIBLE	2, B9, B14, B16, B32, B74, T38, T43, T45
	Methyl picric acid (heavy metal salts of)	Forbidden				
	Methyl propionate	3	UN1248	II	FLAMMABLE LIQUID	B101, T2
	Methyl propyl ether	3	UN2612	II	FLAMMABLE LIQUID	T14
	Methyl propyl ketone	3	UN1249	II	FLAMMABLE LIQUID	T1
	Methyl sulfate, see *Dimethyl sulfate*					
	Methyl sulfide, see *Dimethyl sulfide*					
	Methyl trichloroacetate	6.1	UN2533	III	KEEP AWAY FROM FOOD	T1
	Methyl trimethylol methane trinitrate	Forbidden				
	Methyl vinyl ketone	3	UN1251	II	FLAMMABLE LIQUID	T8
	Methylal	3	UN1234	II	FLAMMABLE LIQUID	T14
	Methylamine, anhydrous	2.1	UN1061		FLAMMABLE GAS	

HAZARDOUS MATERIALS TABLE

Sym.	Descriptions and shipping names	Hazard	ID No.	PG	Label(s)	Special provisions
	Methylamine, aqueous solution	3	UN1235	II	FLAMMABLE LIQUID, CORROSIVE	B1, T8
	Methylamine dinitramine and dry salts thereof	Forbidden				
	Methylamine nitroform	Forbidden				
	Methylamine perchlorate (dry)	Forbidden				
	Methylamyl acetate	3	UN1233	III	FLAMMABLE LIQUID	B1, T1
	N-Methylaniline	6.1	UN2294	III	KEEP AWAY FROM FOOD	T7
	alpha-Methylbenzyl alcohol	6.1	UN2937	III	KEEP AWAY FROM FOOD	T1
	3-Methylbutan-2-one	3	UN2397	II	FLAMMABLE LIQUID	T1
	N-Methylbutylamine	3	UN2945	II	FLAMMABLE LIQUID, CORROSIVE	T8
	Methylchlorosilane	2.3	UN2534		POISON GAS, FLAMMABLE GAS, CORROSIVE	2, A2, A3, A7, B9, B14, N34
	Methylcyclohexane	3	UN2296	II	FLAMMABLE LIQUID	B1, T1
	Methylcyclohexanols, *flammable*	3	UN2617	III	FLAMMABLE LIQUID	B1, T2
	Methylcyclohexanone	3	UN2297	III	FLAMMABLE LIQUID	B1, T1
	Methylcyclopentane	3	UN2298	II	FLAMMABLE LIQUID	T8
	Methyldichloroarsine	6.1	NA1556	I	POISON	2
	Methyldichlorosilane	4.3	UN1242	I	DANGEROUS WHEN WET, CORROSIVE, FLAMMABLE LIQUID	A2, A3, A7, B6, B77, N34, T16, T26
D	*Methylene chloride, see Dichloromethane*					

HAZARDOUS MATERIALS TABLE

Sym.	Descriptions and shipping names	Hazard	ID No.	PG	Label(s)	Special provisions
	Methylene glycol dinitrate	Forbidden				
	2-Methylfuran	3	UN2301	II	FLAMMABLE LIQUID	T7
	a-Methylglucoside tetranitrate	Forbidden				
	a-Methylglycerol trinitrate	Forbidden				
	5-Methylhexan-2-one	3	UN2302	III	FLAMMABLE LIQUID	B1, T1
	Methylhydrazine	6.1	UN1244	I	POISON, FLAMMABLE LIQUID, CORROSIVE	1, B9, B14, B30, B72, B77, N34, T38, T43, T44
	Methylmorpholine	3	UN2535	II	FLAMMABLE LIQUID, CORROSIVE	B6, T8
	Methylpentadienes	3	UN2461	II	FLAMMABLE LIQUID	T7
	2-Methylpentan-2-ol	3	UN2560	III	FLAMMABLE LIQUID	B1, T1
	Methylpentanes, see Hexanes					
	Methylphenyldichlorosilane	8	UN2437	II	CORROSIVE	T8, T26
	1-Methylpiperidine	3	UN2399	II	FLAMMABLE LIQUID, CORROSIVE	T8
	Methyltetrahydrofuran	3	UN2536	II	FLAMMABLE LIQUID	B101, T7
	Methyltrichlorosilane	3	UN1250	I	FLAMMABLE LIQUID, CORROSIVE	A7, B6, B77, N34, T14, T26
	alpha-Methylvaleraldehyde	3	UN2367	II	FLAMMABLE LIQUID	B1, T1
	Mine rescue equipment containing carbon dioxide, see Carbon dioxide					

HAZARDOUS MATERIALS TABLE

Sym.	Descriptions and shipping names	Hazard	ID No.	PG	Label(s)	Special provisions
	Mines with bursting charge	1.1F	UN0136	II	EXPLOSIVE 1.1F	
	Mines with bursting charge	1.1D	UN0137	II	EXPLOSIVE 1.1D	
	Mines with bursting charge	1.2D	UN0138	II	EXPLOSIVE 1.2D	
	Mines with bursting charge	1.2F	UN0294	II	EXPLOSIVE 1.2F	
	Mixed acid, see Nitrating acid, mixtures etc.					
	Mobility aids, see Wheel chair, electric					
D	Model rocket motor	1.4C	NA0276	II	EXPLOSIVE 1.4C	51
D	Model rocket motor	1.4S	NA0323	II	EXPLOSIVE 1.4S	51
	Molybdenum pentachloride	8	UN2508	III	CORROSIVE	T8, T26
	Monochloroacetone (unstabilized)	Forbidden				
	Monochloroethylene, see Vinyl chloride, Inhibited					
	Monoethanolamine, see Ethanolamine, solutions					
	Monoethylamine, see Ethylamine					
	Morpholine	3	UN2054	III	FLAMMABLE LIQUID	B1, T1
	Morpholine, aqueous, mixture, see Corrosive liquids, n.o.s.					
	Motor fuel anti-knock compounds see Motor fuel anti-knock mixtures					
+	Motor fuel anti-knock mixtures	6.1	UN1649	I	POISON, FLAMMABLE LIQUID	14, B9, B12, B90, T26, T39
	Motor spirit, see Gasoline					

HAZARDOUS MATERIALS TABLE

Sym.	Descriptions and shipping names	Hazard	ID No.	PG	Label(s)	Special provisions
	Muriatic acid, see **Hydrochloric acid solution**					
	Musk xylene, see **5-tert-Butyl2,4,6-trinitro-m-xylene**					
	Naphtha see **Petroleum distillate n.o.s.**					
	Naphthalene, crude or **Naphthalene, refined**	4.1	UN1334	III	FLAMMABLE SOLID	A1
	Naphthalene diozonide	Forbidden				
	Naphthalene, molten	4.1	UN2304	III	FLAMMABLE SOLID	A1, T8
	beta-Naphthylamine	6.1	UN1650	II	POISON	T12, T26
	alpha-Naphthylamine	6.1	UN2077	III	KEEP AWAY FROM FOOD	T7
	Naphthylamineperchlorate	Forbidden				
	Naphthylthiourea	6.1	UN1651	II	POISON	
	Naphthylurea	6.1	UN1652	II	POISON	
	Natural gases (with high methane content), see **Methane, etc.** *(UN 1971, UN 1972)*					
	Neohexane, see **Hexanes**					
	Neon, compressed	2.2	UN1065		NONFLAMMABLE GAS	
	Neon, refrigerated liquid *(cryogenic liquid)*	2.2	UN1913		NONFLAMMABLE GAS	
	New explosive or explosive device, see sections 173.51 and 173.56					
	Nickel carbonyl	6.1	UN1259	I	POISON, FLAMMABLE LIQUID	1
	Nickel cyanide	6.1	UN1653	II	POISON	N74, N75

HAZARDOUS MATERIALS TABLE

Sym.	Descriptions and shipping names	Hazard	ID No.	PG	Label(s)	Special provisions
	Nickel nitrate	5.1	UN2725	III	OXIDIZER	A1
	Nickel nitrite	5.1	UN2726	III	OXIDIZER	A1
	Nickel picrate	Forbidden				
	Nicotine	6.1	UN1654	II	POISON	
	Nicotine compounds, liquid, n.o.s. or **Nicotine preparations, liquid, n.o.s.**	6.1	UN3144	I	POISON	A4, T42
				II	POISON	T14
				III	KEEP AWAY FROM FOOD	T7
	Nicotine compounds, solid, n.o.s. or **Nicotine preparations, solid, n.o.s.**	6.1	UN1655	I	POISON	
				II	POISON	
				III	KEEP AWAY FROM FOOD	
	Nicotine hydrochloride or **Nicotine hydrochloride solution**	6.1	UN1656	II	POISON	
	Nicotine salicylate	6.1	UN1657	II	POISON	
	Nicotine sulfate, *solid*	6.1	UN1658	II	POISON	T14
	Nicotine sulfate, *solution*	6.1	UN1658	II	POISON	
	Nicotine tartrate	6.1	UN1659	II	POISON	
	Nitrated paper (unstable)	Forbidden				
	Nitrates, Inorganic, aqueous solution, n.o.s.	5.1	UN3218	II	OXIDIZER	T8
				III	OXIDIZER	T8
	Nitrates, Inorganic, n.o.s.	5.1	UN1477	II	OXIDIZER	
				III	OXIDIZER	

HAZARDOUS MATERIALS TABLE

Sym.	Descriptions and shipping names	Hazard	ID No.	PG	Label(s)	Special provisions
	Nitrates of diazonium compounds	Forbidden				
	Nitrating acid mixtures, spent with more than 50 percent nitric acid	8	UN1826	I	CORROSIVE, OXIDIZER	T12, T27
	Nitrating acid mixtures spent with not more than 50 percent nitric acid	8	UN1826	II	CORROSIVE	B2, B100, T12, T27
	Nitrating acid mixtures with more than 50 percent nitric acid	8	UN1796	I	CORROSIVE, OXIDIZER	T12, T27
	Nitrating acid mixtures with not more than 50 percent nitric acid	8	UN1796	II	CORROSIVE	B2, T12, T27
	Nitric acid other than red fuming, with more than 70 percent nitric acid	8	UN2031	I	CORROSIVE	B12, B47, B53, T9, T27
	Nitric acid other than red fuming, with not more than 70 percent nitric acid	8	UN2031	II	CORROSIVE	B2, B12, B47, B53, T9, T27
+	**Nitric acid, red fuming**	8	UN2032	I	CORROSIVE, OXIDIZER, POISON	2, B9, B32, B74, T38, T43, T45
	Nitric oxide	2.3	UN1660		POISON GAS, OXIDIZER, CORROSIVE	1, 25, B12, B37, B46, B50, B60, B77
	Nitric oxide and dinitrogen tetroxide mixtures or **Nitric oxide and nitrogen dioxide mixtures**	2.3	UN1975		POISON GAS, OXIDIZER, CORROSIVE	1, 25, B7, B9, B12, B14, B45, B46, B61, B66, B67, B77
	Nitriles, flammable, toxic, n.o.s.	3	UN3273	I	FLAMMABLE LIQUID, POISON	
				II	FLAMMABLE LIQUID, POISON	T14

HAZARDOUS MATERIALS TABLE

Sym.	Descriptions and shipping names	Hazard	ID No.	PG	Label(s)	Special provisions
	Nitriles, toxic, flammable, n.o.s.	6.1	UN3275	I	POISON, FLAMMABLE LIQUID	
				II	POISON, FLAMMABLE LIQUID	T14
	Nitriles, toxic, n.o.s.	6.1	UN3276	I	POISON	
				II	POISON	T14
				III	KEEP AWAY FROM FOOD	T7
	Nitrites, inorganic, aqueous solution, n.o.s.	5.1	UN3219	II	OXIDIZER	T8
				III	OXIDIZER	T8
	Nitrites, inorganic, n.o.s.	5.1	UN2627	II	OXIDIZER	33
	3-Nitro-4-chlorobenzotrifluoride	6.1	UN2307	II	POISON	T8
	6-Nitro-4-diazotoluene-3-sulfonic acid (dry)	Forbidden				
	Nitro isobutane triol trinitrate	Forbidden				
	N-Nitro-N-methylglycolamide nitrate	Forbidden				
	2-Nitro-2-methylpropanol nitrate	Forbidden				
	Nitro urea	1.1D	UN0147	II	EXPLOSIVE 1.1D	
	N-Nitroaniline	Forbidden				
	Nitroanilines (o-; m-; p-)	6.1	UN1661	II	POISON	T14
	Nitroanisole	6.1	UN2730	III	KEEP AWAY FROM FOOD	T8
	Nitrobenzene	6.1	UN1662	II	POISON	T14

HAZARDOUS MATERIALS TABLE

Sym.	Descriptions and shipping names	Hazard	ID No.	PG	Label(s)	Special provisions
	m-Nitrobenzene diazonium perchlorate	Forbidden				
	Nitrobenzenesulfonic acid	8	UN2305	II	CORROSIVE	
	Nitrobenzol, see Nitrobenzene					
	5-Nitrobenzotriazol	1.1D	UN0385	II	EXPLOSIVE 1.1D	
	Nitrobenzotrifluorides	6.1	UN2306	II	POISON	T8
	Nitrobromobenzenes liquid	6.1	UN2732	III	KEEP AWAY FROM FOOD	T8, T38
	Nitrobromobenzenes solid	6.1	UN2732	III	KEEP AWAY FROM FOOD	
	Nitrocellulose, dry or wetted with less than 25 percent water (or alcohol), by mass	1.1D	UN0340	II	EXPLOSIVE 1.1D	
	Nitrocellulose membrane filters	4.1	UN3270	II	FLAMMABLE SOLID	43, A1
	Nitrocellulose, plasticized with not less than 18 percent plasticizing substance, by mass	1.3C	UN0343	II	EXPLOSIVE 1.3C	
	Nitrocellulose, solution, flammable with not more than 12.6 percent nitrogen, by mass, and not more than 55 percent nitrocellulose	3	UN2059	II	FLAMMABLE LIQUID	T8, T31
				III	FLAMMABLE LIQUID	B1, T7, T30
	Nitrocellulose, unmodified or plasticized with less than 18 percent plasticizing substance, by mass	1.1D	UN0341	II	EXPLOSIVE 1.1D	
	Nitrocellulose, wetted with not less than 25 percent alcohol, by mass	1.3C	UN0342	II	EXPLOSIVE 1.3C	
	Nitrocellulose with alcohol with not less than 25 percent alcohol by mass, and with not more than 12.6 percent nitrogen, by dry mass	4.1	UN2556	II	FLAMMABLE SOLID	

HAZARDOUS MATERIALS TABLE

Sym.	Descriptions and shipping names	Hazard	ID No.	PG	Label(s)	Special provisions
	Nitrocellulose, with not more than 12.6 percent nitrogen, by dry mass, or Nitrocellulose mixture with pigment or Nitrocellulose mixture with plasticizer or Nitrocellulose mixture with pigment and plasticizer	4.1	UN2557	II	FLAMMABLE SOLID	44
	Nitrocellulose with water with not less than 25 percent water, by mass	4.1	UN2555	II	FLAMMABLE SOLID	
	Nitrochlorobenzene, see **Chloronitrobenzenes** etc.					
	Nitrocresols	6.1	UN2446	III	KEEP AWAY FROM FOOD	
	Nitroethane	3	UN2842	III	FLAMMABLE LIQUID	B1, T8
	Nitroethyl nitrate	Forbidden				
	Nitroethylene polymer	Forbidden				
	Nitrogen, compressed	2.2	UN1066		NONFLAMMABLE GAS	
	Nitrogen dioxide, liquefied see **Dinitrogen tetroxide, liquefied**					
	Nitrogen fertilizer solution, see **Fertilizer ammoniating solution** etc.					
	Nitrogen, mixtures with rare gases, see **Rare gases and nitrogen mixtures**					
	Nitrogen peroxide, see **Dinitrogen tetroxide, liquefied**					
	Nitrogen, refrigerated liquid cryogenic liquid	2.2	UN1977		NONFLAMMABLE GAS	

HAZARDOUS MATERIALS TABLE

Sym.	Descriptions and shipping names	Hazard	ID No.	PG	Label(s)	Special provisions
	Nitrogen tetraoxide and nitric oxide mixtures, see **Nitric oxide and nitrogen tetroxide mixtures**	Forbidden				
	Nitrogen tetroxide, see **Dinitrogen tetroxide, liquefied**					
D	Nitrogen trichloride	2.2	UN2451		NONFLAMMABLE GAS, OXIDIZER	
I	**Nitrogen trifluoride**	2.3	UN2451		POISON GAS, OXIDIZER	
	Nitrogen trifluoride	Forbidden				
	Nitrogen triiodide	Forbidden				
	Nitrogen triiodide monoamine	2.3	UN2421		POISON GAS, OXIDIZER, CORROSIVE	1
	Nitrogen trioxide	1.1D	UN0143	II	EXPLOSIVE 1.1D, POISON	
	Nitroglycerin, desensitized with not less than 40 percent non-volatile water insoluble phlegmatizer, by mass	Forbidden				
	Nitroglycerin, liquid, not desensitized	3	UN3064	II	FLAMMABLE LIQUID	N8
	Nitroglycerin, solution in alcohol, with more than 1 percent but not more than 5 percent nitroglycerin	1.1D	UN0144	II	EXPLOSIVE 1.1D	
	Nitroglycerin, solution in alcohol, with more than 1 percent but not more than 10 percent nitroglycerin					

Sym.	Descriptions and shipping names	Hazard	ID No.	PG	Label(s)	Special provisions
	Nitroglycerin solution in alcohol with not more than 1 percent nitroglycerin	3	UN1204	II	FLAMMABLE LIQUID	N34, T25
	Nitroguanidine nitrate	Forbidden				
	Nitroguanidine or **Picrite**, dry or wetted with less than 20 percent water, by mass	1.1D	UN0282	II	EXPLOSIVE 1.1D	
	Nitroguanidine, wetted or **Picrite, wetted** with not less than 20 percent water, by mass	4.1	UN1336	I	FLAMMABLE SOLID	23, A8, A19, A20, N41
	1-Nitrohydantoin	Forbidden				
	Nitrohydrochloric acid	8	UN1798	I	CORROSIVE	A3, B10, N41, T18, T27
	Nitromannite (dry)	Forbidden				
	Nitromannite, wetted, see Mannitol hexanitrate, etc.					
	Nitromethane	3	UN1261	II	FLAMMABLE LIQUID	T25
	Nitromuriatic acid, see Nitrohydrochloric acid					
	Nitronaphthalene	4.1	UN2538	III	FLAMMABLE SOLID	A1
	Nitrophenols (o-; m-; p-;)	6.1	UN1663	III	KEEP AWAY FROM FOOD	T8, T38
	m-Nitrophenyldinitro methane	Forbidden				
	Nitropropanes	3	UN2608	III	FLAMMABLE LIQUID	B1, T1
	p-Nitrosodimethylaniline	4.2	UN1369	II	SPONTANEOUSLY COMBUSTIBLE	A19, A20, B101, N34
D	**Nitrosoguanidine**	1.1A	NA0473	II	EXPLOSIVE 1.1A	111, 117

HAZARDOUS MATERIALS TABLE

Sym.	Descriptions and shipping names	Hazard	ID No.	PG	Label(s)	Special provisions
	Nitrostarch, *dry or wetted with less than 20 percent water, by mass*	1.1D	UN0146	II	EXPLOSIVE 1.1D	23, A8, A19, A20, N41
	Nitrostarch, *wetted with not less than 20 percent water, by mass*	4.1	UN1337	I	FLAMMABLE SOLID	3, B14
	Nitrosugars *(dry)*	Forbidden				
	Nitrosyl chloride	2.3	UN1069		POISON GAS, CORROSIVE	A3, A6, A7, B2, N34, T9, T27
	Nitrosylsulfuric acid	8	UN2308	II	CORROSIVE	T14
	Nitrotoluenes, *liquid o-; m-; p-;*	6.1	UN1664	II	POISON	T14
	Nitrotoluenes, *solid m-, or p-*	6.1	UN1664	II	POISON	
	Nitrotoluidines *(mono)*	6.1	UN2660	III	KEEP AWAY FROM FOOD	
	Nitrotriazolone or NTO	1.1D	UN0490	II	EXPLOSIVE 1.1D	
	Nitrous oxide and carbon dioxide mixtures, see **Carbon dioxide and nitrous oxide mixtures**					
	Nitrous oxide, compressed	2.2	UN1070		NONFLAMMABLE GAS, OXIDIZER	
	Nitrous oxide, refrigerated liquid	2.2	UN2201		NONFLAMMABLE GAS	B6
	Nitroxylenes, *(o-; m-; p-)*	6.1	UN1665	II	POISON	T14
	Nitroxylol, see **Nitroxylenes**					
	Nonanes	3	UN1920	III	FLAMMABLE LIQUID	B1, T1
	Nonflammable gas, n.o.s., see **Compressed or Liquefied gases**, *etc. (UN 1955, UN 1956).*					

HAZARDOUS MATERIALS TABLE

Sym.	Descriptions and shipping names	Hazard	ID No.	PG	Label(s)	Special provisions
	Nonliquefied gases, see **Compressed gases, etc.**					
	Nonliquefied hydrocarbon gas, see **Hydrocarbon gases, compressed, n.o.s.**					
	Nonyltrichlorosilane	8	UN1799	II	CORROSIVE	A7, B2, B6, N34, T8, T26
	2,5-Norbornadiene or Dicycloheptadiene	3	UN2251	II	FLAMMABLE LIQUID	
	Nordhausen acid, see **Sulfuric acid, fuming etc.**					
	Octadecyltrichlorosilane	8	UN1800	II	CORROSIVE	A7, B2, B6 N34, T8
	Octadiene	3	UN2309	II	FLAMMABLE LIQUID	B1, T1
	1,7-Octadine-3,5-diyne-1,8-dimethoxy-9-octadecynoic acid	Forbidden				
	Octafluorobut-2-ene	2.2	UN2422		NONFLAMMABLE GAS	
	Octafluorocyclobutane, RC318	2.2	UN1976		NONFLAMMABLE GAS	
	Octafluoropropane, R218	2.2	UN2424		NONFLAMMABLE GAS	
	Octanes	3	UN1262	II	FLAMMABLE LIQUID	T1
	Octogen, see **Cyclotetramethylene tetranitramine, etc.**					
	Octolite or Octol, *dry or wetted with less than 15 percent water, by mass*	1.1D	UN0266	II	EXPLOSIVE 1.1D	
	Octonal	1.1D	UN0496		EXPLOSIVE 1.1D	
	Octyl aldehydes, *flammable*	3	UN1191	III	FLAMMABLE LIQUID	B1, T1
+	tert-Octyl mercaptan	6.1	UN3023	II	POISON, FLAMMABLE LIQUID	2, B9, B14, B32, B74, T38, T43, T45

HAZARDOUS MATERIALS TABLE

Sym.	Descriptions and shipping names	Hazard	ID No.	PG	Label(s)	Special provisions
	Octyltrichlorosilane	8	UN1801	II	CORROSIVE	A7, B2, B6, N34, T8, T26
	Oil gas	2.3	UN1071		POISON GAS, FLAMMABLE GAS	6
	Oleum, see Sulfuric acid, fuming					
	Organic peroxide type A, liquid or solid	Forbidden				
	Organic peroxide type B, liquid	5.2	UN3101	II	ORGANIC PEROXIDE, EXPLOSIVE	53
	Organic peroxide type B, liquid, temperature controlled	5.2	UN3111	II	ORGANIC PEROXIDE, EXPLOSIVE	53
	Organic peroxide type B, solid	5.2	UN3102	II	ORGANIC PEROXIDE, EXPLOSIVE	53
	Organic peroxide type B, solid, temperature controlled	5.2	UN3112	II	ORGANIC PEROXIDE, EXPLOSIVE	53
	Organic peroxide type C, liquid	5.2	UN3103	II	ORGANIC PEROXIDE	
	Organic peroxide type C, liquid, temperature controlled	5.2	UN3113	II	ORGANIC PEROXIDE	
	Organic peroxide type C, solid	5.2	UN3104	II	ORGANIC PEROXIDE	
	Organic peroxide type C, solid, temperature controlled	5.2	UN3114	II	ORGANIC PEROXIDE	
	Organic peroxide type D, liquid	5.2	UN3105	II	ORGANIC PEROXIDE	
	Organic peroxide type D, liquid, temperature controlled	5.2	UN3115	II	ORGANIC PEROXIDE	
	Organic peroxide type D, solid	5.2	UN3106	II	ORGANIC PEROXIDE	

HAZARDOUS MATERIALS TABLE

Sym.	Descriptions and shipping names	Hazard	ID No.	PG	Label(s)	Special provisions
	Organic peroxide type D, solid, temperature controlled	5.2	UN3116	II	ORGANIC PEROXIDE	
	Organic peroxide type E, liquid	5.2	UN3107	II	ORGANIC PEROXIDE	
	Organic peroxide type E, liquid, temperature controlled	5.2	UN3117	II	ORGANIC PEROXIDE	
	Organic peroxide type E, solid	5.2	UN3108	II	ORGANIC PEROXIDE	
	Organic peroxide type E, solid, temperature controlled	5.2	UN3118	II	ORGANIC PEROXIDE	
	Organic peroxide type F, liquid	5.2	UN3109	II	ORGANIC PEROXIDE	
	Organic peroxide type F, liquid, temperature controlled	5.2	UN3119	II	ORGANIC PEROXIDE	
	Organic peroxide type F, solid	5.2	UN3110	II	ORGANIC PEROXIDE	T42
	Organic peroxide type F, solid, temperature controlled	5.2	UN3120	II	ORGANIC PEROXIDE	
D	Organic phosphate, mixed with compressed gas or Organic phosphate compound, mixed with compressed gas or Organic phosphorus compound, mixed with compressed gas	2.3	NA1955		POISON GAS	3
	Organoarsenic compound, n.o.s.	6.1	UN3280	I	POISON	
				II	POISON	T14
				III	KEEP AWAY FROM FOOD	T7

HAZARDOUS MATERIALS TABLE

Sym.	Descriptions and shipping names	Hazard	ID No.	PG	Label(s)	Special provisions
	Organochlorine pesticides liquid, flammable, toxic, *flash point less than 23 degrees C*	3	UN2762	I	FLAMMABLE LIQUID, POISON	
				II	FLAMMABLE LIQUID, POISON	
	Organochlorine pesticides, liquid, toxic	6.1	UN2996	I	POISON	T42
				II	POISON	T14
				III	KEEP AWAY FROM FOOD	T14
	Organochlorine pesticides, liquid, toxic, flammable, *flashpoint not less than 23 degrees C*	6.1	UN2995	I	POISON, FLAMMABLE LIQUID	T42
				II	POISON, FLAMMABLE LIQUID	T14
				III	KEEP AWAY FROM FOOD, FLAMMABLE LIQUID	B1, T14
	Organochlorine pesticides, solid toxic	6.1	UN2761	I	POISON	
				II	POISON	
				III	KEEP AWAY FROM FOOD	
	Organometallic compound or Compound solution or Compound dispersion, water-reactive, flammable, n.o.s.	4.3	UN3207	I	DANGEROUS WHEN WET, FLAMMABLE LIQUID	B101, B106
				II	DANGEROUS WHEN WET, FLAMMABLE LIQUID	B106
				III	DANGEROUS WHEN WET, FLAMMABLE LIQUID	

HAZARDOUS MATERIALS TABLE

Sym.	Descriptions and shipping names	Hazard	ID No.	PG	Label(s)	Special provisions
	Organometallic compound, toxic n.o.s.	6.1	UN3282	I	POISON	
				II	POISON	T14
				III	KEEP AWAY FROM FOOD	T7
	Organophosphorus compound, toxic, flammable, n.o.s.	6.1	UN3279	I	POISON, FLAMMABLE LIQUID	
				II	POISON, FLAMMABLE	T14
	Organophosphorus compound, toxic n.o.s.	6.1	UN3278	I	POISON	
				II	POISON	T14
				III	KEEP AWAY FROM FOOD	T7
	Organophosphorus pesticides, liquid, flammable, toxic, *flash point less than 23 degrees C*	3	UN2784	I	FLAMMABLE LIQUID, POISON	
				II	FLAMMABLE LIQUID, POISON	
	Organophosphorus pesticides, liquid, toxic	6.1	UN3018	I	POISON	N76, T42
				II	POISON	N76, T14
				III	KEEP AWAY FROM FOOD	N76, T14
	Organophosphorus pesticides, liquid, toxic, flammable, *flashpoint not less than 23 degrees C*	6.1	UN3017	I	POISON, FLAMMABLE LIQUID	N76, T42
				II	POISON, FLAMMABLE LIQUID.	N76, T14
				III	KEEP AWAY FROM FOOD, FLAMMABLE LIQUID	B1, N76, T14

HAZARDOUS MATERIALS TABLE

Sym.	Descriptions and shipping names	Hazard	ID No.	PG	Label(s)	Special provisions
	Organophosphorus pesticides, solid, toxic	6.1	UN2783	I	POISON	N77
				II	POISON	N77
				III	KEEP AWAY FROM FOOD	N77
	Organotin compounds, liquid, n.o.s.	6.1	UN2788	I	POISON	A3, N33, N34, T42
				II	POISON	A3, N33, N34, T14
				III	KEEP AWAY FROM FOOD	T14
	Organotin compounds, solid, n.o.s.	6.1	UN3146	I	POISON	A5
				II	POISON	
				III	KEEP AWAY FROM FOOD	
	Organotin pesticides, liquid, flammable, toxic, *flash point less than 23 degrees C*	3	UN2787	I	FLAMMABLE LIQUID, POISON	
				II	FLAMMABLE LIQUID, POISON	
	Organotin pesticides, liquid, toxic	6.1	UN3020	I	POISON	T42
				II	POISON	T14
				III	KEEP AWAY FROM FOOD	T14
	Organotin pesticides, liquid, toxic, flammable, *flashpoint not less than 23 degrees C*	6.1	UN3019	I	POISON, FLAMMABLE LIQUID	T42
				II	POISON, FLAMMABLE LIQUID	T14
				III	KEEP AWAY FROM FOOD, FLAMMABLE LIQUID	B1, T14

HAZARDOUS MATERIALS TABLE

Sym.	Descriptions and shipping names	Hazard	ID No.	PG	Label(s)	Special provisions
	Organotin pesticides, solid, toxic	6.1	UN2786	I	POISON	
				II	POISON	
				III	KEEP AWAY FROM FOOD	
	Orthonitroaniline, see Nitroanilines etc.					
	Osmium tetroxide	6.1	UN2471	I	POISON	A8, B100, N33, N34
AD	**Other regulated substances, liquid, n.o.s.**	9	NA3082	III	CLASS 9	
AD	**Other regulated substances, solid, n.o.s.**	9	NA3077	III	CLASS 9	B54
	Oxidizing liquid, corrosive, n.o.s.	5.1	UN3098	I	OXIDIZER, CORROSIVE	
				II	OXIDIZER, CORROSIVE	
				III	OXIDIZER, CORROSIVE	
	Oxidizing liquid, n.o.s.	5.1	UN3139	II	OXIDIZER	A2
				III	OXIDIZER	A2
	Oxidizing liquid, toxic, n.o.s.	5.1	UN3099	I	OXIDIZER, POISON	
				II	OXIDIZER, POISON	
				III	OXIDIZER, KEEP AWAY FROM FOOD	
	Oxidizing solid, corrosive, n.o.s.	5.1	UN3085	I	OXIDIZER, CORROSIVE	
				II	OXIDIZER, CORROSIVE	
				III	OXIDIZER, CORROSIVE	
	Oxidizing solid, flammable, n.o.s.	5.1	UN3137	I	OXIDIZER, FLAMMABLE SOLID	

HAZARDOUS MATERIALS TABLE

Sym.	Descriptions and shipping names	Hazard	ID No.	PG	Label(s)	Special provisions
	Oxidizing solid, n.o.s.	5.1	UN1479	I	OXIDIZER	
		5.1		II	OXIDIZER	
		5.1		III	OXIDIZER	
	Oxidizing solid, self-heating, n.o.s.	5.1	UN3100	II	OXIDIZER, SPONTANEOUSLY COMBUSTIBLE	
	Oxidizing solid, toxic, n.o.s.	5.1	UN3087	I	OXIDIZER, POISON	
		5.1		II	OXIDIZER, POISON	
		5.1		III	OXIDIZER, KEEP AWAY FROM FOOD	
	Oxidizing solid, water-reactive, n.o.s.	5.1	UN3121		OXIDIZER, DANGEROUS WHEN WET	
	Oxygen and carbon dioxide mixtures, see **Carbon dioxide and oxygen mixtures**					
	Oxygen, compressed	2.2	UN1072		NONFLAMMABLE GAS, OXIDIZER	
	Oxygen difluoride	2.3	UN2190		POISON GAS, OXIDIZER, CORROSIVE	1
	Oxygen, mixtures with rare gases, see **Rare gases and oxygen mixtures**					
	Oxygen, refrigerated liquid *(cryogenic liquid)*	2.2	UN1073		NONFLAMMABLE GAS, OXIDIZER	
	Paint including paint, lacquer, enamel, stain, shellac solutions, varnish, polish, liquid filler, and liquid lacquer base	3	UN1263	II	FLAMMABLE LIQUID	B52, T7, T30
		3		III	FLAMMABLE LIQUID	B1, B52, T7, T30

HAZARDOUS MATERIALS TABLE

Sym.	Descriptions and shipping names	Hazard	ID No.	PG	Label(s)	Special provisions
	Paint or Paint related material	8	UN3066	II	CORROSIVE	B2, N71, T14
				III	CORROSIVE	B52, N71, T7
	Paint related material *including paint thinning, drying, removing, or reducing compound*	3	UN1263	II	FLAMMABLE LIQUID	B52, T7, T30
				III	FLAMMABLE LIQUID	B1, B52, T7, T30
	Paper, unsaturated oil treated *incompletely dried (including carbon paper)*	4.2	UN1379	III	SPONTANEOUSLY COMBUSTIBLE	B101, B106
	Paraformaldehyde	4.1	UN2213	III	FLAMMABLE SOLID	A1
	Paraldehyde	3	UN1264	III	FLAMMABLE LIQUID	B1, T1
	Paranitroaniline, solid, see Nitroanilines etc.					
D	**Parathion**	6.1	NA2783	I	POISON	T42
				II	POISON	T14
D	**Parathion and compressed gas mixture**	2.3	NA1967		POISON GAS	3
	Paris green, solid, see Copper acetoarsenite					
A,W	**PCB**, *see Polychlorinated biphenyls*					
+	**Pentaborane**	4.2	UN1380	I	SPONTANEOUSLY COMBUSTIBLE, POISON	1
	Pentachloroethane	6.1	UN1669	II	POISON	T14
	Pentachlorophenol	6.1	UN3155	II	POISON	
	Pentaerythrite tetranitrate (dry)	Forbidden				
	Pentaerythrite tetranitrate or **Pentaerythritol tetranitrate** or **PETN, with not less than 7 percent wax by mass**	1.1D	UN0411	II	EXPLOSIVE 1.1D	

HAZARDOUS MATERIALS TABLE

Sym.	Descriptions and shipping names	Hazard	ID No.	PG	Label(s)	Special provisions
	Pentaerythrite tetranitrate, wetted or **Pentaerythritol tetranitrate, wetted,** or **PETN, wetted** *with not less than 25 percent water, by mass,* or **Pentaerythrite tetranitrate,** or **Pentaerythritol tetranitrate** or **PETN, desensitized** *with not less than 15 percent phlegmatizer by mass*	1.1D	UN0150	II	EXPLOSIVE 1.1D	
	Pentaerythritol tetranitrate, *see* Pentaerythrite tetranitrate, *etc.*					
	Pentafluoroethane	2.2	UN3220		NONFLAMMABLE GAS	
	Pentamethylheptane	3	UN2286	III	FLAMMABLE LIQUID	B1, T1
	Pentan-2,4-dione	3	UN2310	III	FLAMMABLE LIQUID	B1, T1
	Pentanes	3	UN1265	I	FLAMMABLE LIQUID	T20
				II	FLAMMABLE LIQUID	T20
	Pentanitroaniline (dry)	Forbidden				
	1-Pentol	8	UN2705	II	CORROSIVE	B2, T8
	Pentolite, dry or wetted *with less than 15 percent water, by mass*	1.1D	UN0151	II	EXPLOSIVE 1.1D	
	Percarbonates, Inorganic, n.o.s.	5.1	UN3217	III	OXIDIZER	T8
	Perchlorates, inorganic, aqueous solution, n.o.s.	5.1	UN3211	II	OXIDIZER	
	Perchlorates, inorganic, n.o.s.	5.1	UN1481	II	OXIDIZER	
	Perchloric acid, with more than 72 percent acid by mass	Forbidden				

HAZARDOUS MATERIALS TABLE

Sym.	Descriptions and shipping names	Hazard	ID No.	PG	Label(s)	Special provisions
	Perchloric acid with more than 50 percent but not more than 72 percent acid, by mass	5.1	UN1873	I	OXIDIZER, CORROSIVE	A2, A3, N41, T9, T27
	Perchloric acid with not more than 50 percent acid by mass	8	UN1802	II	CORROSIVE, OXIDIZER	N41, T9
	Perchloroethylene, see **Tetrachloroethylene**					
	Perchloromethyl mercaptan	6.1	UN1670	I	POISON	2, A3, A7, B9, B14, B32, B74, N34, T38, T43, T45
	Perchloryl fluoride	2.3	UN3083		POISON GAS, OXIDIZER	2, 25, B9, B12, B14
	Percussion caps, see **Primers, cap type**					
	Perfluoro-2-butene, see **Octafluorobut-2-ene**					
	Perfluoroethyl vinyl ether	2.1	UN3154		FLAMMABLE GAS	
	Perfluoromethyl vinyl ether	2.1	UN3153		FLAMMABLE GAS	
	Perfumery products with flammable solvents	3	UN1266	II	FLAMMABLE LIQUID	T7, T30
				III	FLAMMABLE LIQUID	B1, T7, T30
	Permanganates, Inorganic, aqueous solution, n.o.s.	5.1	UN3214	II	OXIDIZER	26, T8
	Permanganates, Inorganic, n.o.s.	5.1	UN1482	II	OXIDIZER	26, A30
				III	OXIDIZER	26, A30
	Peroxides, Inorganic, n.o.s.	5.1	UN1483	II	OXIDIZER	A7, A20, N34
				III	OXIDIZER	A7, A20, N34
	Peroxyacetic acid, with more than 43 percent and with more than 6 percent hydrogen peroxide	Forbidden				

HAZARDOUS MATERIALS TABLE

Sym.	Descriptions and shipping names	Hazard	ID No.	PG	Label(s)	Special provisions
	Persulfates, inorganic, aqueous solution, n.o.s.	5.1	UN3216	III	OXIDIZER	T2
	Persulfates, inorganic, n.o.s.	5.1	UN3215	III	OXIDIZER	
	Pesticides, liquid, flammable, toxic, *flash-point less than 23 degrees C*	3	UN3021	I	FLAMMABLE LIQUID, POISON	B5
				II	FLAMMABLE LIQUID, POISON	
	Pesticides, liquid, toxic, flammable, n.o.s. *flashpoint not less than 23 degrees C*	6.1	UN2903	I	POISON, FLAMMABLE LIQUID	T42
				II	POISON, FLAMMABLE LIQUID	T14
				III	KEEP AWAY FROM FOOD, FLAMMABLE LIQUID	B1, T14
	Pesticides, liquid, toxic, n.o.s.	6.1	UN2902	I	POISON	T42
				II	POISON	T14
				III	KEEP AWAY FROM FOOD	T14
	Pesticides, solid, toxic, n.o.s.	6.1	UN2588	I	POISON	
				II	POISON	
				III	KEEP AWAY FROM FOOD	
	PETN, see Pentaerythrite tetranitrate					
	PETN/TNT, see Pentolite, etc.					
	Petrol, see Gasoline					

HAZARDOUS MATERIALS TABLE

Sym.	Descriptions and shipping names	Hazard	ID No.	PG	Label(s)	Special provisions
	Petroleum crude oil	3	UN1267	I	FLAMMABLE LIQUID	T8, T31
				II	FLAMMABLE LIQUID	T8, T31
				III	FLAMMABLE LIQUID	B1, T7, T30
	Petroleum distillates, n.o.s. or Petroleum products, n.o.s.	3	UN1268	I	FLAMMABLE LIQUID	T8, T31
				II	FLAMMABLE LIQUID	T8, T31
				III	FLAMMABLE LIQUID	B1, T7, T30
	Petroleum gases, liquefied or Liquefied petroleum gas	2.1	UN1075		FLAMMABLE GAS	
D	Petroleum oil	3	NA1270	I	FLAMMABLE LIQUID	T8, T31
				II	FLAMMABLE LIQUID	T8, T31
				III	FLAMMABLE LIQUID	B1, T7, T30
	Phenacyl bromide	6.1	UN2645	II	POISON	B106
	Phenetidines	6.1	UN2311	III	KEEP AWAY FROM FOOD	T7
	Phenol, molten	6.1	UN2312	II	POISON	B14, B100, T8
	Phenol, solid	6.1	UN1671	II	POISON	N78, T14
	Phenol solutions	6.1	UN2821	II	POISON	T14
				III	KEEP AWAY FROM FOOD	T7
	Phenolsulfonic acid, liquid	8	UN1803	II	CORROSIVE	B2, N41, T8
+	Phenoxy pesticides, liquid, flammable, toxic, *flash point less than 23 degrees C*	3	UN2766	I	FLAMMABLE LIQUID, POISON	
				II	FLAMMABLE LIQUID, POISON	

HAZARDOUS MATERIALS TABLE

Sym.	Descriptions and shipping names	Hazard	ID No.	PG	Label(s)	Special provisions
	Phenoxy pesticides, liquid, toxic	6.1	UN3000	I	POISON	T42
				II	POISON	T14
				III	KEEP AWAY FROM FOOD	T14
	Phenoxy pesticides, liquid, toxic, flammable, *flashpoint not less than 23 degrees C*	6.1	UN2999	I	POISON, FLAMMABLE LIQUID	T42
				II	POISON, FLAMMABLE LIQUID	T14
				III	KEEP AWAY FROM FOOD, FLAMMABLE LIQUID	B1, T14
	Phenoxy pesticides, solid, toxic	6.1	UN2765	I	POISON	
				II	POISON	
				III	KEEP AWAY FROM FOOD	
	Phenyl chloroformate	6.1	UN2746	II	POISON, CORROSIVE	T12
	Phenyl isocyanate	6.1	UN2487	I	POISON	2, A3, B9, B14, B32, B74, B77, N33, N34, T38, T43, T45
	Phenyl mercaptan	6.1	UN2337	I	POISON, FLAMMABLE LIQUID	2, B9, B14, B32, B74, B77, T38, T43, T45
	Phenyl phosphorus dichloride	8	UN2798	II	CORROSIVE	B2, B15, T8, T26
	Phenyl phosphorus thiodichloride	8	UN2799	II	CORROSIVE	B2, B15, T8, T26
+	Phenyl urea pesticides, liquid, flammable, toxic, *flash point less than 23 degrees C*	3	UN2768	I	FLAMMABLE LIQUID, POISON	
				II	FLAMMABLE LIQUID, POISON	

HAZARDOUS MATERIALS TABLE

Sym.	Descriptions and shipping names	Hazard	ID No.	PG	Label(s)	Special provisions
	Phenyl urea pesticides, liquid, toxic	6.1	UN3002	I	POISON	T42
				II	POISON	T14
				III	KEEP AWAY FROM FOOD	T14
	Phenyl urea pesticides, liquid, toxic, flammable, *flash point not less than 23 degrees C*	6.1	UN3001	I	POISON, FLAMMABLE LIQUID	T42
				II	POISON, FLAMMABLE LIQUID	T14
				III	KEEP AWAY FROM FOOD, FLAMMABLE LIQUID	B1, T14
	Phenyl urea pesticides, solid, toxic	6.1	UN2767	I	POISON	
				II	POISON	
				III	KEEP AWAY FROM FOOD	
	Phenylacetonitrile, liquid	6.1	UN2470	III	KEEP AWAY FROM FOOD	T8
	Phenylacetyl chloride	8	UN2577	II	CORROSIVE	B2, T8, B14, B32, B74, T38, T43, T45
	Phenylcarbylamine chloride	6.1	UN1672	I	POISON	2, B9, B14, B32, B74, T38, T43, T45
	m-Phenylene diaminediperchlorate (dry)	Forbidden				
	Phenylenediamines (*o*-; *m*-; *p*-)	6.1	UN1673	III	KEEP AWAY FROM FOOD	
	Phenylhydrazine	6.1	UN2572	II	POISON	T8
	Phenylmercuric acetate	6.1	UN1674	II	POISON	
	Phenylmercuric compounds, n.o.s.	6.1	UN2026	I	POISON	
				II	POISON	
				III	KEEP AWAY FROM FOOD	

HAZARDOUS MATERIALS TABLE

Sym.	Descriptions and shipping names	Hazard	ID No.	PG	Label(s)	Special provisions
	Phenylmercuric hydroxide	6.1	UN1894	II	POISON	
	Phenylmercuric nitrate	6.1	UN1895	II	POISON	
	Phenyltrichlorosilane	8	UN1804	II	CORROSIVE	A7, B6, N34, T8
	Phosgene	2.3	UN1076		POISON GAS, CORROSIVE	1, B7, B46
	9-Phosphabicyclononanes or Cyclooctadiene phosphines	4.2	UN2940	II	SPONTANEOUSLY COMBUSTIBLE	A19
	Phosphine	2.3	UN2199		POISON GAS, FLAMMABLE GAS	1
	Phosphoric acid	8	UN1805	III	CORROSIVE	A7, N34, T7
	Phosphoric acid triethyleneimine, see **Tris-(1-aziridiyl)phosphine oxide, solution**					
	Phosphoric anhydride, see **Phosphorus pentoxide**					
	Phosphorous acid	8	UN2834	III	CORROSIVE	T7
	Phosphorus, amorphous	4.1	UN1338	III	FLAMMABLE SOLID	A1, A19, B1, B9, B12, B26
	Phosphorus bromide, see **Phosphorus tribromide**					
	Phosphorus chloride, see **Phosphorus trichloride**					
	Phosphorus heptasulfide, *free from yellow or white phosphorus*	4.1	UN1339	II	FLAMMABLE SOLID	A20, N34
	Phosphorus oxybromide	8	UN1939	II	CORROSIVE	B8, B106, N41, N43
	Phosphorus oxybromide, molten	8	UN2576	II	CORROSIVE	B2, B8, N41, N43, T8, T27

HAZARDOUS MATERIALS TABLE

Sym.	Descriptions and shipping names	Hazard	ID No.	PG	Label(s)	Special provisions
+	Phosphorus oxychloride	8	UN1810	II	CORROSIVE, POISON	2, A7, B9, B14, B32, B74, B77, N34, T38, T43, T45
	Phosphorus pentabromide	8	UN2691	II	CORROSIVE	A7, B106, N34
	Phosphorus pentachloride	8	UN1806	II	CORROSIVE	A7, B106, N34
	Phosphorus pentafluoride	2.3	UN2198		POISON GAS, CORROSIVE	1
	Phosphorus pentasulfide, free from yellow or white phosphorus	4.3	UN1340	II	DANGEROUS WHEN WET, FLAMMABLE SOLID	A20, B59, B101, B106
	Phosphorus pentoxide	8	UN1807	II	CORROSIVE	A7, N34
	Phosphorus sesquisulfide, free from yellow or white phosphorus	4.1	UN1341	II	FLAMMABLE SOLID	A20, N34
	Phosphorus tribromide	8	UN1808	II	CORROSIVE	A3, A6, A7, B9, B14, B15, B32, B74, B77, N34, N43, T8
+	Phosphorus trichloride	8	UN1809	I	CORROSIVE, POISON	2, A3, A7, B9, B14, B15, B32, B74, B77, N34, T38, T43, T45
	Phosphorus trioxide	8	UN2578	III	CORROSIVE	A20, N34
	Phosphorus trisulfide, free from yellow or white phosphorus	4.1	UN1343	II	FLAMMABLE SOLID	
	Phosphorus, white dry or Phosphorus, white, under water or Phosphorus white, in solution or Phosphorus, yellow dry or Phosphorus, yellow, under water or Phosphorus, yellow, in solution	4.2	UN1381	I	SPONTANEOUSLY COMBUSTIBLE, POISON	B9, B12, B26, N34, T15, T26, T33
	Phosphorus white, molten	4.2	UN2447	I	SPONTANEOUSLY COMBUSTIBLE, POISON	B9, B12, B26, N34, T15, T26, T29

HAZARDOUS MATERIALS TABLE

Sym.	Descriptions and shipping names	Hazard	ID No.	PG	Label(s)	Special provisions
	Phosphorus (white or red) and a chlorate, mixtures of	Forbidden				
	Phosphoryl chloride, see **Phosphorus oxychloride**					
	Phthalic anhydride with more than .05 percent maleic anhydride	8	UN2214	III	CORROSIVE	T7
	Phthalimide derivative pesticides, liquid, flammable, toxic, flash point less than 23 degrees C	3	UN2774	I	FLAMMABLE LIQUID, POISON	T42
				II	FLAMMABLE LIQUID, POISON	T14
	Phthalimide derivative pesticides, liquid, toxic	6.1	UN3008	I	POISON	T42
				II	POISON	T14
				III	KEEP AWAY FROM FOOD	T14
	Phthalimide derivative pesticides, liquid, toxic, flammable flashpoint not less than 23 degrees C	6.1	UN3007	I	POISON, FLAMMABLE LIQUID	T42
				II	POISON, FLAMMABLE LIQUID	T14
				III	KEEP AWAY FROM FOOD, FLAMMABLE LIQUID	T14
	Phthalimide derivative pesticides, solid, toxic	6.1	UN2773	I	POISON	
				II	POISON	
				III	KEEP AWAY FROM FOOD	
	Picolines	3	UN2313	III	FLAMMABLE LIQUID	B1, T8

HAZARDOUS MATERIALS TABLE

Sym.	Descriptions and shipping names	Hazard	ID No.	PG	Label(s)	Special provisions
	Picric acid, see **Trinitrophenol**, etc.					
D	Picric acid, wet, *with not less than 10 percent water*	4.1	NA1344	I	FLAMMABLE SOLID	A19, A20, N41
	Picrite, see **Nitroguanidine**, etc.					
	Picryl chloride, see **Trinitrochlorobenzene**					
	Pine oil	3	UN1272	III	FLAMMABLE LIQUID	B1, T1
	alpha-Pinene	3	UN2368	III	FLAMMABLE LIQUID	B1, T1
	Piperazine	8	UN2579	III	CORROSIVE	T7
	Piperidine	3	UN2401	II	FLAMMABLE LIQUID, CORROSIVE	T2
	Pivaloyl chloride, see **Trimethylacetyl chloride**					
A	Plastic molding material *in dough, sheet or extruded rope form*	9		III	CLASS 9	
	Plastic solvent, n.o.s., see **Flammable liquids**, n.o.s.					
	Plastics, nitrocellulose-based, self-heating, n.o.s.	4.2	UN2006	III	SPONTANEOUSLY COMBUSTIBLE	
	Poisonous gases, n.o.s., see **Compressed or liquefied gases, flammable or toxic, n.o.s.**					
	Polyalkylamines, n.o.s., see **Amines**, etc.					
AW	Polychlorinated biphenyls	9	UN2315	II	CLASS 9	9, 81
	Polyester resin kit	3	UN3269	II	FLAMMABLE LIQUID	40
	Polyhalogenated biphenyls, liquid or Polyhalogenated terphenyls liquid	9	UN3151	II	CLASS 9	

HAZARDOUS MATERIALS TABLE

Sym.	Descriptions and shipping names	Hazard	ID No.	PG	Label(s)	Special provisions
	Polyhalogenated biphenyls, solid or **Polyhalogenated terphenyls, solid**	9	UN3152	II	CLASS 9	
	Polymeric beads, expandable, *evolving flammable vapor.*	9	UN2211	III	None	32
	Potassium	4.3	UN2257	I	DANGEROUS WHEN WET	A19, A20, B27, B100, N6, N34, T15, T26
	Potassium arsenate	6.1	UN1677	II	POISON	
	Potassium arsenite	6.1	UN1678	II	POISON	
	Potassium bisulfite solution, see **Bisulfites, inorganic, aqueous solutions, n.o.s.**					
	Potassium borohydride	4.3	UN1870	I	DANGEROUS WHEN WET	A19, B100, N40
	Potassium bromate	5.1	UN1484	II	OXIDIZER	
	Potassium carbonyl	Forbidden				
	Potassium chlorate	5.1	UN1485	II	OXIDIZER	A9, N34
	Potassium chlorate, aqueous solution	5.1	UN2427	II	OXIDIZER	A2, T8
	Potassium chlorate mixed with mineral oil, see **Explosive, blasting, type C**					
	Potassium cuprocyanide	6.1	UN1679	II	POISON	
	Potassium cyanide	6.1	UN1680	II	POISON	B69, B77, N74, N75, T18, T26
	Potassium dichloro isocyanurate or *Potassium dichloro-s-triazinetrione,* see **Dichloroisocyanuric acid, dry** or **Dichloroisocyanuric acid salts** *etc*					

HAZARDOUS MATERIALS TABLE

Sym.	Descriptions and shipping names	Hazard	ID No.	PG	Label(s)	Special provisions
	Potassium dithionite or Potassium hydrosulfite	4.2	UN1929	II	SPONTANEOUSLY COMBUSTIBLE	A8, A19, A20
	Potassium fluoride	6.1	UN1812	III	KEEP AWAY FROM FOOD	T8
	Potassium fluoroacetate	6.1	UN2628	I	POISON	
	Potassium fluorosilicate	6.1	UN2655	III	KEEP AWAY FROM FOOD	
	Potassium hydrate, see Potassium hydroxide, solid					
	Potassium hydrogen fluoride, see Potassium hydrogen difluoride					
	Potassium hydrogen fluoride solution, see Corrosive liquid, n.o.s.					
	Potassium hydrogen sulfate	8	UN2509	II	CORROSIVE	A7, N34
	Potassium hydrogendifluoride, solid	8	UN1811	II	CORROSIVE, POISON	B106, N3, N34, T8
	Potassium hydrogendifluoride, solution	8	UN1811	II	CORROSIVE, POISON	N3, N34, T8
	Potassium hydrosulfite, see Potassium dithionite					
	Potassium hydroxide, liquid, see Potassium hydroxide solution					
	Potassium hydroxide, solid	8	UN1813	II	CORROSIVE	
	Potassium hydroxide, solution	8	UN1814	II	CORROSIVE	B2, T8
				III	CORROSIVE	T7
	Potassium hypochlorite, solution, see Hypochlorite solutions, etc.					
	Potassium, metal alloys	4.3	UN1420	I	DANGEROUS WHEN WET	A19, A20, B27

HAZARDOUS MATERIALS TABLE

Sym.	Descriptions and shipping names	Hazard	ID No.	PG	Label(s)	Special provisions
	Potassium metal, liquid alloy, see Alkali metal alloys, liquid					
	Potassium metavanadate	6.1	UN2864	II	POISON	
	Potassium monoxide	8	UN2033	II	CORROSIVE	
	Potassium nitrate	5.1	UN1486	III	OXIDIZER	A1, A29
	Potassium nitrate and sodium nitrite mixtures	5.1	UN1487	II	OXIDIZER	B12, B78
	Potassium nitrite	5.1	UN1488	II	OXIDIZER	
	Potassium perchlorate, solid	5.1	UN1489	II	OXIDIZER	
	Potassium perchlorate, solution	5.1	UN1489	II	OXIDIZER	
	Potassium permanganate	5.1	UN1490	II	OXIDIZER	B12
	Potassium peroxide	5.1	UN1491	I	OXIDIZER	A20, N34
	Potassium persulfate	5.1	UN1492	III	OXIDIZER	A1, A29
	Potassium phosphide	4.3	UN2012	I	DANGEROUS WHEN WET, POISON	A19, N40
	Potassium salts of aromatic nitro-derivatives, *explosive*	1.3C	UN0158	II	EXPLOSIVE 1.3C	
	Potassium selenate, see Selenates or Selenites					
	Potassium selenite, see Selenates or Selenites					
	Potassium sodium alloys	4.3	UN1422	I	DANGEROUS WHEN WET	A19, B27, N34, N40, T15, T26

HAZARDOUS MATERIALS TABLE

Sym.	Descriptions and shipping names	Hazard	ID No.	PG	Label(s)	Special provisions
	Potassium sulfide, anhydrous or **Potassium sulfide** with less than 30 percent water of crystallization	4.2	UN1382	II	SPONTANEOUSLY COMBUSTIBLE	A19, A20, B16, B106, N34
	Potassium sulfide, hydrated with not less than 30 percent water of crystallization	8	UN1847	II	CORROSIVE	
	Potassium superoxide	5.1	UN2466	I	OXIDIZER	A20
	Powder cake, wetted or **Powder paste, wetted** with not less than 17 percent alcohol by mass	1.1C	UN0433	II	EXPLOSIVE 1.1C	
	Powder cake, wetted or **Powder paste, wetted** with not less than 25 percent water, by mass	1.3C	UN0159	II	EXPLOSIVE 1.3C	
	Powder paste, see **Powder cake, etc.**					
	Powder, smokeless	1.1C	UN0160	II	EXPLOSIVE 1.1C	
	Powder, smokeless	1.3C	UN0161	II	EXPLOSIVE 1.3C	
	Power device, explosive, see **Cartridges, power device**					
	Primers, cap type	1.4S	UN0044	II	None	
	Primers, cap type	1.1B	UN0377	II	EXPLOSIVE 1.1B	
	Primers, cap type	1.4B	UN0378	II	EXPLOSIVE 1.4B	
	Primers, small arms, see **Primers, cap type**					
	Primers, tubular	1.3G	UN0319	II	EXPLOSIVE 1.3G	
	Primers, tubular	1.4G	UN0320	II	EXPLOSIVE 1.4G	
	Primers, tubular	1.4S	UN0376	II	None	

HAZARDOUS MATERIALS TABLE

Sym.	Descriptions and shipping names	Hazard	ID No.	PG	Label(s)	Special provisions
	Printing Ink, *flammable*	3	UN1210	I	FLAMMABLE LIQUID	T8, T31
				II	FLAMMABLE LIQUID	T7, T30
				III	FLAMMABLE LIQUID	B1, T7, T30
	Projectiles, illuminating, see Ammunition, illuminating, etc.					
	Projectiles, *inert with tracer*	1.4S	UN0345	II	EXPLOSIVE 1.4S	
	Projectiles, *inert, with tracer*	1.3G	UN0424	II	EXPLOSIVE 1.3G	
	Projectiles, *inert, with tracer*	1.4G	UN0425	II	EXPLOSIVE 1.4G	
	Projectiles, *with burster or expelling charge*	1.2D	UN0346	II	EXPLOSIVE 1.2D	
	Projectiles, *with burster or expelling charge*	1.4D	UN0347	II	EXPLOSIVE 1.4D	
	Projectiles, *with burster or expelling charge*	1.2F	UN0426	II	EXPLOSIVE 1.2F	
	Projectiles, *with burster or expelling charge*	1.4F	UN0427	II	EXPLOSIVE 1.4F	
	Projectiles, *with burster or expelling charge*	1.2G	UN0434	II	EXPLOSIVE 1.2G	
	Projectiles, *with burster or expelling charge*	1.4G	UN0435	II	EXPLOSIVE 1.4G	
	Projectiles, *with bursting charge*	1.1F	UN0167	II	EXPLOSIVE 1.1F	
	Projectiles, *with bursting charge*	1.1D	UN0168	II	EXPLOSIVE 1.1D	
	Projectiles, *with bursting charge*	1.2D	UN0169	II	EXPLOSIVE 1.2D	
	Projectiles, *with bursting charge*	1.2F	UN0324	II	EXPLOSIVE 1.2F	
	Projectiles, *with bursting charge*	1.4D	UN0344	II	EXPLOSIVE 1.4D	
	Propadiene, inhibited	2.1	UN2200		FLAMMABLE GAS	
	Propadiene mixed with methyl acetylene, see Methyl acetylene and propadiene mixtures, stabilized					

HAZARDOUS MATERIALS TABLE

Sym.	Descriptions and shipping names	Hazard	ID No.	PG	Label(s)	Special provisions
	Propane or Propane mixtures see also **Petroleum gases, liquefied**	2.1	UN1978		FLAMMABLE GAS	19
	Propanethiols	3	UN2402	II	FLAMMABLE LIQUID	T8
	n-Propanol or Propyl alcohol, normal	3	UN1274	II	FLAMMABLE LIQUID	B1, T1
				III	FLAMMABLE LIQUID	B1, T1
D	Propargyl alcohol	3	NA1986	II	FLAMMABLE LIQUID, POISON	
D	Propellant explosive, liquid	1.1C	NA0474	II	EXPLOSIVE 1.1C	
D	Propellant explosive, liquid	1.3C	NA0477	II	EXPLOSIVE 1.3C	
	Propellant, liquid	1.3C	UN0495		EXPLOSIVE 1.3C	37
	Propellant, liquid	1.1C	UN0497		EXPLOSIVE 1.1C	37
	Propellant, solid	1.1C	UN0498		EXPLOSIVE 1.1C	
	Propellant, solid	1.3C	UN0499		EXPLOSIVE 1.3C	
	Propionaldehyde	3	UN1275	II	FLAMMABLE LIQUID	T14
	Propionic acid	8	UN1848	III	CORROSIVE	T7
	Propionic anhydride	8	UN2496	III	CORROSIVE	T2
	Propionitrile	3	UN2404	II	FLAMMABLE LIQUID, POISON	T14
	Propionyl chloride	3	UN1815	II	FLAMMABLE LIQUID, CORROSIVE	B100, T8, T26
	n-Propyl acetate	3	UN1276	II	FLAMMABLE LIQUID	T1
	Propyl alcohol, see **Propanol**					
	n-Propyl benzene	3	UN2364	III	FLAMMABLE LIQUID	B1, T1

HAZARDOUS MATERIALS TABLE

Sym.	Descriptions and shipping names	Hazard	ID No.	PG	Label(s)	Special provisions
	Propyl chloride	3	UN1278	II	FLAMMABLE LIQUID	N34, T14
	n-Propyl chloroformate	6.1	UN2740	I	POISON, FLAMMABLE LIQUID, CORROSIVE	2, A3, A6, A7, B9, B14, B32, B74, B77, N34, T38, T43, T45
	Propyl formates	3	UN1281	II	FLAMMABLE LIQUID	T8
	n-Propyl isocyanate	6.1	UN2482	I	POISON, FLAMMABLE LIQUID	1, A7, B9, B14, B30, B72, T38, T43, T44
	Propyl mercaptan, see **Propanethiols**					
	n-Propyl nitrate	3	UN1865	II	FLAMMABLE LIQUID	T25
	Propylamine	3	UN1277	II	FLAMMABLE LIQUID, CORROSIVE	N34, T14
	Propylene *see also* **Petroleum gases, liquefied**	2.1	UN1077		FLAMMABLE GAS	19
	Propylene chlorohydrin	6.1	UN2611	II	POISON	T9
	Propylene dichloride	3	UN1279	II	FLAMMABLE LIQUID	N36, T1
	Propylene oxide	3	UN1280	I	FLAMMABLE LIQUID	A3, N34, T20, T29
	Propylene tetramer	3	UN2850	III	FLAMMABLE LIQUID	B1, T1
	1,2-Propylenediamine	8	UN2258	II	CORROSIVE, FLAMMABLE LIQUID	A3, A6, N34, T8
	Propyleneimine, Inhibited	3	UN1921	I	FLAMMABLE LIQUID	A3, N34, T25
	Propyltrichlorosilane	8	UN1816	II	CORROSIVE, FLAMMABLE LIQUID	A7, B2, B6, N34, T8, T26
	Prussic acid, see **Hydrogen cyanide**					
	Pyridine	3	UN1282	II	FLAMMABLE LIQUID	T8

HAZARDOUS MATERIALS TABLE

Sym.	Descriptions and shipping names	Hazard	ID No.	PG	Label(s)	Special provisions
	Pyridine perchlorate	Forbidden				
	Pyrophoric liquid, inorganic, n.o.s.	4.2	UN3194	I	SPONTANEOUSLY COMBUSTIBLE	
	Pyrophoric liquids, organic, n.o.s.	4.2	UN2845	I	SPONTANEOUSLY COMBUSTIBLE	B11, T42
	Pyrophoric metals, n.o.s., or Pyrophoric alloys, n.o.s.	4.2	UN1383	I	SPONTANEOUSLY COMBUSTIBLE	B11
	Pyrophoric organometallic compound, n.o.s.	4.2	UN3203	I	SPONTANEOUSLY COMBUSTIBLE	
	Pyrophoric solid, Inorganic, n.o.s.	4.2	UN3200	I	SPONTANEOUSLY COMBUSTIBLE	
	Pyrophoric solids, organic, n.o.s.	4.2	UN2846	I	SPONTANEOUSLY COMBUSTIBLE	
	Pyrosulfuryl chloride	8	UN1817	II	CORROSIVE	B2, T9, T27
	Pyroxylin solution or solvent, see Nitrocellulose					
	Pyrrolidine	3	UN1922	II	FLAMMABLE LIQUID, CORROSIVE	T1
	Quebrachitol pentanitrate	Forbidden				
	Quicklime, see Calcium oxide					
	Quinoline	6.1	UN2656	III	KEEP AWAY FROM FOOD	T8
	R 114, see Dichlorotetrafluoroethane					
	R 115, see Chloropentafluoroethane					
	R 116, see Hexafluoroethane					

HAZARDOUS MATERIALS TABLE

Sym.	Descriptions and shipping names	Hazard	ID No.	PG	Label(s)	Special provisions
	R 124, see **Chlorotetrafluoroethane**					
	R 133a, see **Chlorotrifluoroethane**					
	R 152a, see **Difluoroethane**					
	R 500, see **Dichlorodifluoromethane and difluoroethane, etc.**					
	R 502, see **Chlorodifluoromethane and chloropentafluoroethane mixture, etc.**					
	R 503, see **Chlorotrifluoromethane and trifluoromethane, etc.**					
	R 12, see **Dichlorodifluoromethane**					
	R 12B1, see **Chlorodifluorobromomethane**					
	R 13, see **Chlorotrifluoromethane**					
	R 13B1, see **Bromotrifluoromethane**					
	R 14, see **Tetrafluoromethane**					
	R 21, see **Dichlorofluoromethane**					
	R 22, see **Chlorodifluoromethane**					
	Radioactive material, excepted package-articles manufactured from natural or depleted uranium or natural thorium	7	UN2910		None	
	Radioactive material, excepted package-empty package or empty packaging	7	UN2910		EMPTY	
	Radioactive material, excepted package-instruments or articles	7	UN2910		None	
	Radioactive material, excepted package-limited quantity of material	7	UN2910		None	

HAZARDOUS MATERIALS TABLE

Sym.	Descriptions and shipping names	Hazard	ID No.	PG	Label(s)	Special provisions
	Radioactive material, fissile, n.o.s.	7	UN2918		RADIOACTIVE	
	Radioactive material, low specific activity or Radioactive material, LSA, n.o.s.	7	UN2912		RADIOACTIVE	
	Radioactive material, n.o.s.	7	UN2982		RADIOACTIVE	
	Radioactive material, special form, n.o.s.	7	UN2974		RADIOACTIVE	
	Radioactive material, surface contaminated object or Radioactive material, SCO	7	UN2913		RADIOACTIVE	
	Railway torpedo, see Signals, railway track, explosive					
	Rare gases and nitrogen mixtures	2.2	UN1981		NONFLAMMABLE GAS	
	Rare gases and oxygen mixtures	2.2	UN1980		NONFLAMMABLE GAS	
	Rare gases, mixtures	2.2	UN1979		NONFLAMMABLE GAS	
	RC 318, see Octafluorocyclobutane					
	RDX and cyclotetramethylenetetranitramine, wetted or desensitized *see* RDX and HMX mixtures, wetted or desensitized					
	RDX and HMX mixtures, wetted *with not less than 15 percent water by mass or* RDX and HMX mixtures, desensitized *with not less than 10 percent phlegmatizer by mass*	1.1D	UN0391	II	EXPLOSIVE 1.1D	
	RDX and Octogen mixtures, wetted or desensitized *see* RDX and HMX mixtures, wetted or desensitized *etc.*					
	RDX, see Cyclotrimethylene trinitramine, *etc.*					

HAZARDOUS MATERIALS TABLE

Sym.	Descriptions and shipping names	Hazard	ID No.	PG	Label(s)	Special provisions
	Receptacles, small, containing gas *flammable, without release device, not refillable and not exceeding 1 L capacity*	2.1	UN2037		FLAMMABLE GAS	
	Receptacles, small, containing gas *non-flammable, without release device, not refillable and not exceeding 1 L capacity*	2.2	UN2037		NONFLAMMABLE GAS	
	Red phosphorus, see **Phosphorus, amorphous**					
	Refrigerant gases, n.o.s.	2.2	UN1078		NONFLAMMABLE GAS	
D	**Refrigerant gases, n.o.s.** *or* **Dispersant gases, n.o.s.**	2.1	NA1954		FLAMMABLE GAS	
D	**Refrigerating machine**	3	NA1993	III	FLAMMABLE LIQUID	
D	**Refrigerating machines,** *containing flammable, non-poisonous, liquefied gas*	2.1	NA1954		FLAMMABLE GAS	
D	**Refrigerating machines,** *containing non-flammable, non-toxic, liquefied gas or ammonia solutions (UN2073)*	2.2	UN2857		NONFLAMMABLE GAS	
D	**Regulated medical waste**	6.2	UN3291	II	INFECTIOUS SUBSTANCE	A13, A14
	Release devices, explosive	1.4S	UN0173	II	EXPLOSIVE 1.4S	
	Resin solution, *flammable*	3	UN1866	II	FLAMMABLE LIQUID	B52, T7, T30
				III	FLAMMABLE LIQUID	B1, B52, T7, T30
	Resorcinol	6.1	UN2876	III	KEEP AWAY FROM FOOD	
	Rifle grenade, see **Grenades,** *hand or rifle, etc.*					
	Rifle powder, see **Powder, smokeless** *(UN 0160)*					
	Rivets, explosive	1.4S	UN0174	II	EXPLOSIVE 1.4S	

HAZARDOUS MATERIALS TABLE

Sym.	Descriptions and shipping names	Hazard	ID No.	PG	Label(s)	Special provisions
	Road asphalt or tar liquid, see Tars, liquid, etc.					
	Rocket motors	1.3C	UN0186	II	EXPLOSIVE 1.3C	109
	Rocket motors	1.1C	UN0280	II	EXPLOSIVE 1.1C	109
	Rocket motors	1.2C	UN0281	II	EXPLOSIVE 1.2C	109
	Rocket motors, liquid fueled	1.2J	UN0395	II	EXPLOSIVE 1.2J	109
	Rocket motors, liquid fueled	1.3J	UN0396	II	EXPLOSIVE 1.3J	109
	Rocket motors with hypergolic liquids with or without an expelling charge	1.3L	UN0250	II	EXPLOSIVE 1.3L	109
	Rocket motors with hypergolic liquids with or without an expelling charge	1.2L	UN0322	II	EXPLOSIVE 1.2L	109
	Rockets, line-throwing	1.2G	UN0238	II	EXPLOSIVE 1.2G	
	Rockets, line-throwing	1.3G	UN0240	II	EXPLOSIVE 1.3G	
	Rockets, line-throwing	1.4G	UN0453	II	EXPLOSIVE 1.4G	
	Rockets, liquid fueled with bursting charge	1.1J	UN0397	II	EXPLOSIVE 1.1J	
	Rockets, liquid fueled with bursting charge	1.2J	UN0398	II	EXPLOSIVE 1.2J	
	Rockets, with bursting charge	1.1F	UN0180	II	EXPLOSIVE 1.1F	
	Rockets, with bursting charge	1.1E	UN0181	II	EXPLOSIVE 1.1E	
	Rockets, with bursting charge	1.2E	UN0182	II	EXPLOSIVE 1.2E	
	Rockets, with bursting charge	1.2F	UN0295	II	EXPLOSIVE 1.2F	
	Rockets, with expelling charge	1.2C	UN0436	II	EXPLOSIVE 1.2C	
	Rockets, with expelling charge	1.3C	UN0437	II	EXPLOSIVE 1.3C	
	Rockets, with expelling charge	1.4C	UN0438	II	EXPLOSIVE 1.4C	

HAZARDOUS MATERIALS TABLE

Sym.	Descriptions and shipping names	Hazard	ID No.	PG	Label(s)	Special provisions
	Rockets, with inert head	1.3C	UN0183	II	EXPLOSIVE 1.3C	
	Rosin oil	3	UN1286	II	FLAMMABLE LIQUID	T7
				III	FLAMMABLE LIQUID	B1, T1
	Rubber solution	3	UN1287	II	FLAMMABLE LIQUID	T7, T30
				III	FLAMMABLE LIQUID	B1, T7, T30
	Rubidium	4.3	UN1423	I	DANGEROUS WHEN WET	22, A7, A19, B100, N34, N40, N45
	Rubidium hydroxide	8	UN2678	II	CORROSIVE	T8
	Rubidium hydroxide solution	8	UN2677	II	CORROSIVE	B2, T8
				III	CORROSIVE	T7
	Safety fuse, see Fuse, safety					
	Samples, explosive, other than initiating explosives		UN0190	II		113
	Sand acid, see Fluorosilicic acid					
	Seed cake, containing vegetable oil solvent extractions and expelled seeds, with not more than 10 percent of oil and when the amount of moisture is higher than 11 percent, with not more than 20 percent of oil and moisture combined.	4.2	UN1386	III	None	N7
I	Seed cake with more than 1.5 percent oil and not more than 11 percent moisture	4.2	UN1386	III	None	N7
I	Seed cake with not more than 1.5 percent oil and not more than 11 percent moisture	4.2	UN2217	III	None	N7
	Selenates or Selenites	6.1	UN2630	I	POISON	

HAZARDOUS MATERIALS TABLE

Sym.	Descriptions and shipping names	Hazard	ID No.	PG	Label(s)	Special provisions
	Selenic acid	8	UN1905	I	CORROSIVE	N34
	Selenium compound, n.o.s.	6.1	UN3283	I	POISON	
				II	POISON	T14
				III	POISON	T7
	Selenium disulfide	6.1	UN2657	II	KEEP AWAY FROM FOOD	
	Selenium hexafluoride	2.3	UN2194		POISON GAS, CORROSIVE	1
	Selenium nitride	Forbidden				
D	Selenium oxide	6.1	NA2811	I	POISON	
	Selenium oxychloride	8	UN2879	I	CORROSIVE, POISON	A3, A6, A7, N34, T12, T27
	Selenium powder	6.1	UN2658	III	KEEP AWAY FROM FOOD	
	Self-heating liquid, corrosive, inorganic, n.o.s.	4.2	UN3188	II	SPONTANEOUSLY COMBUSTIBLE, CORROSIVE	
				III	SPONTANEOUSLY COMBUSTIBLE, CORROSIVE	
	Self-heating liquid, corrosive, organic, n.o.s.	4.2	UN3185	II	SPONTANEOUSLY COMBUSTIBLE, CORROSIVE	
				III	SPONTANEOUSLY COMBUSTIBLE, CORROSIVE	
	Self-heating liquid, inorganic, n.o.s.	4.2	UN3186	II	SPONTANEOUSLY COMBUSTIBLE	
				III	SPONTANEOUSLY COMBUSTIBLE	

HAZARDOUS MATERIALS TABLE

Sym.	Descriptions and shipping names	Hazard	ID No.	PG	Label(s)	Special provisions
	Self-heating liquid, organic, n.o.s.	4.2	UN3183	II	SPONTANEOUSLY COMBUSTIBLE	
				III	SPONTANEOUSLY COMBUSTIBLE	
	Self-heating liquid, toxic, inorganic, n.o.s.	4.2	UN3187	II	SPONTANEOUSLY COMBUSTIBLE, POISON	
				III	SPONTANEOUSLY COMBUSTIBLE, KEEP AWAY FROM FOOD	
	Self-heating liquid, toxic, organic, n.o.s.	4.2	UN3184	II	SPONTANEOUSLY COMBUSTIBLE, POISON	
				III	SPONTANEOUSLY COMBUSTIBLE, KEEP AWAY FROM FOOD	
	Self-heating solid, corrosive, inorganic, n.o.s.	4.2	UN3192	II	SPONTANEOUSLY COMBUSTIBLE, CORROSIVE	
				III	SPONTANEOUSLY COMBUSTIBLE, CORROSIVE	
	Self-heating solid, corrosive, organic, n.o.s.	4.2	UN3126	II	SPONTANEOUSLY COMBUSTIBLE, CORROSIVE	
				III	SPONTANEOUSLY COMBUSTIBLE, CORROSIVE	
	Self-heating solid, inorganic, n.o.s.	4.2	UN3190	II	SPONTANEOUSLY COMBUSTIBLE	
				III	SPONTANEOUSLY COMBUSTIBLE	

HAZARDOUS MATERIALS TABLE

Sym.	Descriptions and shipping names	Hazard	ID No.	PG	Label(s)	Special provisions
	Self-heating, solid, organic, n.o.s.	4.2	UN3088	II	SPONTANEOUSLY COMBUSTIBLE	B101
				III	SPONTANEOUSLY COMBUSTIBLE	B101
	Self-heating, solid, oxidizing, n.o.s.	4.2	UN3127		SPONTANEOUSLY COMBUSTIBLE, OXIDIZER	
	Self-heating solid, toxic, Inorganic, n.o.s.	4.2	UN3191	II	SPONTANEOUSLY COMBUSTIBLE, POISON	
				III	SPONTANEOUSLY COMBUSTIBLE, KEEP AWAY FROM FOOD	
	Self-heating, solid, toxic, organic, n.o.s.	4.2	UN3128	II	SPONTANEOUSLY COMBUSTIBLE, POISON	
				III	SPONTANEOUSLY COMBUSTIBLE, KEEP AWAY FROM FOOD	
	Self-propelled vehicle, see Engines or Batteries etc.					
	Self-reactive liquid type B	4.1	UN3221	II	FLAMMABLE SOLID	53
	Self-reactive liquid type B, temperature controlled	4.1	UN3231	II	FLAMMABLE SOLID	53
	Self-reactive liquid type C	4.1	UN3223	II	FLAMMABLE SOLID	
	Self-reactive liquid type C, temperature controlled	4.1	UN3233	II	FLAMMABLE SOLID	
	Self-reactive liquid type D	4.1	UN3225	II	FLAMMABLE SOLID	
	Self-reactive liquid type D, temperature controlled	4.1	UN3235	II	FLAMMABLE SOLID	

HAZARDOUS MATERIALS TABLE

Sym.	Descriptions and shipping names	Hazard	ID No.	PG	Label(s)	Special provisions
	Self-reactive liquid type E	4.1	UN3227	II	FLAMMABLE SOLID	
	Self-reactive liquid type E, temperature controlled	4.1	UN3237	II	FLAMMABLE SOLID	
	Self-reactive liquid type F	4.1	UN3229	II	FLAMMABLE SOLID	
	Self-reactive liquid type F, temperature controlled	4.1	UN3239	II	FLAMMABLE SOLID	
	Self-reactive solid type B	4.1	UN3222	II	FLAMMABLE SOLID	53
	Self-reactive solid type B, temperature controlled	4.1	UN3232	II	FLAMMABLE SOLID	53
	Self-reactive solid type C	4.1	UN3224	II	FLAMMABLE SOLID	
	Self-reactive solid type C, temperature controlled	4.1	UN3234	II	FLAMMABLE SOLID	
	Self-reactive solid type D	4.1	UN3226	II	FLAMMABLE SOLID	
	Self-reactive solid type D, temperature controlled	4.1	UN3236	II	FLAMMABLE SOLID	
	Self-reactive solid type E	4.1	UN3228	II	FLAMMABLE SOLID	
	Self-reactive solid type E, temperature controlled	4.1	UN3238	II	FLAMMABLE SOLID	
	Self-reactive solid type F	4.1	UN3230	II	FLAMMABLE SOLID	
	Self-reactive solid type F, temperature controlled	4.1	UN3240	II	FLAMMABLE SOLID	
	Shale oil	3	UN1288	I	FLAMMABLE LIQUID	T7
				II	FLAMMABLE LIQUID	T7, T30
				III	FLAMMABLE LIQUID	B1, T7, T30

HAZARDOUS MATERIALS TABLE

Sym.	Descriptions and shipping names	Hazard	ID No.	PG	Label(s)	Special provisions
	Shaped charges, commercial, see Charges, shaped, commercial etc.					
	Signal devices, hand	1.4G	UN0191	II	EXPLOSIVE 1.4G	
	Signal devices, hand	1.4S	UN0373	II	EXPLOSIVE 1.4S	
	Signals, distress, ship	1.1G	UN0194	II	EXPLOSIVE 1.1G	
	Signals, distress, ship	1.3G	UN0195	II	EXPLOSIVE 1.3G	
	Signals, highway, see **Signal devices, hand; Fireworks, type D**					
	Signals, railway track, explosive	1.1G	UN0192	II	EXPLOSIVE 1.1G	
	Signals, railway track, explosive	1.4S	UN0193	II	EXPLOSIVE 1.4S	
	Signals, railway track, explosive	1.3G	UN0492	II	EXPLOSIVE 1.3G	
	Signals, railway track, explosive	1.4G	UN0493	II	EXPLOSIVE 1.4G	
	Signals, ship distress, water-activated, see **Contrivances, water-activated, etc.**					
	Signals, smoke	1.1G	UN0196	II	EXPLOSIVE 1.1G	
	Signals, smoke	1.4G	UN0197	II	EXPLOSIVE 1.4G	
	Signals, smoke	1.2G	UN0313	II	EXPLOSIVE 1.2G	
	Signals, smoke	1.3G	UN0487	II	EXPLOSIVE 1.3G	
	Silane	2.1	UN2203		FLAMMABLE GAS	
	Silicofluoric acid, see **Fluorosilicic acid**					
	Silicon chloride, see **Silicon tetrachloride**					
	Silicon powder, amorphous	4.1	UN1346	III	FLAMMABLE SOLID	A1

HAZARDOUS MATERIALS TABLE

Sym.	Descriptions and shipping names	Hazard	ID No.	PG	Label(s)	Special provisions
	Silicon tetrachloride	8	UN1818	II	CORROSIVE	A3, A6, B2, B6, T18, T26, T29
	Silicon tetrafluoride	2.3	UN1859		POISON GAS, CORROSIVE	2, 25
	Silver acetylide (dry)	Forbidden				
	Silver arsenite (dry)	6.1	UN1683	II	POISON	
	Silver azide (dry)	Forbidden				
	Silver chlorite (dry)	Forbidden				
	Silver cyanide	6.1	UN1684	II	POISON	
	Silver fulminate (dry)	Forbidden				
	Silver nitrate	5.1	UN1493	II	OXIDIZER	
	Silver oxalate (dry)	Forbidden				
	Silver picrate (dry)	Forbidden				
	Silver picrate, wetted with not less than 30 percent water, by mass	4.1	UN1347	I	FLAMMABLE SOLID	
	Sludge, acid	8	UN1906	II	CORROSIVE	A3, A7, B2, N34, T9, T27
D	Smokeless powder for small arms (100 pounds or less)	4.1	NA3178	I	FLAMMABLE SOLID	16
	Soda lime with more than 4 percent sodium hydroxide	8	UN1907	III	CORROSIVE	

HAZARDOUS MATERIALS TABLE

Sym.	Descriptions and shipping names	Hazard	ID No.	PG	Label(s)	Special provisions
	Sodium	4.3	UN1428	I	DANGEROUS WHEN WET	A7, A8, A19, A20, B9, B48, B68, N34, T15, T29, T46
	Sodium aluminate, solid	8	UN2812	III	CORROSIVE	
	Sodium aluminate, solution	8	UN1819	II	CORROSIVE	B2, T8
				III	CORROSIVE	T7
	Sodium aluminum hydride	4.3	UN2835	II	DANGEROUS WHEN WET	A8, A19, A20, B100
	Sodium ammonium vanadate	6.1	UN2863	II	POISON	
	Sodium arsanilate	6.1	UN2473	III	KEEP AWAY FROM FOOD	
	Sodium arsenate	6.1	UN1685	II	POISON	
	Sodium arsenite, aqueous solutions	6.1	UN1686	II	POISON	T15
				III	KEEP AWAY FROM FOOD	T15
	Sodium arsenite, solid	6.1	UN2027	II	POISON	
	Sodium azide	6.1	UN1687	II	POISON	B28
	Sodium bifluoride, solution, see **Sodium hydrogendifluoride**					
	Sodium bisulfite, solution, see **Bisulfites, aqueous solutions, n.o.s.**					
	Sodium borohydride	4.3	UN1426	I	DANGEROUS WHEN WET	B100, N40
	Sodium bromate	5.1	UN1494	II	OXIDIZER	
	Sodium cacodylate	6.1	UN1688	II	POISON	
	Sodium chlorate	5.1	UN1495	II	OXIDIZER	A9, N34, T8
	Sodium chlorate, aqueous solution	5.1	UN2428	II	OXIDIZER	A2, B6, T8

HAZARDOUS MATERIALS TABLE

Sym.	Descriptions and shipping names	Hazard	ID No.	PG	Label(s)	Special provisions
	Sodium chlorate mixed with dinitrotoluene, see **Explosive blasting, type C**					
	Sodium chlorite	5.1	UN1496	II	OXIDIZER	A9, N34, T8
	Sodium chloroacetate	6.1	UN2659	III	KEEP AWAY FROM FOOD	
	Sodium cuprocyanide, solid	6.1	UN2316	I	POISON	T8, T26
	Sodium cuprocyanide, solution	6.1	UN2317	I	POISON	
	Sodium cyanide	6.1	UN1689	I	POISON	B69, B77, N74, N75, T42
	Sodium dichloroisocyanurate or Sodium dichloro-s-triazinetrione, see **Dichloroisocyanuric acid etc.**					
	Sodium dinitro-o-cresolate, dry or wetted with less than 15 percent water, by mass	1.3C	UN0234	II	EXPLOSIVE 1.3C	
	Sodium dinitro-o-cresolate, wetted with not less than 15 percent water, by mass	4.1	UN1348	I	FLAMMABLE SOLID, POISON	23, A8, A19, A20, N41
	Sodium dithionite or **Sodium hydrosulfite**	4.2	UN1384	II	SPONTANEOUSLY COMBUSTIBLE	A19, A20, B106
	Sodium fluoride	6.1	UN1690	III	KEEP AWAY FROM FOOD	T8
	Sodium fluoroacetate	6.1	UN2629	I	POISON	
	Sodium fluorosilicate	6.1	UN2674	III	KEEP AWAY FROM FOOD	
	Sodium hydrate, see **Sodium hydroxide, solid**					
	Sodium hydride	4.3	UN1427	I	DANGEROUS WHEN WET	A19, B100, N40
	Sodium hydrogendifluoride	8	UN2439	II	CORROSIVE	B106, N3, N34
	Sodium hydrogendifluoride solution	8	UN2439	II	CORROSIVE	N3, N34
D	**Sodium hydrosulfide, solution**	8	NA2922	II	CORROSIVE, POISON	B2

HAZARDOUS MATERIALS TABLE

Sym.	Descriptions and shipping names	Hazard	ID No.	PG	Label(s)	Special provisions
	Sodium hydrosulfide, with less than 25 percent water of crystallization	4.2	UN2318	II	SPONTANEOUSLY COMBUSTIBLE	A7, A19, A20
	Sodium hydrosulfide with not less than 25 percent water of crystallization	8	UN2949	II	CORROSIVE	A7
	Sodium hydrosulfite, see Sodium dithionite					
	Sodium hydroxide, solid	8	UN1823	II	CORROSIVE	
	Sodium hydroxide solution	8	UN1824	II	CORROSIVE	B2, N34, T8
				III	CORROSIVE	N34, T7
	Sodium hypochlorite, solution, see Hypochlorite solutions etc.					
	Sodium metal, liquid alloy, see Alkali metal alloys, liquid, n.o.s.					
	Sodium methylate	4.2	UN1431	II	SPONTANEOUSLY COMBUSTIBLE, CORROSIVE	A19
	Sodium methylate solutions in alcohol	3	UN1289	II	FLAMMABLE LIQUID, CORROSIVE	T8, T31
				III	FLAMMABLE LIQUID, CORROSIVE	B1, T7, T30
	Sodium monoxide	8	UN1825	II	CORROSIVE	
	Sodium nitrate	5.1	UN1498	III	OXIDIZER	A1, A29
	Sodium nitrate and potassium nitrate mixtures	5.1	UN1499	III	OXIDIZER	A1, A29
	Sodium nitrite	5.1	UN1500	III	OXIDIZER	A1, A29
	Sodium pentachlorophenate	6.1	UN2567	II	POISON	
	Sodium percarbonate	5.1	UN2467	III	OXIDIZER	27, A1, A29

HAZARDOUS MATERIALS TABLE

Sym.	Descriptions and shipping names	Hazard	ID No.	PG	Label(s)	Special provisions
	Sodium perchlorate	5.1	UN1502	II	OXIDIZER	
	Sodium permanganate	5.1	UN1503	II	OXIDIZER	
	Sodium peroxide	5.1	UN1504	I	OXIDIZER	A20, N34
	Sodium peroxoborate, anhydrous	5.1	UN3247	II	OXIDIZER	
	Sodium persulfate	5.1	UN1505	III	OXIDIZER	A1
	Sodium phosphide	4.3	UN1432	I	DANGEROUS WHEN WET, POISON	A19, N40
	Sodium picramate, dry or wetted with less than 20 percent water, by mass	1.3C	UN0235	II	EXPLOSIVE 1.3C	
	Sodium picramate, wetted with not less than 20 percent water, by mass	4.1	UN1349	I	FLAMMABLE SOLID	23, A8, A19, N41
	Sodium picryl peroxide	Forbidden				
	Sodium potassium alloys, see Potassium sodium alloys					
	Sodium salts of aromatic nitro-derivatives, n.o.s. explosive	1.3C	UN0203	II	EXPLOSIVE 1.3C	
	Sodium selenate, see Selenates or Selenites					
D	Sodium selenite	6.1	NA2630	II	POISON	
	Sodium sulfide, anhydrous or Sodium sulfide with less than 30 percent water of crystallization	4.2	UN1385	II	SPONTANEOUSLY COMBUSTIBLE	A19, A20, B106, N34
	Sodium sulfide, hydrated with not less than 30 percent water	8	UN1849	II	CORROSIVE	T8
	Sodium superoxide	5.1	UN2547	I	OXIDIZER	A20, N34

HAZARDOUS MATERIALS TABLE

Sym.	Descriptions and shipping names	Hazard	ID No.	PG	Label(s)	Special provisions
	Sodium tetranitride	Forbidden				
	Solids containing corrosive liquid, n.o.s.	8	UN3244	II	CORROSIVE	49
	Solids containing flammable liquid, n.o.s.	4.1	UN3175	II	FLAMMABLE SOLID	47
	Solids containing toxic liquid, n.o.s.	6.1	UN3243	II	POISON	48
	Sounding devices, explosive	1.2F	UN0204	II	EXPLOSIVE 1.2F	
	Sounding devices, explosive	1.1F	UN0296	II	EXPLOSIVE 1.1F	
	Sounding devices, explosive	1.1D	UN0374	II	EXPLOSIVE 1.1D	
	Sounding devices, explosive	1.2D	UN0375	II	EXPLOSIVE 1.2D	
	Spirits of salt, see Hydrochloric acid					
	Squibs, see Igniters etc.					
	Stannic chloride, anhydrous	8	UN1827	II	CORROSIVE	B2, T8, T26
	Stannic chloride, pentahydrate	8	UN2440	III	CORROSIVE	
	Stannic phosphide	4.3	UN1433	I	DANGEROUS WHEN WET, POISON	A19, B100, N40
	Steel swarf, see Ferrous metal borings, etc.					
	Stibine	2.3	UN2676		POISON GAS, FLAMMABLE GAS	1
	Storage batteries, wet, see Batteries, wet etc.					
	Strontium arsenite	6.1	UN1691	II	POISON	
	Strontium chlorate	5.1	UN1506	II	OXIDIZER	A1, A9, N34
	Strontium nitrate	5.1	UN1507	III	OXIDIZER	A1, A29
	Strontium perchlorate	5.1	UN1508	II	OXIDIZER	

HAZARDOUS MATERIALS TABLE

Sym.	Descriptions and shipping names	Hazard	ID No.	PG	Label(s)	Special provisions
	Strontium peroxide	5.1	UN1509	II	OXIDIZER	
	Strontium phosphide	4.3	UN2013	I	DANGEROUS WHEN WET, POISON	A19, N40
	Strychnine or Strychnine salts	6.1	UN1692	I	POISON	
	Styphnic acid, see Trinitroresorcinol, etc.					
	Styrene monomer, Inhibited	3	UN2055	III	FLAMMABLE LIQUID	B1, T1
	Substances, explosive, n.o.s.	1.1L	UN0357	II	EXPLOSIVE 1.1L	101
	Substances, explosive, n.o.s.	1.2L	UN0358	II	EXPLOSIVE 1.2L	101
	Substances, explosive, n.o.s.	1.3L	UN0359	II	EXPLOSIVE 1.3L	101
	Substances, explosive, n.o.s.	1.1A	UN0473	II	EXPLOSIVE 1.1A	101, 111
	Substances, explosive, n.o.s.	1.1C	UN0474	II	EXPLOSIVE 1.1C	101
	Substances, explosive, n.o.s.	1.1D	UN0475	II	EXPLOSIVE 1.1D	101
	Substances, explosive, n.o.s.	1.1G	UN0476	II	EXPLOSIVE 1.1G	101
	Substances, explosive, n.o.s.	1.3C	UN0477	II	EXPLOSIVE 1.3C	101
	Substances, explosive, n.o.s.	1.3G	UN0478	II	EXPLOSIVE 1.3G	101
	Substances, explosive, n.o.s.	1.4C	UN0479	II	EXPLOSIVE 1.4C	101
	Substances, explosive, n.o.s.	1.4D	UN0480	II	EXPLOSIVE 1.4D	101
	Substances, explosive, n.o.s.	1.4S	UN0481	II	EXPLOSIVE 1.4S	101
	Substances, explosive, n.o.s.	1.4G	UN0485	II	EXPLOSIVE 1.4G	101
	Substances, explosive, very insensitive, n.o.s., or Substances, EVI, n.o.s.	1.5D	UN0482	II	EXPLOSIVE 1.5D	101

HAZARDOUS MATERIALS TABLE

Sym.	Descriptions and shipping names	Hazard	ID No.	PG	Label(s)	Special provisions
	Substituted nitrophenol pesticides, liquid, flammable, toxic, *flash point less than 23 degrees C*	3	UN2780	I	FLAMMABLE LIQUID, POISON	
				II	FLAMMABLE LIQUID, POISON	
	Substituted nitrophenol pesticides, liquid, toxic	6.1	UN3014	I	POISON	T42
				II	POISON	T14
				III	KEEP AWAY FROM FOOD	T14
	Substituted nitrophenol pesticides, liquid, toxic, flammable *flashpoint not less than 23 degrees C*	6.1	UN3013	I	POISON, FLAMMABLE LIQUID	T42
				II	POISON, FLAMMABLE LIQUID	T14
				III	KEEP AWAY FROM FOOD, FLAMMABLE LIQUID	B1, T14
	Substituted nitrophenol pesticides, solid, toxic	6.1	UN2779	I	POISON	
				II	POISON	
				III	KEEP AWAY FROM FOOD	
	Sucrose octanitrate (dry)	Forbidden				
	Sulfamic acid	8	UN2967	III	CORROSIVE	
D	Sulfur	9	NA1350	III	CLASS 9	30, A1
I	Sulfur	4.1	UN1350	III	FLAMMABLE SOLID	A1, N20, T1

HAZARDOUS MATERIALS TABLE

Sym.	Descriptions and shipping names	Hazard	ID No.	PG	Label(s)	Special provisions
	Sulfur and chlorate, loose mixtures of	Forbidden				
	Sulfur chlorides	8	UN1828	I	CORROSIVE	5, A3, B10, B77, N34, T18, T27
	Sulfur dichloride, see Sulfur chlorides					
	Sulfur dioxide, liquefied	2.3	UN1079		POISON GAS, CORROSIVE	3, B14
	Sulfur dioxide solution, see Sulfurous acid					
	Sulfur hexafluoride	2.2	UN1080		NONFLAMMABLE GAS	
D	**Sulfur, molten**	9	NA2448	III	CLASS 9	T9, T38
I	**Sulfur, molten**	4.1	UN2448	III	FLAMMABLE SOLID	T9, T38
	Sulfur tetrafluoride	2.3	UN2418		POISON GAS, CORROSIVE	1
+	**Sulfur trioxide, Inhibited**	8	UN1829	I	CORROSIVE, POISON	2, A7, B9, B12, B14, B32, B49, B74, B77, N34, T38, T43, T45
+, D	**Sulfur trioxide, uninhibited**	8	NA1829	I	CORROSIVE, POISON	2, A7, B9, B12, B14, B32, B49, B74, B77, N34, T38, T43, T45
	Sulfuretted hydrogen, see Hydrogen sulfide, liquefied					
	Sulfuric acid, *fuming with less than 30 percent free sulfur trioxide*	8	UN1831	I	CORROSIVE	A3, A7, B84, N34, T18, T27
+	**Sulfuric acid,** *fuming with 30 percent or more free sulfur trioxide*	8	UN1831	I	CORROSIVE, POISON	2, A3, A6, A7, B9, B14, B32, B74, B77, B84, N34, T38, T43, T45

HAZARDOUS MATERIALS TABLE

Sym.	Descriptions and shipping names	Hazard	ID No.	PG	Label(s)	Special provisions
	Sulfuric acid, spent	8	UN1832	II	CORROSIVE	A3, A7, B2, B83, BB4, N34, T9, T27
	Sulfuric acid with more than 51 percent acid	8	UN1830	II	CORROSIVE	A3, A7, B3, B83, BB4, N34, T9, T27
	Sulfuric acid with not more than 51% acid	8	UN2796	II	CORROSIVE	A3, A7, B2, B15, N6, N34, T9, T27
	Sulfuric and hydrofluoric acid mixtures, see Hydrofluoric and sulfuric acid mixtures					
	Sulfuric anhydride, see Sulfur trioxide, Inhibited					
	Sulfurous acid	8	UN1833	II	CORROSIVE	B3, T8
+	**Sulfuryl chloride**	8	UN1834	I	CORROSIVE, POISON	1, A3, B6, B9, B10, B14, B30, B74, B77, N34, T38, T43, T44
	Sulfuryl fluoride	2.3	UN2191		POISON GAS	4
	Tars, liquid *including road asphalt and oils, bitumen and cut backs*	3	UN1999	II	FLAMMABLE LIQUID	B13, T7, T30
				III	FLAMMABLE LIQUID	B1, B13, T7, T30
	Tear gas candles	6.1	UN1700	II	POISON, FLAMMABLE SOLID	
	Tear gas cartridges, see Ammunition, tear-producing, etc.					
D	**Tear gas devices with more than 2 percent tear gas substances, by mass**	6.1	NA1693	I	POISON	
				II	POISON	

HAZARDOUS MATERIALS TABLE

Sym.	Descriptions and shipping names	Hazard	ID No.	PG	Label(s)	Special provisions
	Tear gas devices, with not more than 2 percent tear gas substances, by mass, see **Aerosols,** *etc.*					
	Tear gas grenades, see **Tear gas candles**					
	Tear gas substances, liquid, n.o.s.	6.1	UN1693	I	POISON	
		6.1		II	POISON	
	Tear gas substances, solid, n.o.s.	6.1	UN1693	I	POISON	
		6.1		II	POISON	
	Tellurium compound, n.o.s.	6.1	UN3284	I	POISON	T14
		6.1		II	POISON	T7
		6.1		III	KEEP AWAY FROM FOOD	
	Tellurium hexafluoride	2.3	UN2195		POISON GAS, CORROSIVE	1
	Terpene hydrocarbons, n.o.s.	3	UN2319	III	FLAMMABLE LIQUID	B1, T1
	Terpinolene	3	UN2541	III	FLAMMABLE LIQUID	B1, T1
	Tetraazido benzene quinone	Forbidden				
	Tetrabromoethane	6.1	UN2504	III	KEEP AWAY FROM FOOD	T7
	Tetrachloroethane	6.1	UN1702	II	POISON	N36, T14
	Tetrachloroethylene	6.1	UN1897	III	KEEP AWAY FROM FOOD	N36, T1
	Tetraethyl dithiopyrophosphate	6.1	UN1704	I	POISON	
D	Tetraethyl lead, *liquid*	6.1	NA1649	I	POISON, FLAMMABLE LIQUID	
D	Tetraethyl pyrophosphate, *liquid*	6.1	NA3018	I	POISON	

HAZARDOUS MATERIALS TABLE

Sym.	Descriptions and shipping names	Hazard	ID No.	PG	Label(s)	Special provisions
D	**Tetraethyl pyrophosphate solid**	6.1	NA2783	I	POISON	N77
	Tetraethyl silicate	3	UN1292	III	FLAMMABLE LIQUID	B1, T1
	Tetraethylammonium perchlorate (dry)	Forbidden				
	Tetraethylenepentamine	8	UN2320	III	CORROSIVE	
	1,1,1,2-Tetrafluoroethane	2.2	UN3159		NONFLAMMABLE GAS	T2
	Tetrafluoroethylene, inhibited	2.1	UN1081		FLAMMABLE GAS	
	Tetrafluoromethane, R14	2.2	UN1982		NONFLAMMABLE GAS	
	1,2,3,6-Tetrahydrobenzaldehyde	3	UN2498	III	FLAMMABLE LIQUID	B1, T1
	Tetrahydrofuran	3	UN2056	II	FLAMMABLE LIQUID	T8
	Tetrahydrofurfurylamine	3	UN2943	III	FLAMMABLE LIQUID	B1, T1
	Tetrahydrophthalic anhydrides *with more than 0.05 percent of maleic anhydride*	8	UN2698	III	CORROSIVE	
	1,2,3,6-Tetrahydropyridine	3	UN2410	II	FLAMMABLE LIQUID	T8
	Tetrahydrothiophene	3	UN2412	II	FLAMMABLE LIQUID	T7
	Tetramethylammonium hydroxide	8	UN1835	II	CORROSIVE	B2, T3
	Tetramethylene diperoxide dicarbamide	Forbidden				
	Tetramethylsilane	3	UN2749	I	FLAMMABLE LIQUID	T21, T26
	Tetranitro diglycerin	Forbidden				
	Tetranitroaniline	1.1D	UN0207	II	EXPLOSIVE 1.1D	
+	**Tetranitromethane**	5.1	UN1510	I	OXIDIZER, POISON	2, B9, B14, B32, B74, T38, T43, T45

HAZARDOUS MATERIALS TABLE

Sym.	Descriptions and shipping names	Hazard	ID No.	PG	Label(s)	Special provisions
	2,3,4,6-Tetranitrophenol	Forbidden				
	2,3,4,6-Tetranitrophenyl methyl nitramine	Forbidden				
	2,3,4,6-Tetranitrophenylnitramine	Forbidden				
	Tetranitroresorcinol (dry)	Forbidden				
	2,3,5,6-Tetranitroso-1,4-dinitrobenzene	Forbidden				
	2,3,5,6-Tetranitroso nitrobenzene (dry)	Forbidden				
	Tetrapropylorthotitanate	3	UN2413	III	FLAMMABLE LIQUID	B1, T8
	Tetrazene, see Guanyl nitrosaminoguanyltetrazene					
	Tetrazine (dry)	Forbidden				
	Tetrazol-1-acetic acid	1.4C	UN0407	II	EXPLOSIVE 1.4C	
	Tetrazolyl azide (dry)	Forbidden				
	Tetryl, see Trinitrophenylmethylnitramine					
	Thallium chlorate	5.1	UN2573	II	OXIDIZER, POISON	
	Thallium compounds, n.o.s.	6.1	UN1707	II	POISON	
	Thallium nitrate	6.1	UN2727	II	POISON, OXIDIZER	
D	Thallium sulfate, solid	6.1	NA1707	II	POISON	

Sym.	Descriptions and shipping names	Hazard	ID No.	PG	Label(s)	Special provisions
	4-Thiapentanal	6.1	UN2785	III	KEEP AWAY FROM FOOD	T8
	Thioacetic acid	3	UN2436	II	FLAMMABLE LIQUID	T8
	Thiocarbonylchloride, see **Thiophosgene**					
	Thioglycol	6.1	UN2966	II	POISON	T8
	Thioglycolic acid	8	UN1940	II	CORROSIVE	A7, B2, N34, T8
	Thiolactic acid	6.1	UN2936	II	POISON	T8
	Thionyl chloride	8	UN1836	I	CORROSIVE	A7, B6, B10, N34, T18, T27
	Thiophene	3	UN2414	II	FLAMMABLE LIQUID	B101, T2
+	Thiophosgene	6.1	UN2474	II	POISON	2, A7, B9, B14, B32, B74, N33, N34, T38, T43, T45
	Thiophosphoryl chloride	8	UN1837	II	CORROSIVE	A3, A7, B2, B8, B25, B101, N34, T12
	Thorium metal, pyrophoric	7	UN2975		RADIOACTIVE, SPONTA-NEOUSLY COMBUSTIBLE	
	Thorium nitrate, solid	7	UN2976		RADIOACTIVE, OXIDIZER	
	Tin chloride, fuming, see **Stannic chloride, anhydrous**					
	Tin perchloride or Tin tetrachloride, see **Stannic chloride, anhydrous**					
	Tinctures, medicinal	3	UN1293	II	FLAMMABLE LIQUID	T8, T31
				III	FLAMMABLE LIQUID	B1, T7, T30
	Tinning flux, see **Zinc chloride**					

HAZARDOUS MATERIALS TABLE

Sym.	Descriptions and shipping names	Hazard	ID No.	PG	Label(s)	Special provisions
	Titanium disulphide	4.2	UN3174	III	SPONTANEOUSLY COMBUSTIBLE	
	Titanium hydride	4.1	UN1871	II	FLAMMABLE SOLID	A19, A20, N34
	Titanium powder, dry	4.2	UN2546	I	SPONTANEOUSLY COMBUSTIBLE	
				II	SPONTANEOUSLY COMBUSTIBLE	A19, A20, N5, N34
				III	SPONTANEOUSLY COMBUSTIBLE	
	Titanium powder, wetted with not less than 25 percent water (a visible excess of water must be present) (a) mechanically produced, particle size less than 53 microns; (b) chemically produced, particle size less than 840 microns	4.1	UN1352	II	FLAMMABLE SOLID	A19, A20, N34
	Titanium sponge granules or Titanium sponge powders	4.1	UN2878	III	FLAMMABLE SOLID	A1
D	Titanium sulfate solution	8	NA1760	II	CORROSIVE	B2, B15
+	Titanium tetrachloride	8	UN1838	II	CORROSIVE, POISON	2, A3, A6, B7, B9, B14, B32, B74, B77, T38, T43, T45
	Titanium trichloride mixtures	8	UN2869	II	CORROSIVE	A7, B106, N34
				III	CORROSIVE	A7, N34
	Titanium trichloride, pyrophoric or Titanium trichloride mixtures, pyrophoric	4.2	UN2441	I	SPONTANEOUSLY COMBUSTIBLE, CORROSIVE	A7, A8, A19, A20, N34
	TNT mixed with aluminum, see *Tritonal*					
	TNT, see *Trinitrotoluene, etc.*					

HAZARDOUS MATERIALS TABLE

Sym.	Descriptions and shipping names	Hazard	ID No.	PG	Label(s)	Special provisions
	Toluene	3	UN1294	II	FLAMMABLE LIQUID	T1
	Toluene diisocyanate	6.1	UN2078	II	POISON	B110, T14
	Toluene sulfonic acid, see Alkyl, or Aryl sulfonic acid etc.					
	Toluidines *liquid*	6.1	UN1708	II	POISON	T14
	Toluidines *solid*	6.1	UN1708	II	POISON	
	2,4-Toluylenediamine or 2,4-Toluenediamine	6.1	UN1709	III	KEEP AWAY FROM FOOD	T7
	Torpedoes, liquid fueled, with inert head	1.3J	UN0450	II	EXPLOSIVE 1.3J	
	Torpedoes, liquid fueled, with or without bursting charge	1.1J	UN0449	II	EXPLOSIVE 1.1J	
	Torpedoes with bursting charge	1.1E	UN0329	II	EXPLOSIVE 1.1E	
	Torpedoes with bursting charge	1.1F	UN0330	II	EXPLOSIVE 1.1F	
	Torpedoes with bursting charge	1.1D	UN0451	II	EXPLOSIVE 1.1D	
	Toxic liquid, corrosive, inorganic, n.o.s.	6.1	UN3289	I	POISON, CORROSIVE	T42
				II	POISON, CORROSIVE	T14
	Toxic liquid, corrosive, inorganic, n.o.s. *Inhalation Hazard, Packing Group I, Zone A*	6.1	UN3289	I	POISON, CORROSIVE	1, B9, B14, B30, B72, T38, T43, T44
	Toxic liquid, corrosive, inorganic, n.o.s. *Inhalation Hazard, Packing Group I, Zone B*	6.1	UN3289	I	POISON, CORROSIVE	2, B9, B14, B32, B74, T38, T43, T45
	Toxic liquid, inorganic, n.o.s.	6.1	UN3287	I	POISON	T42
				II	POISON	B110, T14
				III	KEEP AWAY FROM FOOD	T7
	Toxic liquid, inorganic, n.o.s. *Inhalation Hazard, Packing Group I, Zone A*	6.1	UN3287	I	POISON	1, B9, B14, B30, B72, T38, T43, T44

HAZARDOUS MATERIALS TABLE

Sym.	Descriptions and shipping names	Hazard	ID No.	PG	Label(s)	Special provisions
	Toxic liquid, inorganic, n.o.s. *Inhalation Hazard, Packing Group I, Zone B*	6.1	UN3287	I	POISON	2, B9, B14, B32, B74, T38, T43, T45
	Toxic liquids, corrosive, organic, n.o.s.	6.1	UN2927	I	POISON, CORROSIVE	T42
		6.1	UN2927	II	POISON, CORROSIVE	T42
	Toxic liquids, corrosive, organic, n.o.s., *Inhalation hazard, Packing Group I, Zone A*	6.1	UN2927	I	POISON, CORROSIVE	1, B9, B14, B30, B72, T38, T43, T44
	Toxic liquids, corrosive, organic, n.o.s., *Inhalation hazard, Packing Group I, Zone B*	6.1	UN2927	I	POISON, CORROSIVE	2, B9, B14, B32, B74, T38, T43, T45
	Toxic liquids, flammable, organic, n.o.s.	6.1	UN2929	I	POISON, FLAMMABLE LIQUID	T42
		6.1	UN2929	II	POISON, FLAMMABLE LIQUID	T15
	Toxic liquids, flammable, organic, n.o.s., *inhalation hazard, Packing Group I, Zone A*	6.1	UN2929	I	POISON, FLAMMABLE LIQUID	1, B9, B14, B30, B72, T38, T43, T44
	Toxic liquids, flammable, organic, n.o.s., *Inhalation hazard, Packing Group I, Zone B*	6.1	UN2929	I	POISON, FLAMMABLE LIQUID	2, B9, B14, B32, B74, T38, T43, T45
	Toxic, liquids, organic, n.o.s.	6.1	UN2810	I	POISON	T42
		6.1	UN2810	II	POISON	B110, T14
		6.1	UN2810	III	KEEP AWAY FROM FOOD	T7
	Toxic, liquids, organic, n.o.s. *Inhalation hazard, Packing Group I, Zone A*	6.1	UN2810	I	POISON	1, B9, B14, B30, B72, T38, T43, T44
	Toxic, liquids, organic, n.o.s. *Inhalation hazard, Packing Group I, Zone B*	6.1	UN2810	I	POISON	2, B9, B14, B32, B74, T38, T43, T45
	Toxic liquids, oxidizing, n.o.s.	6.1	UN3122	I	POISON, OXIDIZER	
		6.1	UN3122	II	POISON, OXIDIZER	A4

HAZARDOUS MATERIALS TABLE

Sym.	Descriptions and shipping names	Hazard	ID No.	PG	Label(s)	Special provisions
	Toxic liquids, oxidizing, n.o.s. *Inhalation hazard, Packing Group I, Zone A*	6.1	UN3122	I	POISON, OXIDIZER	1, B9, B14, B30, B72, T38, T43, T44
	Toxic liquids, oxidizing, n.o.s. *Inhalation Hazard, Packing Group I, Zone B*	6.1	UN3122	I	POISON, OXIDIZER	2, B9, B14, B32, T38, T43, T45
	Toxic liquids, water-reactive, n.o.s.	6.1	UN3123	I	POISON, DANGEROUS WHEN WET	A4
				II	POISON, DANGEROUS WHEN WET	
	Toxic liquids, water-reactive, n.o.s. *Inhalation hazard, packing group I, Zone A*	6.1	UN3123	I	POISON, DANGEROUS WHEN WET	1, B9, B14, B30, B72, T38, T43, T44
	Toxic liquids, water-reactive, n.o.s. *Inhalation hazard, packing group I, Zone B*	6.1	UN3123	I	POISON, DANGEROUS WHEN WET	2, B9, B14, B32, B74, T38, T43, T45
	Toxic solid, corrosive, inorganic, n.o.s.	6.1	UN3290	I	POISON, CORROSIVE	
				II	POISON, CORROSIVE	
	Toxic solid, inorganic, n.o.s.	6.1	UN3288	I	POISON	
				II	POISON	
				III	KEEP AWAY FROM FOOD	
	Toxic solids, corrosive, organic, n.o.s.	6.1	UN2928	I	POISON, CORROSIVE	
				II	POISON, CORROSIVE	
	Toxic solids, flammable, organic, n.o.s.	6.1	UN2930	I	POISON, FLAMMABLE SOLID	B106
				II	POISON, FLAMMABLE SOLID	B106

HAZARDOUS MATERIALS TABLE

Sym.	Descriptions and shipping names	Hazard	ID No.	PG	Label(s)	Special provisions
	Toxic solids, organic, n.o.s.	6.1	UN2811	I	POISON	
				II	POISON	
				III	KEEP AWAY FROM FOOD	
	Toxic solids, oxidizing, n.o.s.	6.1	UN3086	I	POISON, OXIDIZER	
				II	POISON, OXIDIZER	
	Toxic solids, self-heating, n.o.s.	6.1	UN3124	I	POISON, SPONTANEOUSLY COMBUSTIBLE	A5, B100
				II	POISON, SPONTANEOUSLY COMBUSTIBLE	A5, B100
	Toxic solids, water-reactive, n.o.s.	6.1	UN3125	I	POISON, DANGEROUS WHEN WET	
				II	POISON, DANGEROUS WHEN WET	B101
D	Toy Caps	1.4S	NA0337	II	EXPLOSIVE 1.4S	
	Tracers for ammunition	1.3G	UN0212	II	EXPLOSIVE 1.3G	
	Tracers for ammunition	1.4G	UN0306	II	EXPLOSIVE 1.4G	
	Tractors, see Vehicles, self propelled					
	Tri-(b-nitroxyethyl) ammonium nitrate	Forbidden				
	Triallyl borate	6.1	UN2609	III	KEEP AWAY FROM FOOD	
	Triallylamine	3	UN2610	III	FLAMMABLE LIQUID, CORROSIVE	B1, T1

HAZARDOUS MATERIALS TABLE

Sym.	Descriptions and shipping names	Hazard	ID No.	PG	Label(s)	Special provisions
	Triazine pesticides, liquid, flammable, toxic, *flash point less than 23 degrees C*	3	UN2764	I	FLAMMABLE LIQUID, POISON	
				II	FLAMMABLE LIQUID, POISON	T42
	Triazine pesticides, liquid, toxic	6.1	UN2998	I	POISON	T14
				II	POISON	T14
				III	KEEP AWAY FROM FOOD	
	Triazine pesticides, liquid, toxic, flammable, *flashpoint not less than 23 degrees C*	6.1	UN2997	I	POISON, FLAMMABLE LIQUID	T42
				II	POISON, FLAMMABLE LIQUID	T14
				III	KEEP AWAY FROM FOOD, FLAMMABLE LIQUID	
	Triazine pesticides, solid, toxic	6.1	UN2763	I	POISON	
				II	POISON	
				III	KEEP AWAY FROM FOOD	
	Tributylamine	8	UN2542	II	CORROSIVE	T2
	Tributylphosphane	4.2	UN3254	I	SPONTANEOUSLY COMBUSTIBLE	
D	mono-(Trichloro) tetra-(monopotassium dichloro)-penta-s-triazinetrione, dry *(with more than 39 percent available chlorine)*	5.1	NA2468	II	OXIDIZER	

HAZARDOUS MATERIALS TABLE

Sym.	Descriptions and shipping names	Hazard	ID No.	PG	Label(s)	Special provisions
	Trichloro-s-triazinetrione dry, with more than 39 percent available chlorine, see Trichloroisocyanuric acid, dry					
	Trichloroacetic acid	8	UN1839	II	CORROSIVE	A7, N34
+	Trichloroacetic acid, solution	8	UN2564	II	CORROSIVE	A3, A6, A7, B2, N34, T8
	Trichloroacetyl chloride	8	UN2442	II	CORROSIVE, POISON	A3, A6, A7, N34, T7
	Trichlorobenzenes, liquid	6.1	UN2321	III	KEEP AWAY FROM FOOD	2, A3, A7, B9, B14, B32, B74, N34, T38, T43, T45
	Trichlorobutene	6.1	UN2322	II	POISON	T7
	1,1,1-Trichloroethane	6.1	UN2831	III	KEEP AWAY FROM FOOD	T8
	Trichloroethylene	6.1	UN1710	III	KEEP AWAY FROM FOOD	N36, T7
	Trichloroisocyanuric acid, dry	5.1	UN2468	II	OXIDIZER	N36, T1
	Trichloromethyl perchlorate	Forbidden				
	Trichlorosilane	4.3	UN1295	I	DANGEROUS WHEN WET, FLAMMABLE LIQUID, CORROSIVE	A7, N34, T24, T26
	Tricresyl phosphate with more than 3 percent ortho isomer	6.1	UN2574	II	POISON	A3, N33, N34, T8
	Triethyl phosphite	3	UN2323	III	FLAMMABLE LIQUID	B1, T1
	Triethylamine	3	UN1296	II	FLAMMABLE LIQUID, CORROSIVE	B101, T8
	Triethylenetetramine	8	UN2259	II	CORROSIVE	B2, T8

HAZARDOUS MATERIALS TABLE

Sym.	Descriptions and shipping names	Hazard	ID No.	PG	Label(s)	Special provisions
	Trifluoroacetic acid	8	UN2699	I	CORROSIVE	A3, A6, A7, B4, N3, N34, T18, T27
	Trifluoroacetyl chloride	2.3	UN3057		POISON GAS	2, 25, B9, B14
	Trifluorochloroethylene, inhibited, R1113	2.3	UN1082		POISON GAS	3, 25, B14
	Trifluoroethane, compressed, R143	2.1	UN2035		FLAMMABLE GAS	
	Trifluoromethane	2.2	UN1984		NONFLAMMABLE GAS	
D	Trifluoromethane and chlorotrifluoromethane mixture (constant boiling mixture) (R-503). See Refrigerant gases, n.o.s.					
	Trifluoromethane, refrigerated liquid	2.2	UN3136		NONFLAMMABLE GAS	
	2-Trifluoromethylaniline	6.1	UN2942	III	KEEP AWAY FROM FOOD	
	3-Trifluoromethylaniline	6.1	UN2948	II	POISON	T14
	Triformoxime trinitrate	Forbidden				
	Triisobutylene	3	UN2324	III	FLAMMABLE LIQUID	B1, T7, T30
	Triisocyanatoisocyanurate of isophoronediisocyanate, solution, with 70 percent, by mass	3	UN2906	III	FLAMMABLE LIQUID	B1, T1
	Triisopropyl borate	3	UN2616	II	FLAMMABLE LIQUID	T8, T31
		3		III	FLAMMABLE LIQUID	B1, T8, T31
D	Trimethoxysilane	6.1	NA9269	I	POISON, FLAMMABLE LIQUID	2, B9, B14, B32, B74, T38, T43, T45
	Trimethyl borate	3	UN2416	II	FLAMMABLE LIQUID	T14
	Trimethyl phosphite	3	UN2329	III	FLAMMABLE LIQUID	B1, T1

HAZARDOUS MATERIALS TABLE

Sym.	Descriptions and shipping names	Hazard	ID No.	PG	Label(s)	Special provisions
	1,3,5-Trimethyl-2,4,6-trinitrobenzene	Forbidden				
	Trimethylacetyl chloride	6.1	UN2438	I	POISON, CORROSIVE, FLAMMABLE LIQUID	2, A3, A6, A7, B3, B9, B14, B32, B74, N34, T38, T43, T45
	Trimethylamine, anhydrous	2.1	UN1083		FLAMMABLE GAS	T42
	Trimethylamine, aqueous solutions with not more than 50 percent trimethylamine by mass	3	UN1297	I	FLAMMABLE LIQUID, CORROSIVE	
				II	FLAMMABLE LIQUID, CORROSIVE	B1, T14
				III	FLAMMABLE LIQUID, CORROSIVE	B1
	1,3,5-Trimethylbenzene	3	UN2325	III	FLAMMABLE LIQUID	B1, T1
	Trimethylchlorosilane	3	UN1298	II	FLAMMABLE LIQUID, CORROSIVE	A3, A7, B77, N34, T14, T26
	Trimethylcyclohexylamine	8	UN2326	III	CORROSIVE	T2
	Trimethylene glycol diperchlorate	Forbidden				
	Trimethylhexamethylene diisocyanate	6.1	UN2328	III	KEEP AWAY FROM FOOD	T8
	Trimethylhexamethylenediamines	8	UN2327	III	CORROSIVE	T7
	Trimethylol nitromethane trinitrate	Forbidden				
	Trinitro-meta-cresol	1.1D	UN0216	II	EXPLOSIVE 1.1D	
	2,4,6-Trinitro-1,3-diazobenzene	Forbidden				

HAZARDOUS MATERIALS TABLE

Sym.	Descriptions and shipping names	Hazard	ID No.	PG	Label(s)	Special provisions
	2,4,6-Trinitro-1,3,5-triazido benzene (dry)	Forbidden				
	Trinitroacetic acid	Forbidden				
	Trinitroacetonitrile	Forbidden				
	Trinitroamine cobalt	Forbidden				
	Trinitroaniline or **Picramide**	1.1D	UN0153	II	EXPLOSIVE 1.1D	
	Trinitroanisole	1.1D	UN0213	II	EXPLOSIVE 1.1D	
	Trinitrobenzene, *dry or wetted with less than 30 percent water, by mass*	1.1D	UN0214	II	EXPLOSIVE 1.1D	
	Trinitrobenzene, *wetted with not less than 30 percent water, by mass*	4.1	UN1354	I	FLAMMABLE SOLID	23, A2, A8, A19, N41
	Trinitrobenzenesulfonic acid	1.1D	UN0386	II	EXPLOSIVE 1.1D	
	Trinitrobenzoic acid, *dry or wetted with less than 30 percent water, by mass*	1.1D	UN0215	II	EXPLOSIVE 1.1D	
	Trinitrobenzoic acid, *wetted with not less than 30 percent water, by mass*	4.1	UN1355	I	FLAMMABLE SOLID	23, A2, A8, A19, N41
	Trinitrochlorobenzene or **Picryl chloride**	1.1D	UN0155	II	EXPLOSIVE 1.1D	
	Trinitroethanol	Forbidden				
	Trinitroethylnitrate	Forbidden				
	Trinitrofluorenone	1.1D	UN0387	II	EXPLOSIVE 1.1D	

HAZARDOUS MATERIALS TABLE

Sym.	Descriptions and shipping names	Hazard	ID No.	PG	Label(s)	Special provisions
	Trinitromethane	Forbidden				
	1,3,5-Trinitronaphthalene	Forbidden				
	Trinitronaphthalene	1.1D	UN0217	II	EXPLOSIVE 1.1D	
	Trinitrophenetole	1.1D	UN0218	II	EXPLOSIVE 1.1D	
	Trinitrophenol or Picric acid, dry or wetted with less than 30 percent water, by mass	1.1D	UN0154	II	EXPLOSIVE 1.1D	
	Trinitrophenol, wetted with not less than 30 percent water, by mass	4.1	UN1344	I	FLAMMABLE SOLID	23, A8, A19, N41
	2,4,6-Trinitrophenyl guanidine (dry)	Forbidden				
	2,4,6-Trinitrophenyl nitramine	Forbidden				
	2,4,6-Trinitrophenyl trimethylol methyl nitramine trinitrate (dry)	Forbidden				
	Trinitrophenylmethylnitramine or Tetryl	1.1D	UN0208	II	EXPLOSIVE 1.1D	
	Trinitroresorcinol or Styphnic acid, dry or wetted with less than 20 percent water, or mixture of alcohol and water, by mass	1.1D	UN0219	II	EXPLOSIVE 1.1D	
	Trinitroresorcinol, wetted or Styphnic acid, wetted with not less than 20 percent water, or mixture of alcohol and water by mass	1.1D	UN0394	II	EXPLOSIVE 1.1D	
	2,4,6-Trinitroso-3-methyl nitraminoanisole	Forbidden				
	Trinitrotetramine cobalt nitrate	Forbidden				

HAZARDOUS MATERIALS TABLE

Sym.	Descriptions and shipping names	Hazard	ID No.	PG	Label(s)	Special provisions
	Trinitrotoluene and Trinitrobenzene mixtures or TNT and trinitrobenzene mixtures or Trinitrotoluene and hexanitrostilbene mixtures	1.1D	UN0388	II	EXPLOSIVE 1.1D	
	Trinitrotoluene mixtures containing Trinitrobenzene and Hexanitrostilbene or TNT mixtures containing trinitrobenzene and hexanitrostilbene	1.1D	UN0389	II	EXPLOSIVE 1.1D	
	Trinitrotoluene or TNT, dry or wetted with less than 30 percent water, by mass	1.1D	UN0209	II	EXPLOSIVE 1.1D	23, A2, A8, A19, N41
	Trinitrotoluene, wetted with not less than 30 percent water, by mass	4.1	UN1356	I	FLAMMABLE SOLID	
	Tripropylamine	3	UN2260	III	FLAMMABLE LIQUID, CORROSIVE	B1, T8
	Tripropylene	3	UN2057	II	FLAMMABLE LIQUID	T1
				III	FLAMMABLE LIQUID	B1, T1
	Tris-(1-aziridinyl)phosphine oxide, solution	6.1	UN2501	II	POISON	T8
				III	KEEP AWAY FROM FOOD	T7
	Tris, bis-bifluoroamino diethoxy propane (TVOPA)	Forbidden				
	Tritonal	1.1D	UN0390	II	EXPLOSIVE 1.1D	
	Tungsten hexafluoride	2.3	UN2196		POISON GAS, CORROSIVE	3
	Turpentine	3	UN1299	III	FLAMMABLE LIQUID	B1, T1
	Turpentine substitute	3	UN1300	I	FLAMMABLE LIQUID	T1
				II	FLAMMABLE LIQUID	T1

HAZARDOUS MATERIALS TABLE

Sym.	Descriptions and shipping names	Hazard	ID No.	PG	Label(s)	Special provisions
	Undecane	3	UN2330	III	FLAMMABLE LIQUID	B1, T1
	Uranium hexafluoride, *fissile excepted or non-fissile*	7	UN2978		RADIOACTIVE, CORROSIVE	B1, T1
	Uranium hexafluoride, fissile *(with more than 1 percent U-235)*	7	UN2977		RADIOACTIVE, CORROSIVE	
	Uranium metal, pyrophoric	7	UN2979		RADIOACTIVE, SPONTANEOUSLY COMBUSTIBLE	
	Uranyl nitrate hexahydrate solution	7	UN2980		RADIOACTIVE, CORROSIVE	
	Uranyl nitrate, solid	7	UN2981		RADIOACTIVE, OXIDIZER	
	Urea hydrogen peroxide	5.1	UN1511	III	OXIDIZER, CORROSIVE	A1, A7, A29
	Urea nitrate, dry or wetted with less than 20 percent water, by mass	1.1D	UN0220	II	EXPLOSIVE 1.1D	
	Urea nitrate, wetted with not less than 20 percent water, by mass	4.1	UN1357	I	FLAMMABLE SOLID	39, A8, A19, N41
	Urea peroxide, see Urea hydrogen peroxide					
	Valeraldehyde	3	UN2058	II	FLAMMABLE LIQUID	T1
	Valeric acid, see Corrosive liquids, n.o.s.					
	Valeryl chloride	8	UN2502	II	CORROSIVE, FLAMMABLE LIQUID	A3, A6, A7, B2, N34, T8
	Vanadium compound, n.o.s.	6.1	UN3285	I	POISON	T14
				II	POISON	T7
				III	KEEP AWAY FROM FOOD	

HAZARDOUS MATERIALS TABLE

Sym.	Descriptions and shipping names	Hazard	ID No.	PG	Label(s)	Special provisions
	Vanadium oxytrichloride	8	UN2443	II	CORROSIVE	A3, A6, A7, B2, B16, N34, T8, T26
	Vanadium pentoxide, *non-fused form*	6.1	UN2862	II	POISON	
	Vanadium tetrachloride	8	UN2444	I	CORROSIVE	A3, A6, A7, B4, N34, T8, T26
	Vanadium trichloride	8	UN2475	III	CORROSIVE	
	Vanadyl sulfate	6.1	UN2931	II	POISON	
D	*Vehicles, self-propelled including internal combustion engines or other apparatus containing an internal combustion Engine or electric storage battery, see Engines etc. or Battery powered etc. or Wheel chair, electric)*					
	Very signal cartridge, see **Cartridges, signal**					
	Vinyl acetate, inhibited	3	UN1301	II	FLAMMABLE LIQUID	T8
	Vinyl bromide, inhibited	2.1	UN1085		FLAMMABLE GAS	T7
	Vinyl butyrate, inhibited	3	UN2838	II	FLAMMABLE LIQUID	
	Vinyl chloride, inhibited or **Vinyl chloride, stabilized**	2.1	UN1086		FLAMMABLE GAS	21, B44
	Vinyl chloroacetate	6.1	UN2589	II	POISON, FLAMMABLE LIQUID	T14
	Vinyl ethyl ether, inhibited	3	UN1302	I	FLAMMABLE LIQUID	A3, B100, T14
	Vinyl fluoride, inhibited	2.1	UN1860		FLAMMABLE GAS	
	Vinyl isobutyl ether, inhibited	3	UN1304	II	FLAMMABLE LIQUID	T8
	Vinyl methyl ether, inhibited	2.1	UN1087		FLAMMABLE GAS	B44

HAZARDOUS MATERIALS TABLE

Sym.	Descriptions and shipping names	Hazard	ID No.	PG	Label(s)	Special provisions
	Vinyl nitrate polymer	Forbidden				
	Vinyl toluene, Inhibited *mixed isomers*	3	UN2618	III	FLAMMABLE LIQUID	B1, T1
	Vinylidene chloride, Inhibited	3	UN1303	I	FLAMMABLE LIQUID	T23, T29
	Vinylpyridines, Inhibited	6.1	UN3073	II	POISON, FLAMMABLE LIQUID	B100, T8
	Vinyltrichlorosilane	3	UN1305	I	FLAMMABLE LIQUID, CORROSIVE	A3, A7, B6, N34, T14, T26
	Warheads, rocket with burster or expelling charge	1.4D	UN0370	II	EXPLOSIVE 1.4D	
	Warheads, rocket with burster or expelling charge	1.4F	UN0371	II	EXPLOSIVE 1.4F	
	Warheads, rocket with bursting charge	1.1D	UN0286	II	EXPLOSIVE 1.1D	
	Warheads, rocket with bursting charge	1.2D	UN0287	II	EXPLOSIVE 1.2D	
	Warheads, rocket with bursting charge	1.1F	UN0369	II	EXPLOSIVE 1.1F	
	Warheads, torpedo with bursting charge	1.1D	UN0221	II	EXPLOSIVE 1.1D	
	Water-reactive liquid, corrosive, n.o.s.	4.3	UN3129	I	DANGEROUS WHEN WET, CORROSIVE	
				II	DANGEROUS WHEN WET, CORROSIVE	B106
				III	DANGEROUS WHEN WET, CORROSIVE	B106
	Water-reactive liquid, n.o.s.	4.3	UN3148	I	DANGEROUS WHEN WET	
				II	DANGEROUS WHEN WET	B106
				III	DANGEROUS WHEN WET	B106

HAZARDOUS MATERIALS TABLE

Sym.	Descriptions and shipping names	Hazard	ID No.	PG	Label(s)	Special provisions
	Water-reactive liquid, toxic, n.o.s.	4.3	UN3130	I	DANGEROUS WHEN WET, POISON	A4
				II	DANGEROUS WHEN WET, POISON	B106
				III	DANGEROUS WHEN WET, KEEP AWAY FROM FOOD	B106
	Water-reactive solid, corrosive, n.o.s.	4.3	UN3131	I	DANGEROUS WHEN WET, CORROSIVE	B101, B106
				II	DANGEROUS WHEN WET, CORROSIVE	B101, B106
				III	DANGEROUS WHEN WET, CORROSIVE	B105, B106
	Water-reactive solid, flammable, n.o.s.	4.3	UN3132	I	DANGEROUS WHEN WET, FLAMMABLE SOLID	B101, B106, N40
				II	DANGEROUS WHEN WET, FLAMMABLE SOLID	B101, B106
				III	DANGEROUS WHEN WET, FLAMMABLE SOLID	B105, B106
	Water-reactive solid, n.o.s.	4.3	UN2813	I	DANGEROUS WHEN WET	B101, B106, N40
				II	DANGEROUS WHEN WET	B101, B106
				III	DANGEROUS WHEN WET	B105, B106
	Water-reactive solid, oxidizing, n.o.s.	4.3	UN3133	I	DANGEROUS WHEN WET, OXIDIZER	
	Water-reactive solid, self-heating, n.o.s.	4.3	UN3135	I	DANGEROUS WHEN WET, SPONTANEOUSLY COMBUSTIBLE	B100, N40

HAZARDOUS MATERIALS TABLE

Sym.	Descriptions and shipping names	Hazard	ID No.	PG	Label(s)	Special provisions
	Water-reactive solid, toxic, n.o.s.	4.3	UN3134	I	DANGEROUS WHEN WET, SPONTANEOUSLY COMBUSTIBLE	B101, B106
				II	DANGEROUS WHEN WET, SPONTANEOUSLY COMBUSTIBLE	B101, B106
				III	DANGEROUS WHEN WET, SPONTANEOUSLY COMBUSTIBLE	A8, B101, B106, N40
				I	DANGEROUS WHEN WET, POISON	B105, B106
				II	DANGEROUS WHEN WET, POISON	B105, B106
				III	DANGEROUS WHEN WET, KEEP AWAY FROM FOOD	B105, B106
AD	Wheel chair, electric *(spillable or non-spillable type batteries)*	9			CLASS 9	
	White acid, see Hydrofluoric acid mixtures					
I	White asbestos *(chrysotile, actinolite, anthophyllite, tremolite)*	9	UN2590	III	CLASS 9	
	Wood preservatives, liquid	3	UN1306	II	FLAMMABLE LIQUID	T7, T30
				III	FLAMMABLE LIQUID	B1, T7, T30
	Xenon	2.2	UN2036		NONFLAMMABLE GAS	
	Xenon, refrigerated liquid *(cryogenic liquids)*	2.2	UN2591		NONFLAMMABLE GAS	T1
	Xylenes	3	UN1307	II	FLAMMABLE LIQUID	B1, T1
				III	FLAMMABLE LIQUID	
	Xylenols	6.1	UN2261	II	POISON	T8
	Xylidines, solid	6.1	UN1711	II	POISON	T14

HAZARDOUS MATERIALS TABLE

320

Sym.	Descriptions and shipping names	Hazard	ID No.	PG	Label(s)	Special provisions
	Xylidines, solution	6.1	UN1711	II	POISON	T14
	Xylyl bromide	6.1	UN1701	II	POISON	A3, A6, A7, N33
	p-Xylyl diazide	Forbidden				
	Zinc ammonium nitrate	5.1	UN1512	II	OXIDIZER	
	Zinc arsenate or Zinc arsenite or Zinc arsenate and zinc arsenite mixtures.	6.1	UN1712	II	POISON	
	Zinc ashes	4.3	UN1435	III	DANGEROUS WHEN WET	A1, A19, B108
	Zinc bisulfite solution *see* **Bisulfites, Inorganic aqueous solutions, n.o.s.**					
	Zinc bromate	5.1	UN2469	III	OXIDIZER	A1, A29
	Zinc chlorate	5.1	UN1513	II	OXIDIZER	A9, N34
	Zinc chloride, anhydrous	8	UN2331	III	CORROSIVE	
	Zinc chloride, solution	8	UN1840	III	CORROSIVE	T7
	Zinc cyanide	6.1	UN1713	I	POISON	
	Zinc dithionite *or* Zinc hydrosulfite	9	UN1931	III	None	
	Zinc ethyl, see **Diethylzinc**					
	Zinc fluorosilicate	6.1	UN2855	III	KEEP AWAY FROM FOOD	
	Zinc hydrosulfite, *see* **Zinc dithionite**					
	Zinc muriate solution, *see* **Zinc chloride, solution**					
	Zinc nitrate	5.1	UN1514	II	OXIDIZER	
	Zinc permanganate	5.1	UN1515	II	OXIDIZER	

HAZARDOUS MATERIALS TABLE

Sym.	Descriptions and shipping names	Hazard	ID No.	PG	Label(s)	Special provisions
	Zinc peroxide	5.1	UN1516	II	OXIDIZER	A19, N40
	Zinc phosphide	4.3	UN1714	I	DANGEROUS WHEN WET, POISON	A19, B109, N40
	Zinc powder or Zinc dust	4.3	UN1436	I	DANGEROUS WHEN WET, SPONTANEOUSLY COMBUSTIBLE	A19, B109
				II	DANGEROUS WHEN WET, SPONTANEOUSLY COMBUSTIBLE	A19, B109
				III	DANGEROUS WHEN WET, SPONTANEOUSLY COMBUSTIBLE	B108
	Zinc resinate	4.1	UN2714	III	FLAMMABLE SOLID	A1
	Zinc selenate, see **Selenates or Selenites**					
	Zinc selenite, see **Selenates or Selenites**					
	Zinc silicofluoride, see **Zinc fluorosilicate**					
	Zirconium, dry, coiled wire, finished metal sheets, strip (thinner than 254 microns but not thinner than 18 microns)	4.1	UN2858	III	FLAMMABLE SOLID	A1
	Zirconium, dry, finished sheets, strip or coiled wire	4.2	UN2009	III	SPONTANEOUSLY COMBUSTIBLE	A1, A19
	Zirconium hydride	4.1	UN1437	II	FLAMMABLE SOLID	A19, A20, N34
	Zirconium nitrate	5.1	UN2728	III	OXIDIZER	A1, A29
	Zirconium picramate, dry or wetted with less than 20 percent water, by mass	1.3C	UN0236	II	EXPLOSIVE 1.3C	
	Zirconium picramate, wetted with not less than 20 percent water, by mass	4.1	UN1517	I	FLAMMABLE SOLID	23, N41

HAZARDOUS MATERIALS TABLE

Sym.	Descriptions and shipping names	Hazard	ID No.	PG	Label(s)	Special provisions
	Zirconium powder, dry	4.2	UN2008	I	SPONTANEOUSLY COMBUSTIBLE	
				II	SPONTANEOUSLY COMBUSTIBLE	A19, A20, N5, N34
				III	SPONTANEOUSLY COMBUSTIBLE	
	Zirconium powder, wetted with not less than 25 percent water (a visible excess of water must be present) (a) mechanically produced, particle size less than 53 microns; (b) chemically produced, particle size less than 840 microns	4.1	UN1358	II	FLAMMABLE SOLID	A19, A20, N34
	Zirconium scrap	4.2	UN1932	III	SPONTANEOUSLY COMBUSTIBLE	N34
D	Zirconium sulfate	8	NA9163	III	CORROSIVE	N34
	Zirconium suspended in a liquid	3	UN1308	I	FLAMMABLE LIQUID	
				II	FLAMMABLE LIQUID	
				III	FLAMMABLE LIQUID	B1
	Zirconium tetrachloride	8	UN2503	III	CORROSIVE	

Appendix A to §172.101 - List of Hazardous Substances and Reportable Quantities

1. This Appendix lists materials and their corresponding reportable quantities (RQs) that are listed or designated as "hazardous substances" under section 101(14) of the Comprehensive Environmental Response, Compensation, and Liability Act, 42 U.S.C. 9601(14) (CERCLA; 42 U.S.C. 9601 *et seq*). This listing fulfills the requirement of CERCLA, 42 U.S.C. 9656 (a), that all "hazardous substances," as defined in 42 U.S.C. 9601 (14), be listed and regulated as hazardous materials under 49 U.S.C. 5101-5127. That definition includes substances listed under sections 311(b)(2)(A) and 307(a) of the Federal Water Pollution Control Act, 33 U.S.C. 1321(b)(2)(A) and 1317(a), section 3001 of the Solid Waste Disposal Act, 42 U.S.C. 6921, and Section 112 of the Clean Air Act, 42 U.S.C. 7412. In addition, this list contains materials that the Administrator of the Environmental Protection Agency has determined to be hazardous substances in accordance with section 102 of CERCLA, 42 U.S.C. 9602. It should be noted that 42 U.S.C. 9656(b) provides that common and contract carriers may be held liable under laws other than CERCLA for the release of a hazardous substance as defined in that Act, during transportation that commenced before the effective date of the listing and regulating of that substance as a hazardous material under 49 U.S.C. 5101-5127.

2. This Appendix is divided into two TABLES which are entitled "TABLE 1–HAZARDOUS SUBSTANCES OTHER THAN RADIONUCLIDES" and "TABLE 2–RADIONU-

CLIDES." A material listed in this Appendix is regulated as a hazardous material and a hazardous substance under this subchapter if it meets the definition of a hazardous substance in §171.8 of this subchapter.

3. The procedure for selecting a proper shipping name for a hazardous substance is set forth in §172.101(c)(8).

4. Column 1 of TABLE 1, entitled *"Hazardous substance"*, contains the names of those elements and compounds which are hazardous substances. Following the listing of elements and compounds is a listing of waste streams. These waste streams appear on the list in numerical sequence and are referenced by the appropriate "D", "F" or "K" numbers. Column 2 of TABLE 1, entitled *"Synonyms"*, contains the names of synonyms for certain elements and compounds listed in Column 1. No synonyms are listed for waste streams. Synonyms are useful in identifying hazardous substances and in identifying proper shipping names. Column 3 of TABLE 1, entitled *"Reportable quantity (RQ)"*, contains the reportable quantity (RQ), in pounds and kilograms, for each hazardous substance listed in Column 1 of TABLE 1.

5. A series of notes is used throughout TABLE 1 and TABLE 2 to provide additional information concerning certain hazardous substances. These notes are explained at the end of each TABLE.

6. TABLE 2 lists radionuclides that are hazardous substances and their corresponding RQ's. The RQ's in TABLE 2 for radionuclides are expressed in units of curies and terabecquerels, whereas those in TABLE 1 are expressed in units of pounds and kilograms. If a material is listed in both TABLE 1 and TABLE 2, the lower RQ shall ap-

ply. Radionuclides are listed in alphabetical order. The RQs for radionuclides are given in the radiological unit of measure of curie, abbreviated "Ci", followed, in parentheses, by an equivalent unit measured in terabecquerels, abbreviated "TBq".

7. For mixtures of radionuclides, the following requirements shall be used in determining if a package contains an RQ of a hazardous substance: (i) if the identity and quantity (in curies or terabecquerels) of each radionuclide in a mixture or solution is known, the ratio between the quantity per package (in curies or terabecquerels) and the RQ for the radionuclide must be determined for each radionuclide. A package contains an RQ of a hazardous substance when the sum of the ratios for the radionuclides in the mixture or solution is equal to or greater than one; (ii) if the identity of each radionuclide in a mixture or solution is known but the quantity per package (in curies or terabecquerels) of one or more of the radionuclides is unknown, an RQ of a hazardous substance is present in a package when the total quantity (in curies or terabecquerels) of the mixture or solution is equal to or greater than the lowest RQ of any individual radionuclide in the mixture or solution; and (iii) if the identity of one or more radionuclides in a mixture or solution is unknown (or if the identity of a radionuclide by itself is unknown), an RQ of a hazardous substance is present when the total quantity (in curies or terabecquerels) in a package is equal to or greater than either one curie or the lowest RQ of any known individual radionuclide in the mixture or solution, whichever is lower.

TABLE 1

Hazardous Substances	Synonyms	Reportable Quantity (RQ) Pounds (Kilograms)
Acenaphthene		100 (45.4)
Acenaphthylene		5000 (2270)
Acetaldehyde	Ethanal	1000 (454)
Acetaldehyde, chloro-	Chloroacetaldehyde	1000 (454)
Acetaldehyde, trichloro-	Chloral	5000 (2270)
Acetamide		100 (45.4)
Acetamide, N-(aminothioxomethyl)-	1-Acetyl-2-thiourea	1000 (454)
Acetamide, N-(4-ethoxyphenyl)-	Phenacetin	100 (45.4)
Acetamide, N-fluoren-2-yl-	2-Acetylaminofluorene	1 (0.454)
Acetamide, 2-fluoro-	Fluoroacetamide	100 (45.4)
Acetic acid		5000 (2270)
Acetic acid (2,4-dichlorophenoxy)-	2,4-D, salts and esters	100 (45.4)
	2,4-D acid	
Acetic acid, ethyl ester	Ethyl acetate	5000 (2270)
Acetic acid, fluoro-, sodium salt	Fluoroacetic acid, sodium salt	10 (4.54)
Acetic acid, lead (2+)salt	Lead acetate	10 (4.54)
Acetic acid, thallium(I+) salt	Thallium(I) acetate	100 (45.4)
Acetic acid, (2,4,5-trichlorophenoxy)	2,4,5-T	1000 (454)
	2,4,5-T acid	
Acetic anhydride		5000 (2270)
Acetone	2-Propanone	5000 (2270)
Acetone cyanohydrin	Propanenitrile, 2-hydroxy-2-methyl-	10 (4.54)
Acetonitrile	2-Methyllactonitrile	5000 (2270)
Acetophenone	Ethanenitrile	5000 (2270)
2-Acetylaminofluorene	Ethanone, 1-phenyl-	1 (0.454)
Acetyl bromide	Acetamide, N-fluoren-2-yl-	5000 (2270)

TABLE 1

Hazardous Substances	Synonyms	(RQ)
Acetyl chloride	Ethanoyl chloride	5000 (2270)
1-Acetyl-2-thiourea	Acetamide, N-(aminothioxomethyl)-	1000 (454)
Acrolein	2-Propenal	1 (0.454)
Acrylamide	2-Propenamide	5000 (2270)
Acrylic acid	2-Propenoic acid	5000 (2270)
Acrylonitrile	2-Propenenitrile	100 (45.4)
Adipic acid		5000 (2270)
Aldicarb	Propanal, 2-methyl-2-(methylthio)-O-[(methyl-amino) carbonyl]oxime	1 (0.454)
Aldrin	1,2,3,4,10,10-Hexachloro-1,4,4a,5,8,8a-hexahydro-1,4,5,8-endo, exodimethanonaphthalene 1,4,5,8-Dimethanonaphthalene,1,2,3,4,10,10-hexachloro-1,4,4a,5,8,8a-hexahydro-,(1alpha,4alpha,4abeta,5alpha,8a	1 (0.454)
Allyl alcohol	2-Propen-1-ol	100 (45.4)
Allyl chloride		1000 (454)
Aluminum phosphide		100 (45.4)
Aluminum sulfate		5000 (2270)
4-Aminobiphenyl		1 (0.454)
5-(Aminomethyl)-3-isoxazolol	3(2H)-isoxazolone, 5-(aminomethyl)-Muscimol	1000 (454)
4-Aminopyridine	4-Pyridinamine	1000 (454)
Amitrole	1H-1,2,4-Triazol-3-amine	10 (4.54)
Ammonia		100 (45.4)
Ammonium acetate		5000 (2270)
Ammonium benzoate		5000 (2270)
Ammonium bicarbonate		5000 (2270)
Ammonium bichromate	Ammonium dichromate @	10 (4.54)
Ammonium bifluoride		100 (45.4)
Ammonium bisulfite		5000 (2270)
Ammonium carbamate		5000 (2270)

TABLE 1

Hazardous Substances	Synonyms	(RQ)
Ammonium carbonate		5000 (2270)
Ammonium chloride		5000 (2270)
Ammonium chromate		10 (4.54)
Ammonium citrate, dibasic		5000 (2270)
Ammonium dichromate @	Ammonium bichromate	10 (4.54)
Ammonium fluoborate		5000 (2270)
Ammonium fluoride		100 (45.4)
Ammonium hydroxide		1000 (454)
Ammonium oxalate		5000 (2270)
Ammonium picrate	Phenol, 2,4,6-trinitro-, ammonium salt	10 (4.54)
Ammonium silicofluoride		1000 (454)
Ammonium sulfamate		5000 (2270)
Ammonium sulfide		100 (45.4)
Ammonium sulfite		5000 (2270)
Ammonium tartrate		5000 (2270)
Ammonium thiocyanate		5000 (2270)
Ammonium vanadate	Vanadic acid, ammonium salt	1000 (454)
Amyl acetate		5000 (2270)
iso-Amyl acetate		
sec-Amyl acetate		
tert-Amyl acetate		
Aniline	Benzenamine	5000 (2270)
o-Anisidine		100 (45.4)
Anthracene		5000 (2270)
Antimony¢		5000 (2270)
Antimony pentachloride		1000 (454)
Antimony potassium tartrate		100 (45.4)
Antimony tribromide		1000 (454)
Antimony trichloride		1000 (454)

TABLE 1

Hazardous Substances	Synonyms	(RQ)
Antimony trifluoride		1000 (454)
Antimony trioxide		1000 (454)
Argentate(1-), bis(cyano-C)-, potassium	Potassium silver cyanide	1 (0.454)
Aroclor 1016	POLYCHLORINATED BIPHENYLS (PCBs)	1 (0.454)
Aroclor 1221	POLYCHLORINATED BIPHENYLS (PCBs)	1 (0.454)
Aroclor 1232	POLYCHLORINATED BIPHENYLS (PCBs)	1 (0.454)
Aroclor 1242	POLYCHLORINATED BIPHENYLS (PCBs)	1 (0.454)
Aroclor 1248	POLYCHLORINATED BIPHENYLS (PCBs)	1 (0.454)
Aroclor 1254	POLYCHLORINATED BIPHENYLS (PCBs)	1 (0.454)
Aroclor 1260	POLYCHLORINATED BIPHENYLS (PCBs)	1 (0.454)
Arsenic¢		1 (0.454)
Arsenic acid	Arsenic acid H3AsO4	1 (0.454)
Arsenic acid H3AsO4	Arsenic acid	1 (0.454)
Arsenic disulfide		1 (0.454)
Arsenic oxide As2O3	Arsenic trioxide	1 (0.454)
Arsenic oxide As2O5	Arsenic pentoxide	1 (0.454)
Arsenic pentoxide	Arsenic oxide As2O5	1 (0.454)
Arsenic trioxide	Arsenic oxide As2O3	1 (0.454)
Arsenic trisulfide		1 (0.454)
Arsine, diethyl-	Diethylarsine	1 (0.454)
Arsinic acid, dimethyl-	Cacodylic acid	1 (0.454)
Arsonous dichloride, phenyl-	Dichlorophenylarsine Phenyl dichloroarsine @	1 (0.454)
Asbestos ¢¢		1 (0.454)
Auramine	Benzenamine, 4,4'-carbonimidoylbis(N,N-dimethyl-	100 (45.4)
Azaserine	L-Serine, diazoacetate (ester)	1 (0.454)
Aziridine	Ethylenimine	1 (0.454)
Aziridine, 2-methyl-	1,2-Propylenimine	1 (0.454)

TABLE 1

Hazardous Substances	Synonyms	(RQ)
Azirino(2',3':3',4)pyrrolo(1,2-a)indole-4,7-dione, 6-amino-8-[[(aminocarbonyl)oxy]methyl]-1,1a,2,8,8b-hexahydro-8a-methoxy-5-methyl-, [1aS-[1aalpha,8beta,8aalpha,8balpha]]-	Mitomycin C	10 (4.54)
Barium cyanide		10 (4.54)
Benz[j]aceanthrylene, 1,2-dihydro-3-methyl-	3-Methylcholanthrene	10 (4.54)
Benz[c]acridine	3,4-Benzacridine	100 (45.4)
3,4-Benzacridine	Benz[c]acridine	100 (45.4)
Benzal chloride	Benzene, dichloromethyl-	5000 (2270)
Benzamide,3,5-dichloro-N-(1,1-dimethyl-2-propynyl)-	Pronamide	5000 (2270)
Benz[a]anthracene	Benzo[a]anthracene	10 (4.54)
	1,2-Benzanthracene	
1,2-Benzanthracene	Benz[a]anthracene	10 (4.54)
	Benzo[a]anthracene	
Benz[a]anthracene, 7,12-dimethyl-	7,12-Dimethylbenz[a]anthracene	1 (0.454)
Benzenamine	Aniline	5000 (2270)
Benzenamine, 4,4'-carbonimidoylbis(N,N-dimethyl-	Auramine	100 (45.4)
Benzenamine, 4-chloro-	p-Chloroaniline	1000 (454)
Benzenamine, 4-chloro-2-methyl-, hydrochloride	4-Chloro-o-toluidine, hydrochloride	100 (45.4)
Benzenamine, N,N-dimethyl-4-(phenylazo)	p-Dimethylaminoazobenzene	10 (4.54)
Benzenamine, 2-methyl-	o-Toluidine	100 (45.4)
Benzenamine, 4-methyl-	p-Toluidine	100 (45.4)
Benzenamine, 4,4'-methylenebis(2-chloro-	4,4'-Methylenebis(2-chloroaniline)	10 (4.54)
Benzenamine, 2-methyl-, hydrochloride	o-Toluidine hydrochloride	100 (45.4)
Benzenamine, 2-methyl-5-nitro-	5-Nitro-o-toluidine	100 (45.4)
Benzenamine, 4-nitro-	p-Nitroaniline	5000 (2270)
Benzene		10 (4.54)
Benzene, 1-bromo-4-phenoxy-	4-Bromophenyl phenyl ether	100 (45.4)
Benzene, chloro-	Chlorobenzene	100 (45.4)
Benzene, chloromethyl-	Benzyl chloride	100 (45.4)

TABLE 1

Hazardous Substances	Synonyms	(RQ)
Benzene, 1,2-dichloro-	o-Dichlorobenzene	100 (45.4)
	1,2-Dichlorobenzene	100 (45.4)
Benzene, 1,3-dichloro-	m-Dichlorobenzene	100 (45.4)
	1,3-Dichlorobenzene	
Benzene, 1,4-dichloro-	p-Dichlorobenzene	100 (45.4)
	1,4-Dichlorobenzene	
Benzene, 1,1'-(2,2-dichloroethylidene)bis[4-chloro	DDD	1 (0.454)
	TDE	
	4,4'-DDD	
Benzene, dichloromethyl-	Benzal chloride	5000 (2270)
Benzene, 1,3-diisocyanatomethyl-	Toluene diisocyanate	100 (45.4)
Benzene, dimethyl-	Xylene; Xylene (mixed); Xylenes (isomers and mixtures)	1000 (454)
Benzene, m-dimethyl-	m-Xylene	1000 (454)
Benzene, o-dimethyl-	o-Xylene	1000 (454)
Benzene, p-dimethyl-	p-Xylene	100 (45.4)
Benzene, hexachloro-	Hexachlorobenzene	10 (4.54)
Benzene, hexahydro-	Cyclohexane	1000 (454)
Benzene, hydroxy-	Phenol	1000 (454)
Benzene, methyl-	Toluene	1000 (454)
Benzene, 1-methyl-2,4-dinitro-	2,4-Dinitrotoluene	10 (4.54)
Benzene, 2-methyl-1,3-dinitro-	2,6-Dinitrotoluene	100 (45.4)
Benzene, 1-methylethyl-	Cumene	5000 (2270)
Benzene, nitro-	Nitrobenzene	1000 (454)
Benzene, pentachloro-	Pentachlorobenzene	10 (4.54)
Benzene, pentachloronitro-	Pentachloronitrobenzene (PCNB)	1000 (454)
Benzene, 1,2,4,5-tetrachloro-	1,2,4,5-Tetrachlorobenzene	5000 (2270)
Benzene, 1,1'-(2,2,2-trichloroethylidene)bis [4-chloro	DDT	1 (0.454)
	4,4'-DDT	
Benzene, 1,1'-(2,2,2-trichloroethylidene)bis [4-methoxy]	Methoxychlor	1 (0.454)
Benzene, (trichloromethyl)	Benzotrichloride	10 (4.54)

TABLE 1

Hazardous Substances	Synonyms	(RQ)
Benzene, 1,3,5-trinitro-	1,3,5-Trinitrobenzene	10 (4.54)
Benzeneacetic acid, 4-chloro-alpha-(4-chlorophenyl)-alpha-hydroxy-, ethyl ester	Chlorobenzilate	10 (4.54)
Benzenebutanoic acid, 4-[bis(2-chloroethyl)amino]-	Chlorambucil	10 (4.54)
Benzenediamine, ar-methyl-	Toluenediamine	10 (45.4)
1,2-Benzenedicarboxylic acid, [bis(2-ethylhexyl)] ester	Bis(2-ethylhexyl)phthalate Diethylhexyl phthalate	100 (45.4)
1,2-Benzenedicarboxylic acid, dibutyl ester	Di-n-butyl phthalate Dibutyl phthalate n-Butyl phthalate	10 (4.54)
1,2-Benzenedicarboxylic acid, diethyl ester	Diethyl phthalate	1000 (454)
1,2-Benzenedicarboxylic acid, dimethyl ester	Dimethyl phthalate	5000 (2270)
1,2-Benzenedicarboxylic acid, dioctyl ester	Di-n-octyl phthalate	5000 (2270)
1,3-Benzenediol	Resorcinol	5000 (2270)
1,2-Benzenediol,4-[1- hydroxy-2-(methylamino) ethyl]-	Epinephrine	1000 (454)
Benzeneethanamine, alpha,alpha-dimethyl-	alpha,alpha-Dimethylphenethylamine	5000 (2270)
Benzeneethanamine, alpha,alpha-dimethyl-	alpha,alpha-Dimethylphenethylamine	5000 (2270)
Benzenesulfonic acid chloride	Benzenesulfonyl chloride	100 (45.4)
Benzenesulfonyl chloride		100 (45.4)
Benzenethiol	Phenyl mercaptan @ Thiophenol	100 (45.4)
Benzidine	(1,1'-Biphenyl)-4,4' diamine	1 (0.454)
1,2-Benzisothiazol-3(2H)-one,1,1-dioxide	Saccharin and salts	100 (45.4)
Benz[a]anthracene	1,2-Benzanthracene	10 (4.54)
1,3-Benzodioxole, 5-(2-propenyl)-	Safrole	100 (45.4)
1,3-Benzodioxole, 5-(1-propenyl)-	Isosafrole	100 (45.4)
1,3-Benzodioxole, 5-propyl-	Dihydrosafrole	10 (4.54)
Benzo[b]fluoranthene		1 (0.454)
Benzo[k]fluoranthene		5000 (2270)

TABLE 1

Hazardous Substances	Synonyms	(RQ)
Benzo[j,k]fluorene	Fluoranthene	100 (45.4)
Benzoic acid		5000 (2270)
Benzonitrile		5000 (2270)
Benzo[g,h,i]perylene		5000 (2270)
2H-1-Benzopyran-2-one, 4-hydroxy-3-(3-oxo-1-phenyl-butyl), & salts, when present at concentrations greater than 0.3%	Warfarin, & salts, when present at concentrations greater than 0.3%	100 (45.4)
Benzo[a]pyrene	3,4-Benzopyrene	1 (0.454)
3,4-Benzopyrene	Benzo[a]pyrene	1 (0.454)
p-Benzoquinone	2,5-Cyclohexadiene-1,4-dione	10 (4.54)
Benzo[rst]pentaphene	Dibenz[a,i]pyrene	10 (4.54)
Benzotrichloride	Benzene, (trichloromethyl)	10 (4.54)
Benzoyl chloride		1000 (454)
1,2-Benzphenanthrene	Chrysene	100 (45.4)
Benzyl chloride	Benzene, chloromethyl-	100 (45.4)
Beryllium ¢	Beryllium dust ¢	10 (4.54)
Beryllium chloride		1 (0.454)
Beryllium dust ¢	Beryllium ¢	10 (4.54)
Beryllium fluoride		1 (0.454)
Beryllium nitrate		1 (0.454)
alpha - BHC		10 (4.54)
beta - BHC		1 (0.454)
delta - BHC		1 (0.454)
gamma - BHC	Hexachlorocyclohexane (gamma isomer) Lindane,1,2,3,4,5,6-hexachloro- Cyclohexane,(1alpha,2alpha,3beta,4alpha,5alpha,6beta)-	1 (0.454)
2,2'-Bioxirane	1,2-3,4-Diepoxybutane	10 (4.54)
Biphenyl		100 (45.4)
(1,1'-Biphenyl)-4,4' diamine	Benzidine	1 (0.454)
(1,1'-Biphenyl)-4,4' diamine 3,3'dichloro-	3,3'-Dichlorobenzidine	1 (0.454)

TABLE 1

Hazardous Substances	Synonyms	(RQ)
(1,1'-Biphenyl)-4,4' diamine 3,3'-dimethoxy-	3,3'-Dimethoxybenzidine	10 (4.54)
(1,1'-Biphenyl)-4,4' diamine 3,3'-dimethyl-	3,3'-Dimethylbenzidine	10 (4.54)
Bis(2-chloroethoxy) methane	Ethane, 1,1'-[methylenebis (oxy)]bis[2-chloro- Dichloromethoxy ethane	1000 (454)
Bis(2-chloroethyl) ether	Dichloroethyl ether Ethane, 1,1'-oxybis[2-chloro-	10 (4.54)
Bis(2-ethylhexyl)phthalate	1,2-Benzenedicarboxylic acid, [bis(2-ethylhexyl) ester Diethylhexyl phthalate	100 (45.4)
Bromoacetone	2-Propanone, 1-bromo-	1000 (454)
Bromoform	Methane, tribromo-	100 (45.4)
4-Bromophenyl phenyl ether	Benzene, 1-bromo-4- phenoxy-	100 (45.4)
Brucine	Strychnidin-10-one, 2,3-dimethoxy-	100 (45.4)
1,3-Butadiene		10 (4.54)
1,3-Butadiene, 1,1,2,3,4,4-hexachloro-	Hexachlorobutadiene	1 (0.454)
1-Butanamine, N-butyl-N-nitroso-	N-Nitrosodi-n-butylamine	10 (4.54)
1-Butanol	n-Butyl alcohol	5000 (2270)
2-Butanone	Ethyl methyl ketone @ Methyl ethyl ketone(MEK)	5000 (2270)
2-Butanone, 3,3-dimethyl-1-(methylthio)-,O-[(methylamino)carbonyl] oxime	Thiofanox	100 (45.4)
2-Butanone peroxide	Methyl ethyl ketone peroxide	10 (4.54)
2-Butenal	Crotonaldehyde	100 (45.4)
2-Butene, 1,4-dichloro-	1,4-Dichloro-2-butene	1 (0.454)
2-Butenoic acid, 2-methyl-,7[[2,3-dihydroxy-2-(1-methoxyethyl)-3-methyl-1-oxobutoxy]methyl]-2-3,5,7a-tetrahydro-1H-pyrrolizin-1-ylester[1S-[1alpha(Z),7(2S*,3R*),7aalpha]]-	Lasiocarpine	10 (4.54)
Butyl acetate iso- Butyl acetate sec- Butyl acetate tert- Butyl acetate		5000 (2270)
n-Butyl alcohol	1-Butanol	5000 (2270)

TABLE 1

Hazardous Substances	Synonyms	(RQ)
Butylamine		1000 (454)
iso-Butylamine		
sec-Butylamine		
tert-Butylamine		
Butyl benzyl phthalate		100 (45.4)
n-Butyl phthalate		10 (4.54)
	Di-n-butyl phthalate	
	Dibutyl phthalate	
	1,2-Benzenedicarboxylic acid,dibutyl ether	
Butyric acid		5000 (2270)
iso-Butyric acid		
Cacodylic acid	Arsenic acid, dimethyl-	1 (0.454)
Cadmium ¢		10 (4.54)
Cadmium acetate		10 (4.54)
Cadmium bromide		10 (4.54)
Cadmium chloride		10 (4.54)
Calcium arsenate		1 (0.454)
Calcium arsenite		1 (0.454)
Calcium carbide		10 (4.54)
Calcium chromate	Chromic acid H2CrO4, calcium salt	10 (4.54)
Calcium cyanamide		1000 (454)
Calcium cyanide	Calcium Cyanide Ca(CN)2	10 (4.54)
Calcium cyanide Ca(CN)2	Calcium cyanide	10 (4.54)
Calcium dodecylbenzene sulfonate		1000 (454)
Calcium hypochlorite		10 (4.54)
Camphene, octachloro-	Toxaphene	1 (0.454)
Caprolactam		5000 (2270)
Captan		10 (4.54)
Carbamic acid, ethyl ester	Ethyl carbamate (Urethan)	100 (45.4)
Carbamic acid, methylnitroso-, ethyl ester	N-Nitroso-N-methylurethane	1 (0.454)
Carbamic chloride, dimethyl-	Dimethylcarbamoyl chloride	1 (0.454)
Carbamide, thio-	Thiourea	10 (4.54)

TABLE 1

Hazardous Substances	Synonyms	(RQ)
Carbamimidoselenoic acid	Selenourea	1000 (454)
Carbamothioic acid, bis (1-methylethyl)-, S-(2,3-dichloro-2) ester	Diallate	100 (45.4)
Carbaryl		100 (45.4)
Carbofuran		10 (4.54)
Carbon bisulfide	Carbon disulfide	100 (45.4)
Carbon disulfide	Carbon bisulfide	100 (45.4)
Carbonic acid, dithallium (l+)	Thallium(I) carbonate	100 (45.4)
Carbonic dichloride	Phosgene	10 (4.54)
Carbonic difluoride	Carbon oxyfluoride	1000 (454)
Carbonochloridic acid, methyl ester	Methyl chlorocarbonate Methyl chloroformate	1000 (454)
Carbonyl sulfide		100 (45.4)
Carbon oxyfluoride	Carbonic difluoride	1000 (454)
Carbon tetrachloride	Methane, tetrachloro-	10 (4.54)
Catechol		100 (45.4)
Chloral	Acetaldehyde, trichloro-	5000 (2270)
Chloramben		100 (45.4)
Chlorambucil	Benzenebutanoic acid, 4-[bis(2-chloroethyl)amino]-	10 (4.54)
Chlordane	Chlordane, technical 4,7-Methano-1H-indene, 1,2,4,5,6,7,8,8-octachloro-2,3,3a,4,7,7a-hexahydro- Chlordane, alpha & gamma isomers	1 (0.454)
Chlordane, alpha & gamma isomers	Chlordane, technical Chlordane 4,7-Methano-1H-indene, 1,2,4,5,6,7,8,8-octachloro-2,3,3a,4,7,7a-hexahydro-	1 (0.454)
Chlordane, technical	Chlordane 4,7-Methano-1H-indene, 1,2,4,5,6,7,8,8-octachloro-2,3,3a,4,7,7a-hexahydro- Chlordane, alpha & gamma isomers	1 (0.454)
Chlorine		10 (4.54)
Chlornaphazine	Naphthylamine, N,N'-bis (2-chloroethyl)-	100 (45.4)

TABLE 1

Hazardous Substances	Synonyms	(RQ)
Chloroacetaldehyde	Acetaldehyde, chloro-	1000 (454)
Chloroacetic acid		100 (45.4)
2-Chloroacetophenone		100 (45.4)
p-Chloroaniline	Benzenamine, 4-chloro-	1000 (454)
Chlorobenzene	Benzene, chloro-	100 (45.4)
Chlorobenzilate	Benzeneacetic acid, 4-chloro-alpha-(4-chlorophenyl)-alpha-hydroxy, ethyl ester	10 (4.54)
4-Chloro-m-cresol	p-Chloro-m-cresol Phenol, 4-chloro-3-methyl-	5000 (2270)
p-Chloro-m-cresol	Phenol, 4-chloro-3-methyl- 4-Chloro-m-cresol	5000 (2270)
Chlorodibromomethane		100 (45.4)
Chloroethane	Ethyl chloride @	100 (45.4)
2-Chloroethyl vinyl ether	Ethene, 2-chloroethoxy-	1000 (454)
Chloroform	Methane, trichloro-	10 (4.54)
Chloromethane	Methane, chloro Methyl chloride	100 (45.4)
Chloromethyl methyl ether	Methane, chloromethroxy- Methylchloromethyl ether @	1 (0.454)
beta-Chloronaphthalene	Naphthalene, 2-chloro- 2-Chloronaphthalene	5000 (2270)
2-Chloronaphthalene	beta-Chloronaphthalene Naphthalene, 2-chloro-	5000 (2270)
2-Chlorophenol	o-Chlorophenol Phenol, 2-chloro-	100 (45.4)
o-Chlorophenol	Phenol, 2-chloro- 2-Chlorophenol	100 (45.4)
4-Chlorophenyl phenyl ether		5000 (2270)
1-(o-Chlorophenyl)thiourea	Thiourea, (2-chlorophenyl)-	100 (45.4)
Chloroprene		100 (45.4)
3-Chloropropionitrile	Propanenitrile, 3-chloro-	1000 (454)

TABLE 1

Hazardous Substances	Synonyms	(RQ)
Chlorosulfonic acid		1000 (454)
4-Chloro-o-toluidine, hydrochloride	Benzenamine, 4-chloro-2-methyl hydrochloride	100 (45.4)
Chlorpyrifos		1 (0.454)
Chromic acetate		1000 (454)
Chromic acid		10 (4.54)
Chromic acid H2CrO4, calcium salt	Calcium chromate	10 (4.54)
Chromic sulfate		1000 (454)
Chromium ¢		5000 (2270)
Chromous chloride		1000 (454)
Chrysene	1,2-Benzphenanthrene	100 (45.4)
Cobaltous bromide		1000 (454)
Cobaltous formate		1000 (454)
Cobaltous sulfamate		1000 (454)
Coke Oven Emissions		1 (0.454)
Copper ¢		5000 (2270)
Copper chloride @	Cupric chloride	10 (4.54)
Copper cyanide	Copper cyanide CuCN	10 (4.54)
Copper cyanide CuCN	Copper cyanide	10 (4.54)
Coumaphos		1 (0.454)
Creosote		100 (45.4)
Cresols (isomers and mixture)	Cresylic acid (isomers and mixture) Phenol, methyl-	100 (45.4)
m-Cresol	m-Cresylic acid	100 (45.4)
o-Cresol	o-Cresylic acid	100 (45.4)
p-Cresol	p-Cresylic acid	100 (45.4)
Cresylic acid (isomers and mixture)	Cresols (isomers and mixture); phenol, methyl-	100 (45.4)
m-Cresylic acid	m-Cresol	100 (45.4)
o-Cresylic acid	o-Cresol	100 (45.4)
p-Cresylic acid	p-Cresol	100 (45.4)
Crotonaldehyde	2-Butenal	100 (45.4)

Hazardous Substances	Synonyms	(RQ)
Cumene	Benzene, 1-methylethyl-	5000 (2270)
Cupric acetate		100 (45.4)
Cupric acetoarsenite		1 (0.454)
Cupric chloride	Copper chloride@	10 (4.54)
Cupric nitrate		100 (45.4)
Cupric oxalate		100 (45.4)
Cupric sulfate		10 (4.54)
Cupric sulfate ammoniated		100 (45.4)
Cupric tartrate		100 (45.4)
Cyanides (soluble salts and complexes) not otherwise specified		10 (4.54)
Cyanogen	Ethanedinitrile	100 (45.4)
Cyanogen bromide	Cyanogen bromide (CN)Br	1000 (454)
Cyanogen bromide (CN)Br	Cyanogen bromide	1000 (454)
Cyanogen chloride	Cyanogen Chloride (CN)Cl	10 (4.54)
Cyanogen chloride (CN)Cl	Cyanogen chloride	10 (4.54)
2,5-Cyclohexadiene-1,4-dione	p-Benzoquinone	10 (4.54)
Cyclohexane	Benzene, hexahydro-	1000 (454)
Cyclohexane, 1,2,3,4,5,6-hexachloro-,(1alpha,2alpha,3beta,4alpha, 5alpha,6beta)-	gamma-BHC Hexachlorocyclohexane(gamma isomer) Lindane	1 (0.454)
Cyclohexanone		5000 (2270)
2-Cyclohexyl-4,6-dinitrophenol	Phenol,2-cyclohexyl-4,6-dinitro-	100 (45.4)
1,3-Cyclopentadiene, 1,2,3,4,5,5-hexachloro-	Hexachlorocyclopentadiene	10 (4.54)
Cyclophosphamide	2H-1,3,2-Oxazaphosphorin, 2-amine,N,N-bis(2-chloroethyl) tetrahydro-)2-oxide	
2,4-D Acid	2,4-D, salts and esters Acetic acid (2,4-dichlorophenoxy)-	100 (45.4)
2,4-D Ester		100 (45.4)

TABLE 1

Hazardous Substances	Synonyms	(RQ)
Daunomycin	5,12-Naphthacenedione, 8-acetyl-10-[3-amino,2,3,6-trideoxy-alpha-L-lyxo-hexopyranosyl)oxy]-7,8,9,10-tetrahydro-6,8,11-trihydroxy-1-methoxy-(8S-cis)-	10 (4.54)
DDD	Benzene,1,1'-(2,2-dichloroethylidene)bis[4-chloro-	1 (0.454)
	TDE	
	4,4'-DDD	
4,4'-DDD	DDD	1 (0.454)
	Dichlorodiphenyl dichloroethane	
	TDE	
DDE	4,4'-DDE	5000 (2270)
4,4'-DDE	DDE	5000 (2270)
DDT	Benzene, 1,1'-(2,2,2-trichloroethylidene)bis[4-chloro-	1 (0.454)
	4,4'-DDT	
4,4'-DDT	DDT	1 (0.454)
	Benzene,1,1'-(2,2,2-trichloroethylidene)bis[4-chloro-	
Diallate	Carbamothioic acid, bis (1-methylethyl)-, S-(2,3-dichloro-2-propenyl) ester	100 (45.4)
Diamine	Hydrazine	1 (0.454)
Diazinon		1 (0.454)
Diazomethane		100 (45.4)
Dibenz[a,h]anthracene		1 (0.454)
1,2,5,6-Dibenzanthracene	Dibenzo[a,h]anthracene	1 (0.454)
	1,2:5,6-Dibenzanthracene	
Dibenz[a,h]anthracene	Dibenz[a,h]anthracene	1 (0.454)
	Dibenzo[a,h]anthracene	
	1,2:5,6-Dibenzanthracene	
Dibenzofuran		100 (45.4)
Dibenz[a,j]pyrene	Benzo [rst]pentaphene	10 (4.54)
1,2-Dibromo-3-chloropropane	Propane, 1,2-dibromo-3-chloro-	1 (0.454)
Dibutyl phthalate	Di-n-butyl phthalate	10 (4.54)
	n-Butyl phthalate	
	1,2-Benzenedicarboxylic acid, dibutyl ester	

TABLE 1

Hazardous Substances	Synonyms	(RQ)
Di-n-butylphthalate	Dibutyl phthalate	10 (4.54)
	n-butyl phthalate	
	1,2-Benzenedicarboxylic acid, dibutyl ester	
Dicamba		1000 (454)
Dichlobenil		100 (45.4)
Dichlone		1 (0.454)
Dichlorobenzene		100 (45.4)
1,2-Dichlorobenzene	Benzene, 1,2-dichloro-	100 (45.4)
	o-Dichlorobenzene	
1,3-Dichlorobenzene	Benzene, 1,3-dichloro-	100 (45.4)
	m-Dichlorobenzene	
1,4-Dichlorobenzene	Benzene, 1,4-cichloro-	100 (45.4)
	p-Dichlorobenzene	
m-Dichlorobenzene	Benzene, 1,3-cichloro-	100 (45.4)
	1,3-Dichlorobenzene	
o-Dichlorobenzene	Benzene, 1,2-cichloro-	100 (45.4)
	1,2-Dichlorobenzene	
p-Dichlorobenzene	Benzene, 1,4-cichloro-	100 (45.4)
	1,4-Dichlorobenzene	
3,3'-Dichlorobenzidine	(1,1'-Biphenyl)-4,4'diamine,3,3' dichloro-	1 (0.454)
Dichlorobromomethane		5000 (2270)
1,4-Dichloro-2-butene	2-Butene, 1,4-dichloro-	1 (0.454)
Dichlorodifluoromethane	Methane, dichlorotifluoro-	5000 (2270)
1,1-Dichloroethane	Ethane, 1,1-dichloro-	1000 (454)
	Ethylidene dichloride	
1,2-Dichloroethane	Ethane, 1,2-dichloro-	100 (45.4)
	Ethylene dichloride	
1,1-Dichloroethylene	Ethane, 1,1-dichloro-	100 (45.4)
	Vinylidene chloride	
1,2-Dichloroethylene	Ethene, 1,2-dichloro-(E)	1000 (454)
Dichloroethyl ether	Bis (2-chloroethyl) ether	10 (4.54)
	Ethane, 1,1'-oxybis(2-chloro-	

TABLE 1

Hazardous Substances	Synonyms	(RQ)
Dichloroisopropyl ether	Propane,2,2'-oxybis [2-chloro-	1000 (454)
Dichloromethane@	Methane, dichloro-	1000 (454)
	Methylene chloride	
Dichloromethoxy ethane	Bis(2-chloroethoxy)methane	1000 (454)
	Ethane, 1,1'-[methylenebis(oxy)]bis[2-chloro-	
Dichloromethyl ether	Methane, oxybis(chloro-	1 (0.454)
2,4-Dichlorophenol	Phenol, 2,4-dichloro-	100 (45.4)
2,6-Dichlorophenol	Phenol, 2,6-dichloro-	100 (45.4)
Dichlorophenylarsine	Arsonous dichloride, phenyl-	1 (0.454)
Dichloropropane		1000 (454)
1,1-Dichloropropane		
1,3-Dichloropropane		
1,2-Dichloropropane	Propylene dichloride	1000 (454)
	Propane, 1,2-dichloro-	
Dichloropropane Dichloropropene (mixture)		100 (45.4)
Dichloropropene		100 (45.4)
2,3-Dichloropropene		
1,3-Dichloropropene	1-Propene, 1,3-dichloro-	100 (45.4)
2,2-Dichloropropionic acid		5000 (2270)
Dichlorvos		10 (4.54)
Dicofol		10 (4.54)
Dieldrin	2,7:3,6-Dimethanonaphth[2,3-b]oxirene,3,4,5,6,9,9-hexachloro-1a,2,2a,3,6,6a,7,7a-octahydro-(1alpha, 2beta,2aalpha,3beta,6beta,6aalpha,7beta,7aalpha)-	1 (0.454)
1,2,3,4-Diepoxybutane		10 (4.54)
Diethanolamine	2,2'-Bioxirane	100 (45.4)
Diethylamine		100 (45.4)
N,N-diethylaniline		1000 (454)
Diethylarsine	Arsine, diethyl-	1 (0.454)
1,4-Diethylenedioxide	1,4-Dioxane	100 (45.4)

TABLE 1

Hazardous Substances	Synonyms	(RQ)
Diethylhexyl phthalate	1,2-Benzenedicarboxylic acid, [bis(2-ethylhexyl) ester	100 (45.4)
	Bis(2-ethylhexyl)phthalate	
N,N-Diethylhydrazine	Hydrazine, 1,2-diethyl-	10 (4.54)
O,O-Diethyl S-methyl dithiophosphate	Phosphorodithioic acid, O,O-diethyl S-methyl ester	5000 (2270)
Diethyl-p-nitrophenyl phosphate	Phosphoric acid, diethyl 4-nitrophenyl ester	100 (45.4)
Diethyl phthalate	1,2-Benzenedicarboxylic acid, diethyl ester	1000 (454)
O,O-Diethyl O-pyrazinyl phosphorothioate	Phosperothioic acid, O,O-diethyl O-pyrazinyl ester	100 (45.4)
Diethyl sulfate		10 (4.54)
Diethylstilbestrol	Phenol, 4,4'-(1,2-diethyl-1,2-ethenediyl)bis-,(E)	1 (0.454)
Dihydrosafrole	Benzene, 1,2- methylenedioxy-4-propyl-	10 (4.54)
Diisopropyl fluorophosphate	Phosphorofluoridic acid, bis(1-methylethyl) ester	100 (45.4)
1,4,5,8-Dimethanonaphthalene,1,2,3,4,10,10-hexachloro-1,4,4a,5,8,8a-hexahydro-,(1alpha,4alpha,4abeta,5alpha,8beta,8abeta)-	Isodrin	1 (0.454)
1,4,5,8-Dimethanonaphthalene,1,2,3,4,10,10-hexachloro-1,4,4a,5,8,8a-hexahydro-, (1alpha,4alpha,4abeta,5alpha,8abeta,8abeta)-	Aldrin	1 (0.454)
	1,2,3,4,10,10-Hexachloro-1,4,4a,5,8,8a-hexahydro-1,4,5,8-endo,exo-dimethanonapthalene	
2,7:3,6-Dimetthanonaphth[2,3-b]oxirene,3,4,5,6,9,9-hexachloro-1a,2,2a,3,6, 6a,7,7a-octahydro- (1aalpha,2beta,2abeta,3alpha,6alpha,6abeta,7beta, 7aalpha)-	Endrin	1 (0.454)
	Endrin, and metabolites	
2,7:3,6-Dimethanonaphth[2,3-b]oxirene, 3,4,5,6,9,9-hexachloro- 1a,2,2a,3,6,6a, 7,7a-octahydro-, (1aalpha,2beta,2aalpha,3beta, 6alpha,6aalpha,7beta, 7aalpha)-	Dieldrin	1 (0.454)
Dimethoate	Phosphorodithioic acid, O,O- dimethyl S-[2(methyl amino)-2-oxoethyl]ester	10 (4.54)
3,3'-Dimethoxybenzidine	(1,1'-Biphenyl)-4,4' diamine,3,3' dimethoxy-	10 (4.54)
Dimethylamine	Methanamine, N-methyl-	1000 (454)
p-Dimethylaminoazobenzene	Benzenamine, N,N-dimethyl-4-(phenylazo)-	10 (4.54)
N,N-dimethylaniline		100 (45.4)
7,12-Dimethylbenz[a]anthracene	Benz[a]anthracene,7,12-dimethyl-	1 (0.454)
3,3'-Dimethylbenzidine	(1,1' Biphenyl)-4,4'-diamine,3,3'-dimethyl-	10 (4.54)

TABLE 1

Hazardous Substances	Synonyms	(RQ)
alpha,alpha-Dimethylbenzyl-hydroperoxide	Hydroperoxide,1-methyl-1-phenylethyl-	10 (4.54)
Dimethylcarbamoyl chloride	Carbamic chloride, dimethyl-	1 (0.454)
Dimethylformamide		100 (45.4)
1,1-Dimethylhydrazine	Dimethylhydrazine, unsymmetrical @ Hydrazine, 1,1-dimethyl-	10 (4.54)
1,2-Dimethylhydrazine	Hydrazine, 1,2-dimethyl-	1 (0.454)
Dimethylhydrazine,unsymmetrical@	1,1-Dimethylhydrazine Hydrazine,1,1-dimethyl-	10 (4.54)
alpha,alpha-Dimethylphenethylamine	Benzeneethanamine, alpha,alpha-dimethyl-	5000 (2270)
2,4-Dimethylphenol	Phenol, 2,4-dimethyl-	100 (45.4)
Dimethyl phthalate	1,2-Benzenedicarboxylic acid, dimethyl ester	5000 (2270)
Dimethyl sulfate	Sulfuric acid, dimethyl ester	100 (45.4)
Dinitrobenzene (mixed) m- Dinitrobenzene o- Dinitrobenzene p- Dinitrobenzene		
4,6-Dinitro-o-cresol and salts	Phenol,2-methyl-4,6-dinitro-	10 (4.54)
Dinitrogen tetroxide @	Nitrogen dioxide Nitrogen oxide NO2	10 (4.54)
Dinitrophenol 2,5- Dinitrophenol 2,6- Dinitrophenol		10 (4.54)
2,4-Dinitrophenol	Phenol, 2,4-dinitro-	10 (4.54)
Dinitrotoluene 3,4-Dinitrotoluene		10 (4.54)
2,4-Dinitrotoluene	Benzene, 1-methyl-2,4-dinitro-	10 (4.54)
2,6-Dinitrotoluene	Benzene, 2-methyl-1,3-dinitro-	100 (45.4)
Di-n-octyl phthalate	1,2-Benzenedicarboxylic acid, dioctyl ester	1000 (454)
Dioxane	1,2-Benzenedicarboxylic acid, 2-(1-methylpropyl)-4,6-dinitro	5000 (2270)
1,4-Dioxane	1,4-Diethylene dioxide	100 (45.4)|†
1,2-Diphenylhydrazine	Hydrazine, 1,2-diphenyl-	10 (4.54)

TABLE 1

Hazardous Substances	Synonyms	(RQ)
Diphosphoramide, octamethyl-	Octamethylpyrophosphoramide	100 (45.4)
Diphosphoric acid,tetraethyl ester	Tetraethyl pyrophosphate	10 (4.54)
Dipropylamine	1-Propanamine, N-propyl-	5000 (2270)
Di-n-propylnitrosamine	1-Propanamine, N-nitroso-N-propyl-	10 (4.54)
Diquat		1000 (454)
Disulfoton	Phosphorodithioic acid, O,O-diethyl S-[2-(ethylthio) ethyl]ester	1 (0.454)
Dithiobiuret	Thioimidodicarbonic diamide[(H2N)C(S)]2NH	100 (45.4)
Diuron		100 (45.4)
Dodecylbenzenesulfonic acid		1000 (454)
2,4-D, salts and esters	2,4-D Acid Acetic acid (2,4-dichloro-phenoxy)-	100 (45.4)
Endosulfan	6-9-Methano-2,4,3-benzodioxathiepin,6,7,8,9,10,10-hexachloro-1,5,5a,6,9,9a-hexahydro-, 3-oxide.	1 (0.454)
alpha - Endosulfan		1 (0.454)
beta - Endosulfan		1 (0.454)
Endosulfan sulfate		1 (0.454)
Endothall	7-Oxabicyclo[2,2,1]heptane-2,3-dicarboxylic acid	1000 (454)
Endrin	2,7,3,6-Dimethanonaphth[2,3-b]oxirene, 3,4,5,6,9,9-hexachloro-1,a,2,2a,3,6,6a,7,7a-octa-hydro-, (1aalpha,2beta,2abeta,3alpha,6alpha,6abeta,7beta, 7aalpha)-	1 (0.454)
Endrin, & metabolites	Endrin 2,7,3,6-Dimethanonaphth[2,3-b]oxirene, 3,4,5,6,9,9-hexachloro-1a,2,2a,3,6,6a,7,7a oct-hydro-, (1aalpha,2beta,2abeta,3alpha,6alpha,6abeta,7beta, 7aalpha)-	1 (0.454)
Endrin aldehyde		1 (0.454)
Epichlorohydrin	Oxirane, (chloromethyl)-	1000 (454)
Epinephrine	1,2-Benzenediol, 4-[1-hydroxy-2-(methylamino)ethyl]	100 (45.4)
1,2-Epoxybutane		

TABLE 1

Hazardous Substances	Synonyms	(RQ)
Ethanal	Acetaldehyde	1000 (454)
Ethanamine, N-ethyl-N-nitroso-	N-Nitrosodiethylamine	1 (0.454)
Ethane, 1,2-dibromo-	Ethylene dibromide	1 (0.454)
Ethane, 1,1-dichloro-	Ethylidene dichloride 1,1-Dichloroethane	1000 (454)
Ethane, 1,2-dichloro-	Ethylene dichloride 1,2-Dichloroethane	100 (45.4)
Ethane, hexachloro-	Hexachloroethane	100 (45.4)
Ethane, 1,1'-[methylenebis(oxy)]bis(2-chloro-	Bis(2-chloroethoxy)methane Dichloromethoxy ethane	1000 (454)
Ethane, 1,1'-oxybis-	Ethyl ether	100 (45.4)
Ethane, 1,1'-oxybis(2-chloro-	Bis (2-chloroethyl) ether Dichloroethyl ether	10 (4.54)
Ethane, pentachloro-	Pentachloroethane	10 (4.54)
Ethane, 1,1,1,2-tetrachloro-	1,1,1,2-Tetrachloroethane- Tetrachloroethane@	100 (45.4)
Ethane, 1,1,2,2-tetrachloro-	1,1,2,2-Tetrachloroethane- Tetrachlorethane@	100 (45.4)
Ethane, 1,1,2-trichloro-	1,1,2-Trichloroethane	100 (45.4)
Ethane, 1,1,1-trichloro-	Methyl chloroform 1,1,1-Trichloroethane	1000 (454)
1,2-Ethanediamine, N,N-dimethyl-N'-2-pyridinyl-N'-(2-thienyl- methyl)-	Methapyrilene	5000 (2270)
Ethanedinitrile	Cyanogen	100 (45.4)
Ethanenitrile	Acetonitrile	5000 (2270)
Ethanethioamide	Thioacetamide	10 (4.54)
Ethanimidothioic acid, N-[[(methylamino)carbonyl] oxy]-, methyl ester	Methomyl	100 (45.4)
Ethanol, 2-ethoxy-	Ethylene glycol monoethyl ether	1000 (454)
Ethanol, 2,2'-(nitrosoimino)bis-	N-Nitrosodiethanolamine	1 (0.454)
Ethanone, 1-phenyl-	Acetophenone	5000 (2270)
Ethanoyl chloride	Acetyl chloride	5000 (2270)
Ethene, chloro-	Vinyl chloride	1 (0.454)

TABLE 1

Hazardous Substances	Synonyms	(RQ)
Ethane, 2-chloroethoxy	2-Chloroethyl vinyl ether	1000 (454)
Ethane, 1,1-dichloro-	Vinylidene chloride 1,1-Dichloroethylene	100 (45.4)
Ethane, 1,2-dichloro-(E)	1,2-Dichloroethylene	1000 (454)
Ethane, tetrachloro-	Perchloroethylene Tetrachloroethene Tetrachloroethylene	100 (45.4)
Ethene, trichloro-	Trichloroethene Trichloroethylene	100 (45.4)
Ethion		10 (4.54)
Ethyl acetate	Acetic acid, ethyl ester	5000 (2270)
Ethyl acrylate	2-Propenoicacid, ethyl ester	1000 (454)
Ethylbenzene		1000 (454)
Ethyl carbamate (Urethan)	Carbamic acid, ethyl ester	100 (45.4)
Ethyl chloride @	Chloroethane	100 (45.4)
Ethyl cyanide	Propanenitrile	10 (4.54)
Ethylene dibromide	Ethane, 1,2-dibromo-	1 (0.454)
Ethylene dichloride	1,2-Dichloroethane Ethane, 1,2-dichloro-	100 (45.4)
Ethylene glycol		5000 (2270)
Ethylene glycol monoethyl ether	Ethanol, 2-ethoxy	1000 (454)
Ethylene oxide	Oxirane	10 (4.54)
Ethylenebisdithiocarbamic acid	Ethylenebisdithiocarbamic acid, salts and esters	5000 (2270)
Ethylenebisdithiocarbamic acid, salts and esters	Ethylenebisdithiocarbamic acid	5000 (2270)
Ethylenediamine		5000 (2270)
Ethylenediamine tetraacetic acid (EDTA)		5000 (2270)
Ethylenethiourea	2-Imidazolidinethione	10 (4.54)
Ethylenimine	Aziridine	1 (0.454)
Ethyl ether	Ethane, 1,1'-oxybis-	100 (45.4)
Ethylidene dichloride	Ethane, 1,1-dichloro- 1,1-Dichloroethane	1000 (454)

TABLE 1

Hazardous Substances	Synonyms	(RQ)
Ethyl methacrylate	2-Propenoic acid, 2-methyl-, ethyl ester	1000 (454)
Ethyl methanesulfonate	Methanesulfonic acid, ethyl ester	1 (0.454)
Ethyl methyl ketone @	2-Butanone Methyl ethyl ketone(MEK)	5000 (2270)
Famphur	Phosphorothioic acid, O,[4-[(dimethylamino)-sulfonyl] phenyl] O,O-dimethylester	1000 (454)
Ferric ammonium citrate		1000 (454)
Ferric ammonium oxalate		1000 (454)
Ferric chloride		1000 (454)
Ferric fluoride		100 (45.4)
Ferric nitrate		1000 (454)
Ferric sulfate		1000 (454)
Ferrous ammonium sulfate		1000 (454)
Ferrous chloride		100 (45.4)
Ferrous sulfate		1000 (454)
Fluoranthene	Benzo[j,k]fluorene	100 (45.4)
Fluorene		5000 (2270)
Fluorine		10 (4.54)
Fluoroacetamide	Acetamide, 2-fluoro-	100 (45.4)
Fluoroacetic acid, sodium salt	Acetic acid, fluoro-, sodium salt	10 (4.54)
Formaldehyde	Methylene oxide	100 (45.4)
Formic acid	Methanoic acid	5000 (2270)
Fulminic acid, mercury(2+)salt	Mercury fulminate	10 (4.54)
Fumaric acid		5000 (2270)
Furan	Fururan	100 (45.4)
Furan, tetrahydro-	Tetrahydrofuran	1000 (454)
2-Furancarboxaldehyde	Furfural	5000 (2270)
2,5-Furandione	Maleic anhydride	5000 (2270)
Furfural	2-Furancarboxaldehyde	5000 (2270)
Furfuran	Furan	100 (45.4)

TABLE 1

Hazardous Substances	Synonyms	(RQ)
Glucopyranose, 2-deoxy-2-(3-methyl-3-nitrosoureido)-	Streptozotocin D-Glucose, 2-deoxy-2[[(methylnitrosoamino)-carbonyl]amino]-	1 (0.454)
D-Glucose, 2-deoxy-2-[[(methylnitrosoamino)-carbonyl]amino]-	Streptozotocin Glucopyranose, 2-deoxy-2-(3-methyl-3-nitrosoureido)-	1 (0.454)
Glycidylaldehyde	Oxiranecarboxyaldehyde	10 (4.54)
Guanidine, N-methyl-N'-nitro-N-nitroso-	MNNG	10 (4.54)
Guthion		1 (0.454)
Heptachlor	4,7-Methano-1H-indene,1,4,5,6,7,8,8-heptachloro-3a,4,7,7a-tetrahydro-	1 (0.454)
Heptachlor epoxide		1 (0.454)
Hexachlorobenzene	Benzene, hexachloro-	10 (4.54)
Hexachlorobutadiene	1,3-Butadiene, 1,1,2,3,4,4- hexachloro-	1 (0.454)
Hexachlorocyclohexane (gamma isomer)	gamma - BHC Lindane Cyclohexane,1,2,3,4,5,6-hexachloro-, (1alpha,2alpha,3beta,4alpha,5alpha,6beta)-	1 (0.454)
Hexachlorocyclopentadiene	1,3-Cyclopentadiene, 1,2,3,4,5,5-hexachloro-	10 (4.54)
Hexachloroethane	Ethane, hexachloro-	100 (45.4)
1,2,3,4,10,10-Hexachloro-1,4,4a,5,8,8a-hexahydro-1,4:5,8-endo,exo-dimethanonaphthalene	Aldrin 1,4,5,8-Dimethanonaphthalene,1,2,3,4,10,10-hexachloro-1,4,4a,5,8,8a-hexahydro-,(1alpha,4alpha,4abeta,5alpha,8alpha,8abeta)-	1 (0.454)
Hexachlorophene	Phenol,2,2'-methylenebis[3,4,6-trichloro	100 (45.4)
Hexachloropropene	1-Propene, 1,1,2,3,3,3-hexachloro-	1000 (454)
Hexaethyl tetraphosphate	Tetraphosphoric acid, hexaethyl ester	100 (45.4)
Hexamethylene-1,6-diisocyanate		100 (45.4)
Hexamethylphosphoramide		1 (0.454)
Hexane		5000 (2270)
Hydrazine	Diamine	1 (0.454)

TABLE 1

Hazardous Substances	Synonyms	(RQ)
Hydrazine, 1,2-diethyl-	N,N-Diethylhydrazine	10 (4.54)
Hydrazine, 1,1-dimethyl-	1,1-Dimethylhydrazine; Dimethylhydrazine, unsymmetrical @	10 (4.54)
Hydrazine, 1,2-dimethyl-	1,2-Dimethylhydrazine	1 (0.454)
Hydrazine, 1,2-diphenyl-	1,2-Diphenylhydrazine	10 (4.54)
Hydrazine, methyl-	Methyl hydrazine	10 (4.54)
Hydrazinecarbothioamide	Thiosemicarbazide	100 (45.4)
Hydrochloric acid	Hydrogen chloride	5000 (2270)
Hydrocyanic acid	Hydrogen cyanide	10 (4.54)
Hydrofluoric acid	Hydrogen fluoride	100 (45.4)
Hydrogen chloride	Hydrochloric acid	5000 (2270)
Hydrogen cyanide	Hydrocyanic acid	10 (4.54)
Hydrogen fluoride	Hydrofluoric acid	100 (45.4)
Hydrogen phosphide	Phosphine	100 (45.4)
Hydrogen sulfide	Hydrogen sulfide H2S	100 (45.4)
Hydrogen sulfide H2S	Hydrogen sulfide	100 (45.4)
Hydroperoxide, 1-methyl-1-phenylethyl-	alpha,alpha-Dimethylbenzylhydroperoxide	10 (4.54)
Hydroquinone		100 (45.4)
2-Imidazolidinethione	Ethylenethiourea	10 (4.54)
Indeno(1,2,3-cd)pyrene	1,10-(1,2-Phenylene)pyrene	100 (45.4)
1,3-Isobenzofurandione	Phthalic anhydride	5000 (2270)
Isobutyl alcohol	1-Propanol, 2-methyl-	5000 (2270)
Isodrin	1,4,5,8-Dimethanonaphthalene, 1,2,3,4,10,10-hexachloro-1,4,4a,5,8,8a-hexahydro,(1alpha,4alpha,4abeta,5beta,8beta,8abeta)-	1 (0.454)
Isophorone		5000 (2270)
Isoprene		100 (45.4)
Isopropanolamine dodecylbenzene sulfonate		1000 (454)
Isosafrole	1,3-Benzodioxole,5-(-1propenyl)-	100 (45.4)

TABLE 1

Hazardous Substances	Synonyms	(RQ)
3(2H)-Isoxazolone, 5-(aminomethyl)-	5-(Aminomethyl)-3-isoxazolol Muscimol	1000 (454)
Kepone	1,3,4-Metheno-2H-cyclobuta[cd]-pentalen-2-one,1,1a,3,3a,4,5,5,5a,5b,6-decachloro-tahydro-	1 (0.454)
Lasiocarpine	2-Butenoic acid, 2-methyl-,7[[2,3-dihydroxy-2-(1-methoxyethyl)-3-methyl-1-oxobutoxy]methyl]-2,3,5,7a-tetrahydro-1H-pyrrolizin-1-ylester[1S-[1alpha(Z),7(2S*,3R*),7aalpha]]-	
Lead ¢		10 (4.54)
Lead acetate	Acetic acid, lead (2+)salt	10 (4.54)
Lead arsenate		1 (0.454)
Lead,bis(acetato-O)tetrahydroxytri	Lead subacetate	10 (4.54)
Lead chloride		10 (4.54)
Lead fluoborate		10 (4.54)
Lead fluoride		10 (4.54)
Lead iodide		10 (4.54)
Lead nitrate		10 (4.54)
Lead phosphate	Phosphoric acid, lead(2+) salt (2:3)	10 (4.54)
Lead stearate		10 (4.54)
Lead subacetate	Lead,bis(acetato-O)tetrahydroxytri	10 (4.54)
Lead sulfate		10 (4.54)
Lead sulfide		10 (4.54)
Lead thiocyanate		10 (4.54)
Lindane	gamma - BHC Hexachlorocyclohexane (gamma isomer) Cyclohexene, 1,2,3,4,5,6-hexachloro-,(1alpha,2alpha,3beta,4alpha,5alpha,6beta)-	1 (0.454)
Lithium chromate		10 (4.54)
Malathion		100 (45.4)
Maleic acid		5000 (2270)
Maleic anhydride	2,5-Furandione	5000 (2270)

TABLE 1

Hazardous Substances	Synonyms	(RQ)
Maleic hydrazide	3,6-Pyridazinedione, 1,2-dihydro-	5000 (2270)
Malononitrile	Propanedinitrile	1000 (454)
MDI	Methylene diphenyl diisocyanate	5000 (2270)
Melphalan	L-Phenylalanine, 4-[bis(2-chloroethyl)amino]-	1 (0.454)
Mercaptodimethur		10 (4.54)
Mercuric cyanide		1 (0.454)
Mercuric nitrate		10 (4.54)
Mercuric sulfate		10 (4.54)
Mercuric thiocyanate		10 (4.54)
Mercurous nitrate		10 (4.54)
Mercury,		1 (0.454)
Mercury, (acetato-O)phenyl-	Phenylmercuric acetate	100 (45.4)
Mercury fulminate	Fulminic acid, mercury(2+)salt	10 (4.54)
Methacrylonitrile	2-Propenenitrile, 2-methyl-	1000 (454)
Methanamine, N-methyl-	Dimethylamine	1000 (454)
Methanamine, N-methyl-N-nitroso-	N-Nitrosodimethylamine	10 (4.54)
Methane, bromo-	Methyl bromide	1000 (454)
Methane, chloro-	Chloromethane Methyl chloride	100 (45.4)
Methane, chloromethoxy-	Chloromethyl methyl ether Methylchloromethyl ether @	1 (0.454)
Methane, dibromo-	Methylene bromide	1000 (454)
Methane, dichloro-	Methylene chloride Dichloromethane@	1000 (454)
Methane, dichlorodifluoro-	Dichlorodifluoromethane	5000 (2270)
Methane, iodo-	Methyl iodide	100 (45.4)
Methane, isocyanato-	Methyl isocyanate	1 (0.454)
Methane, oxybis(chloro-	Dichloromethyl ether	1 (0.454)
Methane, tetrachloro-	Carbon tetrachloride	10 (4.54)
Methane, tetranitro-	Tetranitromethane	10 (4.54)

TABLE 1

Hazardous Substances	Synonyms	(RQ)
Methane, tribromo-	Bromoform	100 (45.4)
Methane, trichloro-	Chloroform	10 (4.54)
Methane, trichlorofluoro-	Trichloromonofluoromethane	5000 (2270)
Methenesulfenyl chloride, trichloro	Perchloromethyl mercaptan @ / Trichloromethanesulfenyl chloride	100 (45.4)
Methanesulfonic acid, ethyl ester	Ethyl methanesulfonate	1 (0.454)
Methanethiol	Methyl mercaptan / Thiomethanol	100 (45.4)
6,9-Methano-2,4,3-benzodioxathiepin,6,7,8,9,10-hexachloro-1,5,5a,6,9,9a-hexahydro-, 3-oxide	Endosulfan	1 (0.454)
Methanoic acid	Formic acid	5000 (2270)
4,7-Methano-1H-indene,1,4,5,6,7,8,8-heptachloro-3a,4,7,7a-tetrahydro-	Heptachlor	1 (0.454)
4,7-Methano-1H-indene,1,4,5,6,7,8,8-octachloro-2,3,3a,4,7,7a-hexahydro-	Chlordane, / Chlordane, technical / Chlordane,alpha & gamma isomers	1 (0.454)
Methanol	Methyl alcohol	5000 (2270)
Methapyrilene	1,2-Ethanediamine,N,N-dimethyl-N'-2-pyridinyl-N'-(2-thienylmethyl)-	5000 (2270)
1,3,4-Metheno-2H-cyclobuta[cd]-pentalen-2-one,1,1a,3,3a,4,5,5a,5b,6-decachloroctahydro-	Kepone	1 (0.454)
Methomyl	Ethanimidothioic acid, N-[[(methylamino)carbonyl]oxy]-, methyl ester	100 (45.4)
Methoxychlor	Benzene, 1,1'-(2,2,2-trichloroethylidene) bis[4-methoxy-	1 (0.454)
Methyl alcohol	Methanol	5000 (2270)
Methylamine @	Monomethylamine	100 (45.4)
Methyl bromide	Methane, bromo-	1000 (454)
1-Methylbutadiene	1,3-Pentadiene	100 (45.4)
Methyl chloride	Chloromethane / Methane, chloro-	100 (45.4)
Methyl chlorocarbonate	Carbonochloridic acid, methyl ester / Methyl chloroformate	1000 (414)

TABLE 1

Hazardous Substances	Synonyms	(RQ)
Methyl chloroform	1,1,1-Trichloroethane Ethane, 1,1,1-trichloro-	1000 (454)
Methyl chloroformate	Carbonochloridic acid, methyl ester Methyl chlorocarbonate	1000 (454)
Methylchloromethyl ether @	Chloromethyl methyl ether Methane, chloromethoxy-	1 (0.454)
3-Methylcholanthrene	Benz[j]aceanthrylene, 1,2-dihydro-3-methyl-	10 (4.54)
4,4'-Methylenebis(2-chloroaniline)	Benzenamine, 4,4'-methylenebis(2-chloro-	10 (4.54)
Methylene bromide	Methane, dibromo-	1000 (454)
Methylene chloride	Methane, dichloro- Dichloromethane@	1000 (454)
4,4'-Methylenedianiline		10 (4.54)
Methylene diphenyl diisocyanate	MDI	5000 (2270)
Methylene oxide	Formaldehyde	100 (45.4)
Methyl ethyl ketone(MEK)	2-Butanone Ethyl methyl ketone @	5000 (2270)
Methyl ethyl ketone peroxide	2-Butanone peroxide	10 (4.54)
Methyl hydrazine	Hydrazine, methyl-	10 (4.54)
Methyl iodide	Methane, iodo-	100 (45.4)
Methyl isobutyl ketone	4-Methyl-2-pentanone	5000 (2270)
Methyl isocyanate	Methane, isocyanato-	10 (4.54)
2-Methyllactonitrile	Acetone cyanohydrin Propanenitrile, 2-hydroxy-2-methyl-	10 (4.54)
Methyl mercaptan	Methanethiol Thiomethanol	100 (45.4)
Methyl methacrylate	2-Propenoic acid, 2-methyl-, methyl ester	1000 (454)
Methyl parathion	Phosphorothioic acid, O,O-dimethyl O-(4-nitrophenyl) ester	100 (45.4)
4-Methyl-2-pentanone	Methyl Isobutyl ketone	5000 (2270)
Methyl tert-butyl ether		1000 (454)
Methylthiouracil	4(1H)-Pyrimidinone, 2,3-dihydro-6-methyl-2-thioxo-	10 (4.54)

TABLE 1

Hazardous Substances	Synonyms	(RQ)
Mevinphos		10 (4.54)
Mexacarbate		1000 (454)
Mitomycin C	Aziriono[2',3':3,4]pyrrolo [1,2-a]indole-4,7-dione,6-amino-8-[[(amino-carbonyl)oxy]methyl]-1,1a,2,8,8a,8b-hexahydro-8a-methoxy-5-methyl-, [1aS-(1aalpha,8beta,8aalpha,8balpha)]-	
MNNG	Guanidine, N-methyl-N'-nitro-N-nitroso-	10 (4.54)
Monoethylamine		10 (4.54)
Monomethylamine		100 (45.4)
Muscimol	5-(Aminomethyl)-3-isoxazolol 3(2H)-isoxazolone, 5-(aminomethyl)-	1000 (454)
Naled		10 (4.54)
5,12-Naphthacenedione, 8-acetyl-10-[3-amino-2,3,6-trideoxy-alpha-L-lyxo-hexopyranosyl)oxy]-7,8,9,10-tetrahdro-6,8,11-trihydroxy-1-methoxy-,(8S-cis)-	Daunomycin	10 (4.54)
Naphthalenamine,N,N-bis(2-chloroethyl)-	Chlornaphazine	100 (45.4)
Naphthalene		100 (45.4)
Naphthalene, 2-chloro-	beta-Chloronaphthalene 2-Chloronaphthalene	5000 (2270)
1,4-Naphthalenedione	1,4-Naphthoquinone	5000 (2270)
2,7-Naphthalenedisulfonic acid,3,3'-[(3,3'-dimethyl-(1,1'-biphenyl)-4,4'-diyl)-bis(azo)]bis(5-amino-4-hydroxy)-tetrasodium salt.	Trypan blue	10 (4.54)
Naphthenic acid		100 (45.4)
1,4-Naphthoquinone	1,4-Naphthalenedione	5000 (2270)
alpha-Naphthylamine	1-Naphthylamine	100 (45.4)
beta-Naphthylamine	2-Naphthylamine	1 (0.454)
1-Naphthylamine	alpha-Naphthylamine	100 (45.4)
2-Naphthylamine	beta-Naphthylamine	1 (0.454)
alpha-Naphthylthiourea	Thiourea, 1-naphthalenyl-	100 (45.4)
Nickel ϵ		100 (45.4)
Nickel ammonium sulfate		100 (45.4)
Nickel carbonyl	Nickel carbonyl Ni(CO)4,(T-4)	10 (4.54)

TABLE 1

Hazardous Substances	Synonyms	(RQ)
Nickel carbonyl Ni(CO)4,(T-4)-	Nickel carbonyl	10 (4.54)
Nickel chloride		100 (45.4)
Nickel cyanide	Nickel cyanide Ni(CN)2	10 (4.54)
Nickel cyanide Ni(CN)2	Nickel cyanide	10 (4.54)
Nickel hydroxide		10 (4.54)
Nickel nitrate		100 (45.4)
Nickel sulfate		100 (45.4)
Nicotine and salts	Pyridine, 3-(1-methyl-2-pyrrolidinyl)-,(S)-	100 (45.4)
Nitric acid		1000 (454)
Nitric acid, thallium(1+)salt	Thallium(I) nitrate	100 (45.4)
Nitric oxide	Nitrogen oxide NO	10 (4.54)
p-Nitroaniline	Benzenamine, 4-nitro-	5000 (2270)
Nitrobenzene	Benzene, nitro-	1000 (454)
4-nitrobiphenyl		10 (4.54)
Nitrogen dioxide	Nitrogen oxide NO2 Dinitrogen tetroxide @	10 (4.54)
Nitrogen oxide NO	Nitric oxide	10 (4.54)
Nitrogen oxide NO2	Nitrogen dioxide Dinitrogen tetroxide @	10 (4.54)
Nitroglycerine	1,2,3-Propanetriol, trinitrate-	10 (4.54)
Nitrophenol (mixed)	2-Nitrophenol 4-Nitrophenol Phenol, 4-nitro-	100 (45.4)
m-	2-Nitrophenol	
o-	4-Nitrophenol	
p-	Phenol, 4-nitro-	
o-Nitrophenol	Phenol, 4-nitro-	100 (45.4)
p-Nitrophenol	4-Nitrophenol	100 (45.4)
2-Nitrophenol	o-Nitrophenol	100 (45.4)
4-Nitrophenol	p-Nitrophenol Phenol, 4-nitro-	100 (45.4)

TABLE 1

Hazardous Substances	Synonyms	(RQ)
2-Nitropropane	Propane, 2-nitro-	10 (4.54)
N-Nitrosodi-n-butylamine	1-Butanamine, N-butyl-N-nitroso-	10 (4.54)
N-Nitrosodiethanolamine	Ethanol, 2,2'- (nitrosimino)bis-	1 (0.454)
N-Nitrosodiethylamine	Ethanamine, N-ethyl-N-nitroso-	1 (0.454)
N-Nitrosodimethylamine	Methanamine, N-methyl-N-nitroso-	10 (4.54)
N-Nitrosodiphenylamine		100 (45.4)
N-Nitroso-N-ethylurea	Urea, N-ethyl-N-nitroso-	1 (0.454)
N-Nitroso-N-methylurea	Urea, N-methyl-N-nitroso-	1 (0.454)
N-Nitroso-N-methylurethane	Carbamic acid, methylnitroso-, ethyl ester	1 (0.454)
N-Nitrosomethylvinylamine	Vinylamine, N-methyl-N-nitroso-	10 (4.54)
n-Nitrosomorpholine		1 (0.454)
N-Nitrosopiperidine	Piperidine, 1-nitroso-	10 (4.54)
N-Nitrosopyrrolidine	Pyrrolidine, 1-nitroso-	1 (0.454)
Nitrotoluene		1000 (454)
m-Nitrotoluene		
o-Nitrotoluene		
p-Nitrotoluene		
5-Nitro-o-toluidine	Benzenamine, 2-methyl-5-nitro-	100 (45.4)
Octamethylpyrophosphoramide	Diphosphoramide, octamethyl-	100 (45.4)
Osmium oxide OsO4 (T4)	Osmium tetroxide	1000 (454)
Osmium tetroxide	Osmium oxide OsO4(T4)-	1000 (454)
7-Oxabicyclo[2.2.1]heptane-2,3-dicarboxylic acid	Endothall	1000 (454)
1,2-Oxathiolane, 2,2-dioxide	1,3-Propane sultone	10 (4.54)
2H-1,3,2-Oxazaphosphorin-2-amine, N,N-bis(2-chloroethyl)tetrahydro-2-oxide	Cyclophosphamide	10 (4.54)
Oxirane	Ethylene oxide	10 (4.54)
Oxiranecarboxyaldehyde	Glycidylaldehyde	10 (4.54)
Oxirane, (chloromethyl)-	Epichlorohydrin	100 (45.4)
Paraformaldehyde		1000 (454)
Paraldehyde	1,3,5-Trioxane, 2,4,6- trimethyl-	1000 (454)

TABLE 1

Hazardous Substances	Synonyms	(RQ)
Parathion	Phosphorothioic acid, O,O-diethyl O-(4-nitrophenyl) ester	10 (4.54)
Pentachlorobenzene	Benzene, pentachloro-	10 (4.54)
Pentachloroethane	Ethane, pentachloro-	10 (4.54)
Pentachloronitrobenzene (PCNB)	Benzene, pentachloronitro-	100 (45.4)
Pentachlorophenol	Phenol, pentachloro-	10 (4.54)
1,3-Pentadiene	1-Methylbutadiene	100 (45.4)
Perchloroethylene	Ethene, tetrachloro Tetrachloroethene Tetrachloroethylene	100 (45.4)
Perchloromethyl mercaptan @	Methanesulfenyl chloride, trichloro- Trichloromethanesulfenyl chloride	100 (45.4)
Phenacetin	Acetamide, N-(4-ethoxyphenyl)-	100 (45.4)
Phenanthrene		5000 (2270)
Phenol	Benzene, hydroxy-	1000 (454)
Phenol, 2-chloro-	o-Chlorophenol 2-Chlorophenol	100 (45.4)
Phenol, 4-chloro-3-methyl-	p-Chloro-m-cresol 4-Chloro-m-cresol	5000 (2270)
Phenol, 2-cyclohexyl-4,6-dinitro-	2-Cyclohexyl-4,6-dinitrophenol	100 (45.4)
Phenol, 2,4-dichloro-	2,4-Dichlorophenol	100 (45.4)
Phenol, 2,6-dichloro-	2,6-Dichlorophenol	100 (45.4)
Phenol, 4,4'-(1,2-diethyl-1,2-ethenediyl)bis-, (E)	Diethylstilbestrol	1 (0.454)
Phenol, 2,4-dimethyl-	2,4-Dimethylphenol	100 (45.4)
Phenol, 2,4-dinitro-	2,4-Dinitrophenol	10 (4.54)
Phenol, methyl-	Cresols (isomers and mixture); Cresylic acid (isomers and mixture).	100 (45.4)
Phenol, 2-methyl-4,6-dinitro-	4,6-Dinitro-o-cresol and salts	10 (4.54)
Phenol, 2,2'-methylenebis[3,4,6-trichloro-	Hexachlorophene	100 (45.4)
Phenol, 2-(1-methylpropyl)-4,6-dinitro-	Dinoseb	1000 (454)

TABLE 1

Hazardous Substances	Synonyms	(RQ)
Phenol, 4-nitro-	p-Nitrophenol	100 (45.4)
	4-Nitrophenol	
Phenol, pentachloro-	Pentachlorophenol	10 (4.54)
Phenol, 2,3,4,6-tetrachloro-	2,3,4,6-Tetrachlorophenol	10 (4.54)
Phenol, 2,4,5-trichloro-	2,4,5-Trichlorophenol	10 (4.54)
Phenol, 2,4,6-trichloro-	2,4,6-Trichlorophenol	10 (4.54)
Phenol, 2,4,6-trinitro-, ammonium salt	Ammonium picrate	1 (0.454)
L-Phenylalanine, 4-[bis(2-chloroethyl)amino]-	Melphalan	5000 (2270)
p-Phenylenediamine		100 (45.4)
1,10-(1,2-Phenylene)pyrene	Indeno(1,2,3-cd)pyrene	100 (45.4)
Phenyl mercaptan @	Benzenethiol	100 (45.4)
	Thiophenol	
Phenylmercuric acetate	Mercury, (acetato-O) phenyl-	100 (45.4)
Phenylthiourea	Thiourea, phenyl-	100 (45.4)
Phorate	Phosphorodithioic acid, O,O- diethyl S-(ethylthio), methylester	10 (4.54)
Phosgene	Carbonic dichloride	10 (4.54)
Phosphine	Hydrogen phosphide	100 (45.4)
Phosphoric acid		5000 (2270)
Phosphoric acid, diethyl 4-nitrophenyl ester	Diethyl-p-nitrophenyl phosphate	100 (45.4)
Phosphoric acid, lead(2+) salt(2:3)	Lead phosphate	100 (45.4)
Phosphorodithioic acid, O,O-diethyl S-[2-(ethylthio) ethyl]ester	Disulfoton	1 (0.454)
Phosphorodithioic acid, O,O-diethyl S-(ethylthio), methyl ester	Phorate	10 (4.54)
Phosphorodithioic acid, O,O-diethyl S-methyl ester	O,O-Diethyl S-methyl dithiophosphate	5000 (2270)
Phosphorodithioic acid, O,O-dimethyl S-[2(methyl amino)-2-oxoethyl] ester	Dimethoate	10 (4.54)
Phosphorofluoridic acid, bis(1-methylethyl) ester	Diisopropyl fluorophosphate	100 (45.4)
Phosphorothioic acid, O,O-diethyl O-(4-nitrophenyl) ester	Parathion	10 (4.54)
Phosphorothioic acid, O,O-diethyl O-pyrazinyl ester	O,O-Diethyl O-pyrazinyl phosphorothioate	10 (4.54)
Phosphorothioic acid, O,O-dimethyl O-(4-nitrophenyl) ester	Methyl parathion	100 (45.4)

TABLE 1

Hazardous Substances	Synonyms	(RQ)
Phosphorothioic acid, O,[4-[(dimethylamino)sulfonyl] phenyl] O,O-di-methyl ester.	Famphur	1000 (454)
Phosphorus		1 (0.454)
Phosphorus oxychloride		1000 (454)
Phosphorus pentasulfide	Phosphorous sulfide, Sulfur phosphide	100 (45.4)
Phosphorus sulfide	Phosphorous pentasulfide, Sulfur phosphide	100 (45.4)
Phosphorus trichloride		1000 (454)
Phthalate anhydride	1,3-Isobenzofurandione	5000 (2270)
2-Picoline	Pyridine,2-methyl-	5000 (2270)
Piperidine, 1-nitroso-	N-Nitrosopiperidine	10 (4.54)
Plumbane, tetraethyl-	Tetraethyl lead	10 (4.54)
POLYCHLORINATED BIPHENYLS (PCBs)	Aroclor 1016	1 (0.454)
	Aroclor 1221	
	Aroclor 1232	
	Aroclor 1242	
	Aroclor 1248	
	Aroclor 1254	
	Aroclor 1260	
Potassium arsenate		1 (0.454)
Potassium arsenite		1 (0.454)
Potassium bichromate		10 (4.54)
Potassium chromate		10 (4.54)
Potassium cyanide	Potassium cyanide K(CN)	10 (4.54)
Potassium cyanide K(CN)	Potassium cyanide	10 (4.54)
Potassium hydroxide		1000 (454)
Potassium permanganate		100 (45.4)
Potassium silver cyanide	Argentate(1-), bis(cyano-C), potassium	1 (0.454)
Pronamide	Benzamide, 3,5-dichloro-N-(1,1-dimethyl-2-propynyl)-	5000 (2270)
Propanal, 2-methyl-2-(methylthio)-,O-[(methylamino) carbonyl]oxime	Aldicarb	1 (0.454)

Hazardous Substances	Synonyms	(RQ)
1-Propanamine	n-Propylamine	5000 (2270)
1-Propanamine, N-nitroso-N-propyl-	Di-n-propylnitrosamine	10 (4.54)
1-Propanamine, N-propyl-	Dipropylamine	5000 (2270)
Propane, 1,2-dibromo-3-chloro-	1,2-Dibromo-3-chloropropane	1 (0.454)
Propane, 1,2-dichloro-	1,2-Dichloropropane	1000 (454)
	Propylene dichloride	
Propane, 2-nitro-	2-Nitropropane	10 (4.54)
Propane, 2,2'-oxybis[2-chloro-	Dichloroisopropyl ether	1000 (454)
1,3-Propane sultone	1,2-Oxathiolane, 2,2-dioxide	10 (4.54)
Propanedinitrile	Malononitrile	1000 (454)
Propanenitrile	Ethyl cyanide	10 (4.54)
Propanenitrile, 3-chloro-	3-Chloropropionitrile	1000 (454)
Propanenitrile, 2-hydroxy-2-methyl-	Acetone cyanohydrin	10 (4.54)
	2-Methyllactonitrile	
1,2,3-Propanetriol, trinitrate-	Nitroglycerine	10 (4.54)
1-Propanol, 2,3-dibromo-, phosphate (3:1)	Tris(2,3-dibromopropyl) phosphate	10 (4.54)
1-Propanol, 2-methyl-	Isobutyl alcohol	5000 (2270)
2-Propanone	Acetone	5000 (2270)
2-Propanone, 1-bromo-	Bromoacetone	1000 (454)
Propargite		10 (4.54)
Propargyl alcohol	2-Propyn-1-ol	1000 (454)
2-Propenal	Acrolein	1 (0.454)
2-Propenamide	Acrylamide	5000 (2270)
1-Propene, 1,3-dichloro-	1,3-Dichloropropene	100 (45.4)
1-Propene, 1,1,2,3,3,3-hexachloro-	Hexachloropropene	1000 (454)
2-Propenenitrile	Acrylonitrile	100 (45.4)
2-Propenenitrile, 2-methyl-	Methacrylonitrile	1000 (454)
2-Propenoic acid	Acrylic acid	5000 (2270)
2-Propenoic acid, ethyl ester	Ethyl acrylate	1000 (454)
2-Propenoic acid, 2-methyl-, ethyl ester	Ethyl methacrylate	1000 (454)

TABLE 1

Hazardous Substances	Synonyms	(RQ)
2-Propenoic acid, 2-methyl-, methyl ester	Methyl methacrylate	1000 (454)
2-Propen-1-ol	Allyl alcohol	100 (45.4)
beta-Propioaldehyde		1000 (454)
Propionic acid		5000 (2270)
Propionic acid, 2-(2,4,5-trichlorophenoxy)-	Silvex (2,4,5-TP), 2,4,5-TP acid	100 (45.4)
Propionic anhydride		5000 (2270)
Propoxur (baygon)		100 (45.4)
n-Propylamine	1-Propanamine	5000 (2270)
Propylene dichloride	1,2-Dichloropropane Propane, 1,2-dichloro-	1000 (454)
Propylene oxide		100 (45.4)
1,2-Propylenimine	Aziridine, 2-methyl-	1 (0.454)
2-Propyn-1-ol	Propargyl alcohol	1000 (454)
Pyrene		5000 (2270)
Pyrethrins		1 (0.454)
3,6-Pyridazinedione, 1,2-dihydro-	Maleic hydrazide	5000 (2270)
4-Pyridinamine	4-Aminopyridine	1000 (454)
Pyridine		1000 (454)
Pyridine, 2-methyl-	2-Picoline	5000 (2270)
Pyridine, 3-(1-methyl-2-pyrrolidinyl)-, (S)	Nicotine and salts	100 (45.4)
2,4-(1H,3H)-Pyrimidinedione, 5-[bis(2-chloroethyl) amino]-	Uracil mustard	10 (4.54)
4(1H)-Pyrimidinone, 2,3-dihydro-6-methyl-2-thioxo-	Methylthiouracil	10 (4.54)
Pyrrolidine, 1-nitroso-	N-Nitrosopyrrolidine	1 (0.454)
Quinoline		5000 (2270)
RADIONUCLIDES		See Table 2
Reserpine	Yohimban-16-carboxylic acid,11,17-dimethoxy-18-[(3,4,5-trimethoxy-benzoyl)oxy-, methyl ester-(3beta, 16beta,17alpha,18beta,20alpha)	5000 (2270)
Resorcinol	1,3-Benzenediol	5000 (2270)

TABLE 1

Hazardous Substances	Synonyms	(RQ)
Saccharin and salts	1,2-Benzisothiazol-3(2H)-one,1,1-dioxide	100 (45.4)
Safrole	1,3-Benzodioxole,5-(2-propenyl)-	100 (45.4)
Selenious acid		10 (4.54)
Selenious acid, dithallium(1+)salt	Thallium selenite	1000 (454)
Selenium ¢		100 (45.4)
Selenium dioxide	Selenium oxide	10 (4.54)
Selenium oxide	Selenium dioxide	10 (4.54)
Selenium sulfide	Selenium sulfide SeS2	10 (4.54)
Selenium sulfide SeS2	Selenium sulfide	10 (4.54)
Selenourea	Carbamimidoselenoic acid	1000 (454)
L-Serine, diazoacetate (ester)	Azaserine	1 (0.454)
Silver ¢		1000 (454)
Silver cyanide	Silver cyanide Ag(CN)	1 (0.454)
Silver cyanide Ag(CN)	Silver cyanide	1 (0.454)
Silver nitrate		1 (0.454)
Silvex(2,4,5-TP)	Propionic acid, 2-(2,4,5-trichlorophenoxy)- 2,4,5-TP acid	100 (45.4)
Sodium		10 (4.54)
Sodium arsenate		1 (0.454)
Sodium arsenite		1 (0.454)
Sodium azide		1000 (454)
Sodium bichromate		10 (4.54)
Sodium bifluoride		100 (45.4)
Sodium bisulfite		5000 (2270)
Sodium chromate		10 (4.54)
Sodium cyanide	Sodium cyanide Na(CN)	10 (4.54)
Sodium cyanide Na(CN)	Sodium cyanide	10 (4.54)
Sodium dodecylbenzene sulfonate		1000 (454)
Sodium fluoride		1000 (454)

TABLE 1

Hazardous Substances	Synonyms	(RQ)
Sodium hydrosulfide		5000 (2270)
Sodium hydroxide		1000 (454)
Sodium hypochlorite		100 (45.4)
Sodium methylate		1000 (454)
Sodium nitrite		100 (45.4)
Sodium phosphate, dibasic		5000 (2270)
Sodium phosphate, tribasic		5000 (2270)
Sodium selenite		100 (45.4)
Streptozotocin	Glucopyranose, 2-deoxy-2-(3-methyl-3-nitrosoureido)- D-Glucose, 2-deoxy-2-[[(methylnitrosoamino)- carbonyl]amino]-	1 (0.454)
Strontium chromate		10 (4.54)
Strychnidin-10-one	Strychnine and salts	10 (4.54)
Strychnidin-10-one, 2,3-dimethoxy-	Brucine	100 (45.4)
Strychnine and salts	Strychnidin-10-one	10 (4.54)
Styrene		1000 (454)
Styrene oxide		100 (45.4)
Sulfur chloride@	Sulfur monochloride	1000 (454)
Sulfur monochloride	Sulfur chloride@	1000 (454)
Sulfur phosphide	Phosphorus pentasulfide Phosphorus sulfide	100 (45.4)
Sulfuric acid		1000 (454)
Sulfuric acid, dimethyl ester	Dimethyl sulfate	100 (45.4)
Sulfuric acid, dithallium(I+)salt	Thallium(I) sulfate	100 (45.4)
2,4,5-T	2,4,5-T acid, Acetic acid, (2,4,5-trichlorophenoxy)	1000 (454)
2,4,5-T acid	2,4,5-T Acetic acid, (2,4,5-trichlorophenoxy)	1000 (454)
2,4,5-T amines		5000 (2270)
2,4,5-T esters		1000 (454)

TABLE 1

Hazardous Substances	Synonyms	(RQ)
2,4,5-T salts	DDD	1000 (454)
TDE	Benzene, 1,1'-(2,2-dichloroethylidene)bis[4-chloro-4,4'-DDD	1 (0.454)
1,2,4,5-Tetrachlorobenzene	Benzene, 1,2,4,5-tetrachloro-	5000 (2270)
2,3,7,8-Tetrachlorodibenzo-p-dioxin (TCDD)		1 (0.454)
1,1,1,2-Tetrachloroethane	Ethane,1,1,1,2-tetrachloro- Tetrachloroethane@	100 (45.4)
1,1,2,2-Tetrachloroethane	Ethane,1,1,2,2-tetrachloro- Tetrachloroethane@	100 (45.4)
Tetrachloroethane@	Ethane, 1,1,1,2-tetrachloro- Ethane, 1,1,2,2-tetrachloro- 1,1,1,2-Tetrachloroethane 1,1,2,2-Tetrachloroethane	100 (45.4)
Tetrachloroethene	Ethene, tetrachloro- Perchloroethylene Tetrachloroethylene	100 (45.4)
Tetrachloroethylene	Ethene, tetrachloro- Perchloroethylene Tetrachloroethene	100 (45.4)
2,3,4,6-Tetrachlorophenol	Phenol, 2,3,4,6-tetrachloro-	10 (4.54)
Tetraethyl lead	Plumbane, tetraethyl-	10 (4.54)
Tetraethyl pyrophosphate	Diphosphoric acid, tetraethyl ester	10 (4.54)
Tetraethyldithiopyrophosphate	Thiodiphosphoric acid, tetraethyl ester	100 (45.4)
Tetrahydrofuran	Furan, tetrahydro-	1000 (454)
Tetranitromethane	Methane, tetranitro-	10 (4.54)
Tetraphosphoricacid, hexaethyl ester	Hexaethyl tetraphosphate	100 (45.4)
Thallic oxide	Thallium oxide Tl2O3	100 (45.4)
Thallium ¢		1000 (454)
Thallium(I) acetate	Acetic acid, thallium(I+)salt	100 (45.4)
Thallium(I) carbonate	Carbonic acid, dithallium (I+)	100 (45.4)
Thallium(I) chloride	Thallium chloride TlCl	100 (45.4)

Hazardous Substances	Synonyms	(RQ)
Thallium chloride TlCl	Thallium(I) chloride	100 (45.4)
Thallium(I) nitrate	Nitric acid, thallium(1+) salt	100 (45.4)
Thallium oxide Tl2O3	Thallic oxide	100 (45.4)
Thallium selenite	Selenious acid, dithallium(1+) salt	1000 (454)
Thallium(I) sulfate	Sulfuric acid, dithallium(1+) salt	100 (45.4)
Thioacetamide	Ethanethioamide	10 (4.54)
Thiodiphosphoric acid, tetraethyl ester	Tetraethyldithiopyrophosphate	100 (45.4)
Thiofanox	2-butanone, 3,3-Dimethyl-1-(methylthio)-, O-[(methylamino)carbonyl]oxime	100 (45.4)
Thioimidodicarbonic diamide [(H2N)C(S)]2NH	Dithiobiuret	100 (45.4)
Thiomethanol	Methanethiol Methyl mercaptan	100 (45.4)
Thioperoxydicarbonic diamide [((H2N)C(S)]2S2, tetramethyl-	Thiram	10 (4.54)
Thiophenol	Benzenethiol Phenyl mercaptan @	100 (45.4)
Thiosemicarbazide	Hydrazinecarbothioamide	100 (45.4)
Thiourea	Carbamide, thio-	10 (4.54)
Thiourea, (2-chlorophenyl)-	1-(o-Chlorophenyl)thiourea	100 (45.4)
Thiourea, 1-naphthalenyl-	alpha-Naphthylthiourea	100 (45.4)
Thiourea, phenyl-	Phenylthiourea	100 (45.4)
Thiram	Thioperoxydicarbonic diamide [(H2N)C(S)]2S2, tetramethyl-	10 (4.54)
Titanium tetrachloride		1000 (454)
Toluene	Benzene, methyl-	1000 (454)
Toluenediamine	Benzenediamine, ar-methyl-	10 (4.54)
Toluene diisocyanate	Benzene, 1,3- diisocyanatomethyl	100 (45.4)
o-Toluidine	2-Amino-1-methyl benzene	100 (45.4)
p-Toluidine	Benzenamine, 4-methyl-	100 (45.4)
o-Toluidine hydrochloride	Benzenamine, 2-methyl-, hydrochloride	100 (45.4)
Toxaphene	Camphene, octachloro-	1 (0.454)

TABLE 1

Hazardous Substances	Synonyms	(RQ)
2,4,5-TP acid	Propionic acid, 2-(2,4,5-trichlorophenoxy)- Silvex (2,4,5-TP)	100 (45.4)
2,4,5-TP acid esters		100 (45.4)
1H-1,2,4-Triazol-3-amine	Amitrole	10 (4.54)
Trichlorfon		100 (45.4)
1,2,4-Trichlorobenzene		100 (45.4)
1,1,1-Trichloroethane	Methyl chloroform Ethane, 1,1,1-trichloro-	1000 (454)
1,1,2-Trichloroethane	Ethane, 1,1,2-trichloro-	100 (45.4)
Trichloroethene	Trichloroethylene Ethene, trichloro-	100 (45.4)
Trichloroethylene	Trichloroethene Ethene, trichloro-	100 (45.4)
Trichloromethanesulfenyl chloride	Methanesulfenyl chloride, trichloro- Perchloromethyl mercaptan @	100 (45.4)
Trichloromonofluoromethane	Methane, trichlorofluoro-	5000 (2270)
Trichlorophenol		10 (4.54)
2,3,4-Trichlorophenol		
2,3,5-Trichlorophenol		
2,3,6-Trichlorophenol		
2,4,5-Trichlorophenol		
2,4,6-Trichlorophenol		
3,4,5-Trichlorophenol		
2,4,5-Trichlorophenol	Phenol, 2,4,5-trichloro-	10 (4.54)
2,4,6-Trichlorophenol	Phenol, 2,4,6-trichloro-	10 (4.54)
Triethanolamine dodecylbenzene sulfonate		1000 (454)
Triethylamine		5000 (2270)
Trifluralin		10 (4.54)
Trimethylamine		100 (45.4)
2,2,4-Trimethylpentane		1000 (454)
1,3,5-Trinitrobenzene	Benzene, 1,3,5-trinitro-	10 (4.54)
1,3,5-Trioxane, 2,4,6-trimethyl-	Paraldehyde	1000 (454)

TABLE 1

Hazardous Substances	Synonyms	(RQ)
Tris(2,3-dibromopropyl) phosphate	1-Propanol, 2,3-dibromo-, phosphate(3:1)	10 (4.54)
Trypan blue	2,7-Naphthalenedisulfonic acid, 3,3'-[(3,3'-dimethyl-(1,1'-biphenyl)-4,4'-diyl)-bis(azo)]bis (5-amino-4-hydroxy)-tetrasodium salt	
Uracil mustard	2,4-(1H,3H)-Pyrimidinedione, 5-[bis(2-chloroethyl)amino]-	10 (4.54)
Uranyl acetate		100 (45.4)
Uranyl nitrate		100 (45.4)
Urea, N-ethyl-N-nitroso-	N-Nitroso-N-ethylurea	1 (0.454)
Urea, N-methyl-N-nitroso-	N-Nitroso-N-methylurea	1 (0.454)
Vanadic acid, ammonium salt	Ammonium vanadate	1000 (454)
Vanadium oxide V2O5	Vanadium pentoxide	1000 (454)
Vanadium pentoxide	Vanadium oxide V2O5	1000 (454)
Vanadyl sulfate		1000 (454)
Vinyl acetate	Vinyl acetate monomer	5000 (2270)
Vinyl acetate monomer	Vinyl acetate	5000 (2270)
Vinylamine, N-methyl-N-nitroso-	N-nitrosomethylvinylamine	10 (4.54)
Vinyl bromide		100 (45.4)
Vinyl chloride	Ethene, chloro-	1 (0.454)
Vinylidene chloride	Ethene, 1,1-dichloro- 1,1-Dichloroethylene	100 (45.4)
Warfarin, & salts, when present at concentrations greater than 0.3%	2H-1-Benzopyran-2-one, 4-hydroxy-3-(3oxo-1-phenylbutyl)-, & salts, when present at concentrations greater than 0.3%	100 (45.4)
Xylene	Benzene dimethyl-; Xylene (mixed); Xylenes (isomers and mixture)	100 (45.4)
m-Xylene	Benzene, m-dimethyl-	1000 (454)
o-Xylene	Benzene, o-dimethyl-	1000 (454)
p-Xylene	Benzene, p-dimethyl-	100 (45.4)
Xylene (mixed)	Benzene, dimethyl-; Xylene; Xylenes (isomers and mixture)	100 (45.4)

TABLE 1

Hazardous Substances	Synonyms	(RQ)
Xylenes (isomers and mixture)	Benzene, dimethyl-; Xylene; Xylenes (mixed)	100 (45.4)
Xylenol		1000 (454)
Yohimban-16-carboxylic acid, 11,17-dimethoxy-18-[(3,4,5-trimethoxybenzoyl)oxy]-, methyl ester (3beta,16beta,17alpha,18beta,20alpha)	Reserpine	5000 (2270)
Zinc ¢		1000 (454)
Zinc acetate		1000 (454)
Zinc ammonium chloride		1000 (454)
Zinc borate		1000 (454)
Zinc bromide		1000 (454)
Zinc carbonate		1000 (454)
Zinc chloride		1000 (454)
Zinc cyanide	Zinc cyanide Zn(CN)2	10 (4.54)
Zinc cyanide Zn(CN)2	Zinc cyanide	10 (4.54)
Zinc fluoride		1000 (454)
Zinc formate		1000 (454)
Zinc hydrosulfite		1000 (454)
Zinc nitrate		1000 (454)
Zinc phenolsulfonate		5000 (2270)
Zinc phosphide	Zinc phosphide Zn3P2, when present at concentrations greater than 10%	100 (45.4)
Zinc phosphide Zn3P2, when present at concentrations greater than 10%	Zinc phosphide	100 (45.4)
Zinc silicofluoride		5000 (2270)
Zinc sulfate		1000 (454)
Zirconium nitrate		5000 (2270)
Zirconium potassium fluoride		1000 (454)
Zirconium sulfate		5000 (2270)
Zirconium tetrachloride		5000 (2270)
D001 Unlisted Hazardous Wastes Characteristic of Ignitability		100 (45.4)
D002 Unlisted Hazardous Wastes Characteristic of Corrosivity		100 (45.4)
D003 Unlisted Hazardous Wastes Characteristic of Reactivity		100 (45.4)

TABLE 1

Hazardous Substances	Synonyms	(RQ)
D004–D043 Unlisted Hazardous Wastes Characteristic of Toxicity		
D004 Arsenic		1 (0.454)
D005 Barium		1000 (454)
D006 Cadmium		10 (4.54)
D007 Chromium		10 (4.54)
D008 Lead		10 (4.54)
D009 Mercury		1 (0.454)
D010 Selenium		10 (4.54)
D011 Silver		1 (0.454)
D012 Endrin		1 (0.454)
D013 Lindane		1 (0.454)
D014 Methoxyclor		1 (0.454)
D015 Toxaphene		1 (0.454)
D016 2,4-D		100 (45.4)
D017 2,4,5-TP		100 (45.4)
D018 Benzene		10 (4.54)
D019 Carbon tetrachloride		10 (4.54)
D020 Chlordane		1 (0.454)
D021 Chlorobenzene		100 (45.4)
D022 Chloroform		10 (4.54)
D023 o-Cresol		100 (45.4)
D024 m-Cresol		100 (45.4)
D025 p-Cresol		100 (45.4)
D026 Cresol		100 (45.4)
D027 1,4-Dichlorobenzene		100 (45.4)
D028 1,2-Dichloroethane		100 (45.4)
D029 1,1-Dichloroethylene		100 (45.4)
D030 2,4-Dinitrotoluene		10 (4.54)
D031 Heptachlor (and hydroxide)		1 (0.454)

TABLE 1

Hazardous Substances	Synonyms	(RQ)
D032 Hexachlorobenzene	....	10 (4.54)
D033 Hexachlorobutadiene	....	1 (0.454)
D034 Hexachloroethane	....	100 (45.4)
D035 Methyl ethyl ketone	....	5000 (2270)
D036 Nitrobenzene	....	1000 (454)
D037 Pentachlorophenol	....	10 (4.54)
D038 Pyridine	....	1000 (454)
D039 Tetrachloroethylene	....	100 (45.4)
D040 Trichloroethylene	....	100 (45.4)
D041 2,4,5-Trichlorophenol	....	10 (4.54)
D042 2,4,6-Trichlorophenol	....	10 (4.54)
D043 Vinyl Chloride	....	1 (0.454)
F001 .... The following spent halogenated solvents used in degreasing: all spent, solvent mixtures/blends used in degreasing containing, before use, a total of ten percent or more (by volume) of one or more of the below listed halogenated solvents or those solvents listed in F002, F004, and F005; and stillbottoms from the recovery of these spent solvents and spent solvent mixtures.	....	10 (4.54)
(a) Tetrachloroethylene		
(b) Trichloroethylene	....	100 (45.4)
(c) Methylene chloride	....	100 (45.4)
(d) 1,1,1-Trichloroethane	....	1000 (454)
(e) Carbon tetrachloride	....	1000 (454)
(f) Chlorinated fluorocarbons	....	10 (4.54)
	....	5000 (2270)

TABLE 1

Hazardous Substances	Synonyms	(RQ)
F002 ... The following spent halogenated solvents; all spent solvent mixtures/blends containing, before use, a total of ten percent or more (by volume) of one or more of the below listed halogenated solvents or those listed in F001, F004, F005; and stillbottoms from the recovery of these spent solvents and spent solvent mixtures.		
(a) Tetrachloroethylene	....	100 (45.4)
(b) Methylene chloride	....	1000 (454)
(c) Trichloroethylene	....	100 (45.4)
(d) 1,1,1-Trichloroethane	....	1000 (454)
(e) Chlorobenzene	....	100 (45.4)
(f) 1,1,2-Trichloro-1,2,2- trifluoroethane	....	5000 (2270)
(g) o-Dichlorobenzene	....	100 (45.4)
(h) Trichlorofluoromethane	....	5000 (2270)
(i) 1,1,2 Trichloroethane	....	100 (45.4)
F003 ... The following spent non-halogenated solvents and solvents:		
(a) Xylene	....	1000 (454)
(b) Acetone	....	5000 (2270)
(c) Ethyl acetate	....	5000 (2270)
(d) Ethylbenzene	....	1000 (454)
(e) Ethyl ether	....	100 (45.4)
(f) Methyl isobutyl ketone	....	5000 (2270)
(g) n-Butyl alcohol	....	5000 (2270)
(h) Cyclohexanone	....	5000 (2270)
(i) Methanol	....	5000 (2270)
F004 ... The following spent non-halogenated solvents and the stillbottoms from the recovery of these solvents:		
(a) Cresols/Cresylic acid	....	1000 (454)
(b) Nitrobenzene	....	100 (45.4)
F005 ... The following spent non-halogenated solvents and the stillbottoms from the recovery of these solvents:		
(a) Toluene	....	1000 (454)
(b) Methyl ethyl ketone	....	5000 (2270)
(c) Carbon disulfide	....	100 (45.4)
(d) Isobutanol	....	5000 (2270)
(e) Pyridine	....	1000 (454)

TABLE 1

Hazardous Substances	Synonyms	(RQ)
F006 Wastewater treatment sludges from electroplating operations except from the following processes: (1) sulfuric acid anodizing of aluminum; (2) tin plating on carbon steel; (3) zinc plating (segregated basis) on carbon steel; (4) aluminum or zinc-aluminum plating on carbon steel; (5) cleaning/stripping associated with tin, zinc and aluminum plating on carbon steel; and (6) chemical etching and milling of aluminum.	....	10 (4.54)
F007 Spent cyanide plating bath solutions from electroplating operations.	....	10 (4.54)
F008 Plating bath residues from the bottom of plating baths from electroplating operations where cyanides are used in the process.	....	10 (4.54)
F009 Spent stripping and cleaning bath solutions from electroplating operations where cyanides are used in the process.	....	10 (4.54)
F010 Quenching bath residues from oil baths from metal heat treating operations where cyanides are used in the process.	....	10 (4.54)
F011 Spent cyanide solutions from salt bath pot cleaning from metal heat treating operations (except for precious metals heat treating spent cyanide solutions from salt bath pot cleaning).	....	10 (4.54)
F012 Quenching wastewater treatment sludges from metal heat treating operations where cyanides are used in the process.	....	10 (4.54)
F019 Wastewater treatment sludges from the chemical conversion coating of aluminum—except from zirconium phosphating in aluminum can washing when such phosphating is an exclusive conversion coating process.	....	10 (4.54)

TABLE 1

Hazardous Substances	Synonyms	(RQ)
F020 Wastes (except wastewater and spent carbon from hydrogen chloride purification) from the production or manufacturing use (as a reactant, chemical intermediate, or component in a formulating process) of tri- or tetrachlorophenol, or of intermediates used to produce their pesticide derivatives. (This listing does not include wastes from the production of hexachlorophene from highly purified 2,4,5-trichlorophenol.)	....	1 (0.454)
F021 Wastes (except wastewater and spent carbon from hydrogen chloride purification) from the production or manufacturing use (as a reactant, chemical intermediate, or component in a formulating process) of pentachlorophenol, or of intermediates used to produce its derivatives.	....	1 (0.454)
F022 Wastes (except wastewater and spent carbon from hydrogen chloride purification) from the manufacturing use (as a reactant, chemical intermediate, or component in a formulating process) of tetra-, penta-, or hexachlorobenzenes under alkaline conditions.	....	1 (0.454)
F023 Wastes (except wastewater and spent carbon from hydrogen chloride purification) from the production of materials on equipment previously used for the production or manufacturing use (as a reactant, chemical intermediate, or component in a formulating process) of tri- and tetrachlorophenols. (This listing does not include wastes from equipment used only for the production or use of hexachlorophene from highly purified 2,4, 5-trichlorophenol.)	....	1 (0.454)
F024 Wastes, including but not limited to distillation residues, heavy ends, tars, and reactor cleanout wastes, from the production of chlorinated aliphatic hydrocarbons, having carbon content from one to five, utilizing free radical catalyzed processes. (This listing does not include light ends, spent filters and filter aids, spent dessicants(sic), wastewater, wastewater treatment sludges, spent catalysts, and wastes listed in Section 261.32.)	....	1 (0.454)
F025 Condensed light ends, spent filters and filter aids, and spent desiccant wastes from the production of certain chlorinated aliphatic hydrocarbons, by free radical catalyzed processes. These chlorinated aliphatic hydrocarbons are those having carbon chain lengths ranging from one to and including five, with varying amounts and positions of chlorine substitution.	....	1 (0.454)

TABLE 1

Hazardous Substances	Synonyms	(RQ)
F026 ... Wastes (except wastewater and spent carbon from hydrogen chloride purification) from the production of materials on equipment previously used for the manufacturing use (as a reactant, chemical intermediate, or component in a formulating process) of tetra-, penta-, or hexachlorobenzene under alkaline conditions.		1 (0.454)
F027 ... Discarded unused formulations containing tri-, tetra-, or pentachlorophenol or discarded unused formulations containing compounds derived from these chlorophenols. (This listing does not include formulations containing hexachlorophene synthesized from prepurified 2,4,5-trichlorophenol as the sole component.)		1 (0.454)
F028 ... Residues resulting from the incineration or thermal treatment of soil contaminated with EPA Hazardous Waste Nos. F020, F021, F022, F023, F026, and F027.		1 (0.454)
F032 ...		1 (0.454)
F034 ...		1 (0.454)
F035 ...		1 (0.454)
F037 ...		1 (0.454)
F038 ...		1 (0.454)
F039 ... Multi source leachate.		1 (0.454)
K001 ... Bottom sediment sludge from the treatment of wastewaters from wood preserving processes that use creosote and/or pentachlorophenol.		10 (4.54)
K002 ... Wastewater treatment sludge from the production of chrome yellow and orange pigments.		10 (4.54)
K003 ... Wastewater treatment sludge from the production of molybdate orange pigments.		10 (4.54)
K004 ... Wastewater treatment sludge from the production of zinc yellow pigments.		10 (4.54)
K005 ... Wastewater treatment sludge from the production of chrome green pigments.		10 (4.54)

TABLE 1

Hazardous Substances	Synonyms	(RQ)
K006 Wastewater treatment sludge from the production of chrome oxide green pigments (anhydrous and hydrated).	....	10 (4.54)
K007 Wastewater treatment sludge from the production of iron blue pigments.	....	10 (4.54)
K008 Oven residue from the production of chrome oxide green pigments.	....	10 (4.54)
K009 Distillation bottoms from the production of acetaldehyde from ethylene.	....	10 (4.54)
K010 Distillation side cuts from the production of acetaldehyde from ethylene.	....	10 (4.54)
K011 Bottom stream from the wastewater stripper in the production of acrylonitrile.	....	10 (4.54)
K013 Bottom stream from the acetonitrile column in the production of acrylonitrile.	....	10 (4.54)
K014 Bottoms from the acetonitrile purification column in the production of acrylonitrile.	....	5000 (2270)
K015 Still bottoms from the distillation of benzyl chloride.	....	10 (4.54)
K016 Heavy ends or distillation residues from the production of carbon tetrachloride.	....	1 (0.454)
K017 Heavy ends (still bottoms) from the purification column in the production of epichlorohydrin.	....	10 (4.54)
K018 Heavy ends from the fractionation column in ethyl chloride production.	....	1 (0.454)
K019 Heavy ends from the distillation of ethylene dichloride in ethylene dichloride production.	....	1 (0.454)
K020 Heavy ends from the distillation of vinyl chloride in vinyl chloride monomer production.	....	1 (0.454)

TABLE 1

Hazardous Substances	Synonyms	(RQ)
K021 Aqueous spent antimony catalyst waste from fluoromethanes production.		10 (4.54)
K022 Distillation bottom tars from the production of phenol/acetone from cumene.		1 (0.454)
K023 Distillation light ends from the production of phthalic anhydride from naphthalene.		5000 (2270)
K024 Distillation bottoms from the production of phthalic anhydride from naphthalene.		5000 (2270)
K025 Distillation bottoms from the production of nitrobenzene by the nitration of benzene.		10 (4.54)
K026 Stripping still tails from the production of methyl ethyl pyridines.		1000 (454)
K027 Centrifuge and distillation residues from toluene diisocyanate production.		10 (4.54)
K028 Spent catalyst from the hydrochlorinator reactor in the production of 1,1,1-trichloroethane.		1 (0.454)
K029 Waste from the product steam stripper in the production of 1,1,1-trichloroethane.		1 (0.454)
K030 Column bottoms or heavy ends from the combined production of trichloroethylene and perchloroethylene.		1 (0.454)
K031 By-product salts generated in the production of MSMA and cacodylic acid.		1 (0.454)
K032 Wastewater treatment sludge from the production of chlordane.		10 (4.54)
K033 Wastewater and scrub water from the chlorination of cyclopentadiene in the production of chlordane.		10 (4.54)
K034 Filter solids from the filtration of hexachlorocyclopentadiene in the production of chlordane.		10 (4.54)

TABLE 1

Hazardous Substances	Synonyms	(RQ)
K035 Wastewater treatment sludges generated in the production of creosote.		1 (0.454)
K036 Still bottoms from toluene reclamation distillation in the production of disulfoton.		1 (0.454)
K037 Wastewater treatment sludges from the production of disulfoton.		1 (0.454)
K038 Wastewater from the washing and stripping of phorate production.		10 (4.54)
K039 Filter cake from the filtration of diethylphosphorodithioic acid in the production of phorate.		10 (4.54)
K040 Wastewater treatment sludge from the production of phorate.		10 (4.54)
K041 Wastewater treatment sludge from the production of toxaphene.		1 (0.454)
K042 Heavy ends or distillation residues from the distillation of tetrachlorobenzene in the production of 2,4,5-T.		10 (4.54)
K043 2,6-Dichlorophenol waste from the production of 2,4-D.		10 (4.54)
K044 Wastewater treatment sludges from the manufacturing and processing of explosives.		10 (4.54)
K045 Spent carbon from the treatment of wastewater containing explosives.		10 (4.54)
K046 Wastewater treatment sludges from the manufacturing, formulation and loading of lead-based initiating compounds.		10 (4.54)
K047 Pink/red water from TNT operations.		10 (4.54)
K048 Dissolved air flotation (DAF) float from the petroleum refining industry.		10 (4.54)

TABLE 1

Hazardous Substances	Synonyms	(RQ)
K049 Slop oil emulsion solids from the petroleum refining industry.		10 (4.54)
K050 Heat exchanger bundle cleaning sludge from the petroleum refining industry.		10 (4.54)
K051 API separator sludge from the petroleum refining industry.		10 (4.54)
K052 Tank bottoms (leaded) from the petroleum refining industry.		10 (4.54)
K060 Ammonia still lime sludge from coking operations.		1 (0.454)
K061 Emission control dust/sludge from the primary production of steel in electric furnaces.		10 (4.54)
K062 Spent pickle liquor generated by steel finishing operations of facilities within the iron and steel industry.		10 (4.54)
K064 Acid plant blowdown slurry/sludge resulting from thickening of blowdown slurry from primary copper production.		10 (4.54)
K065 Surface impoundment solids contained in and dredged from surface impoundments at primary lead smelting facilities.		10 (4.54)
K066 Sludge from treatment of process wastewater and/or acid plant blowdown from primary zinc production.		10 (4.54)
K069 Emission control dust/sludge from secondary lead smelting.		10 (4.54)
K071 Brine purification muds from the mercury cell process in chlorine production, where separately prepurified brine is not used.		1 (0.454)
K073 Chlorinated hydrocarbon waste from the purification step of the diaphragm cell process using graphite anodes in chlorine production.		10 (4.54)

TABLE 1

Hazardous Substances	Synonyms	(RQ)
K083 Distillation bottoms from aniline extraction.		100 (45.4)
K084 Wastewater treatment sludges generated during the production of veterinary pharmaceuticals from arsenic or organo-arsenic compounds.		1 (0.454)
K085 Distillation or fractionation column bottoms from the production of chlorobenzenes.		10 (4.54)
K086 Solvent washes and sludges, caustic washes and sludges, or water washes and sludges from cleaning tubs and equipment used in the formulation of ink from pigments, driers, soaps, and stabilizers containing chromium and lead.		10 (4.54)
K087 Decanter tank tar sludge from coking operations.		100 (45.4)
K088 Spent potliners from primary aluminum reduction.		10 (4.54)
K090 Emission control dust or sludge from ferrochromiumsilicon production.		10 (4.54)
K091 Emission control dust or sludge from ferrochromium production.		10 (4.54)
K093 Distillation light ends from the production of phthalic anhydride from ortho-xylene.		5000 (2270)
K094 Distillation bottoms from the production of phthalic anhydride from ortho-xylene.		5000 (2270)
K095 Distillation bottoms from the production of 1,1,1-trichloroethane.		100 (45.4)
K096 Heavy ends from the heavy ends column from the production of 1,1,1-trichloroethane.		100 (45.4)
K097 Vacuum stripper discharge from the chlordane chlorinator in the production of chlordane.		1 (0.454)

TABLE 1

Hazardous Substances	Synonyms	(RQ)
K098 Untreated process wastewater from the production of toxaphene.		1 (0.454)
K099 Untreated wastewater from the production of 2,4-D.		10 (4.54)
K100 Waste leaching solution from acid leaching of emission control dust/sludge from secondary lead smelting.		10 (4.54)
K101 Distillation tar residues from the distillation of aniline-based compounds in the production of veterinary pharmaceuticals from arsenic or organo-arsenic compounds.		1 (0.454)
K102 Residue from the use of activated carbon for decolorization in the production of veterinary pharmaceuticals from arsenic or organo-arsenic compounds.		1 (0.454)
K103 Process residues from aniline extraction from the production of aniline.		100 (45.4)
K104 Combined wastewater streams generated from nitrobenzene/aniline chlorobenzenes.		10 (4.54)
K105 Separated aqueous stream from the reactor product washing step in the production of chlorobenzenes.		10 (4.54)
K106 Wastewater treatment sludge from the mercury cell process in chlorine production.		1 (0.454)
K107 Column bottoms from product seperation from the production of 1,1-dimethylhydrazine (UDMH) from carboxylic acid hydrazines.		10 (4.54)
K108 Condensed column overheads from product seperation and condensed reactor vent gases from the production of 1,1-dimethylhydrazine (UDMH) from carboxylic acid hydrazides.		10 (4.54)
K109 Spent filter cartridges from product purification from the production of 1,1-dimethylhydrazine (UDMH), from carboxylic acid hydrazides.		10 (4.54)

TABLE 1

Hazardous Substances	Synonyms	(RQ)
K110 Condensed column overheads from intermediate separation from the production of 1,1-dimethylhydrazines (UDMH) from carboxylic acid hydrazides.		10 (4.54)
K111 Product washwaters from the production of dinitrotoluene via nitration of toluene.		10 (4.54)
K112 Reaction by-product water from the drying column in the production of toluenediamine via hydrogenation of dinitrotoluene.		10 (4.54)
K113 Condensed liquid light ends from the purification of toluenediamine in the production of toluenediamine via hydrogenation of dinitrotoluene.		10 (4.54)
K114 Vicinals from the purification of toluenediamine in the production of toluenediamine via hydrogenation of dinitrotoluene.		10 (4.54)
K115 Heavy ends from the purification of toluenediamine in the production of toluenediamine via hydrogenation of dinitrotoluene.		10 (4.54)
K116 Organic condensate from the solvent recovery column in the production of toluene diisocyanate via phosgenation of toluenediamine.		10 (4.54)
K117 Wastewater from the reaction vent gas scrubber in the production of ethylene bromide via bromination of ethene.		1 (0.454)
K118 Spent absorbent solids from purification of ethylene dibromide in the production of ethylene dibromide.		1 (0.454)
K123 Process wastewater (including supernates, filtrates, and washwaters) from the production of ethylenebisdithiocarbamic acid and its salts.		10 (4.54)
K124 Reactor vent scrubber water from the production of ethylenebisdithiocarbamic acid and it salts.		10 (4.54)
K125 Filtration, evaporation, and centrifugation solids from the production of ethylenebisdithiocarbamic acid and its salts.		10 (4.54)

TABLE 1

Hazardous Substances	Synonyms	(RQ)
K126 Baghouse dust and floor sweepings in milling and packaging operations from the production or formulation of ethylenebisdithiocarbamic acid and its salts.	....	10 (4.54)
K131 Waste water from the reactor and spent sulfuric acid from the acid dryer in the production of methyl bromide.	....	100 (45.4)
K132 Spent absorbent and wastewater solids from the production of methyl bromide.	....	1000 (454)
K136 Still bottoms from the purification of ethylene dibromide in the production of ethylene dibromide via bromination of ethene.	....	1 (0.454)
K141	....	1 (0.454)
K142	....	1 (0.454)
K143	....	1 (0.454)
K144	....	1 (0.454)
K145	....	1 (0.454)
K147	....	1 (0.454)
K148	....	1 (0.454)
K149	....	10 (4.54)
K150	....	10 (4.54)
K151	....	10 (4.54)

Footnotes:

¢ the RQ for these hazardous substances is limited to those pieces of the metal having a diameter smaller than 100 micrometers (0.004 inches)

¢¢ the RQ for asbestos is limited to friable forms only

@ indicates that the name was added by RSPA because (1) the name is a synonym for a specific hazardous substance and (2) the name appears in the Hazardous Materials Table as a proper shipping name.

TABLE 2

(1)-Radionuclides	(2)-Atomic number	(3)-Reportable quantity (RQ) Ci (TBq)
Actinium-224	89	100 (3.7)
Actinium-225	89	1 (.037)
Actinium-226	89	10 (.37)
Actinium-227	89	0.001 (.000037)
Actinium-228	89	10 (.37)
Aluminum-26	13	10 (.37)
Americium-237	95	1000 (37)
Americium-238	95	100 (3.7)
Americium-239	95	100 (3.7)
Americium-240	95	10 (.37)
Americium-241	95	0.01 (.00037)
Americium-242	95	100 (3.7)
Americium-242m	95	0.01 (.00037)
Americium-243	95	0.01 (.00037)
Americium-244	95	10 (.37)
Americium-244m	95	1000 (37)
Americium-245	95	1000 (37)
Americium-246	95	1000 (37)
Americium-246m	95	1000 (37)
Antimony-115	51	1000 (37)
Antimony-116	51	1000 (37)
Antimony-116m	51	100 (3.7)
Antimony-117	51	1000 (37)
Antimony-118m	51	10 (.37)
Antimony-119	51	1000 (37)
Antimony-120 (16 min)	51	1000 (37)
Antimony-120 (5.76 day)	51	10 (.37)
Antimony-122	51	10 (.37)
Antimony-124	51	10 (.37)
Antimony-124m	51	1000 (37)

TABLE 2

(1)-Radionuclides	(2)-Atomic number	(3)-Reportable quantity (RQ) Ci (TBq)
Antimony-125	51	10 (.37)
Antimony-126	51	10 (.37)
Antimony-126m	51	1000 (37)
Antimony-127	51	10 (.37)
Antimony-128 (10.4 min)	51	1000 (37)
Antimony-128 (9.01 hr)	51	10 (.37)
Antimony-129	51	100 (3.7)
Antimony-130	51	100 (3.7)
Antimony-131	51	1000 (37)
Argon-39	18	1000 (37)
Argon-41	18	10 (.37)
Arsenic-69	33	1000 (37)
Arsenic-70	33	100 (3.7)
Arsenic-71	33	100 (3.7)
Arsenic-72	33	10 (.37)
Arsenic-73	33	100 (3.7)
Arsenic-74	33	10 (.37)
Arsenic-76	33	100 (3.7)
Arsenic-77	33	1000 (37)
Arsenic-78	33	100 (3.7)
Astatine-207	85	100 (3.7)
Astatine-211	85	100 (3.7)
Barium-126	56	1000 (37)
Barium-128	56	10 (.37)
Barium-131	56	10 (.37)
Barium-131m	56	1000 (37)
Barium-133	56	10 (.37)
Barium-133m	56	100 (3.7)
Barium-135m	56	1000 (37)
Barium-139	56	1000 (37)

TABLE 2

(1)-Radionuclides	(2)-Atomic number	(3)-Reportable quantity (RQ) Ci (TBq)
Barium-140	56	10 (.37)
Barium-141	56	1000 (37)
Barium-142	56	1000 (37)
Berkelium-245	97	100 (3.7)
Berkelium-246	97	10 (.37)
Berkelium-247	97	0.01 (.00037)
Berkelium-249	97	1 (.037)
Berkelium-250	97	100 (3.7)
Beryllium-7	4	100 (3.7)
Beryllium-10	4	1 (.037)
Bismuth-200	83	100 (3.7)
Bismuth-201	83	100 (3.7)
Bismuth-202	83	1000 (37)
Bismuth-203	83	10 (.37)
Bismuth-205	83	10 (.37)
Bismuth-206	83	10 (.37)
Bismuth-207	83	10 (.37)
Bismuth-210	83	10 (.37)
Bismuth-210m	83	0.1 (.0037)
Bismuth-212	83	100 (3.7)
Bismuth-213	83	100 (3.7)
Bismuth-214	83	100 (3.7)
Bromine-74	35	100 (3.7)
Bromine-74m	35	100 (3.7)
Bromine-75	35	100 (3.7)
Bromine-76	35	10 (.37)
Bromine-77	35	100 (3.7)
Bromine-80m	35	1000 (37)
Bromine-80	35	1000 (37)
Bromine-82	35	10 (.37)

TABLE 2

(1)-Radionuclides	(2)-Atomic number	(3)-Reportable quantity (RQ) Ci (TBq)
Bromine-83	35	1000 (37)
Bromine-84	35	100 (3.7)
Cadmium-104	48	1000 (37)
Cadmium-107	48	1000 (37)
Cadmium-109	48	1 (.037)
Cadmium-113	48	0.1 (.0037)
Cadmium-113m	48	0.1 (.0037)
Cadmium-115	48	100 (3.7)
Cadmium-115m	48	10 (.37)
Cadmium-117	48	100 (3.7)
Cadmium-117m	48	10 (.37)
Calcium-41	20	10 (.37)
Calcium-45	20	10 (.37)
Calcium-47	20	10 (.37)
Californium-244	98	1000 (37)
Californium-246	98	10 (.37)
Californium-248	98	0.1 (.0037)
Californium-249	98	0.01 (.00037)
Californium-250	98	0.01 (.00037)
Californium-251	98	0.01 (.00037)
Californium-252	98	0.1 (.0037)
Californium-253	98	10 (.37)
Californium-254	98	0.1 (.0037)
Carbon-11	6	1000 (37)
Carbon-14	6	10 (.37)
Cerium-134	58	10 (.37)
Cerium-135	58	10 (.37)
Cerium-137	58	1000 (37)
Cerium-137m	58	100 (3.7)
Cerium-139	58	100 (3.7)

(1)-Radionuclides	(2)-Atomic number	(3)-Reportable quantity (RQ) Ci (TBq)
Cerium-141	58	10 (.37)
Cerium-143	58	100 (3.7)
Cerium-144	58	1 (.037)
Cesium-125	55	1000 (37)
Cesium-127	55	100 (3.7)
Cesium-129	55	100 (3.7)
Cesium-130	55	1000 (37)
Cesium-131	55	1000 (37)
Cesium-132	55	10 (.37)
Cesium-134	55	1 (.037)
Cesium-134m	55	1000 (37)
Cesium-135	55	10 (.37)
Cesium-135m	55	100 (3.7)
Cesium-136	55	10 (.37)
Cesium-137	55	1 (.037)
Cesium-138	55	100 (3.7)
Chlorine-36	17	10 (.37)
Chlorine-38	17	100 (3.7)
Chlorine-39	17	100 (3.7)
Chromium-48	24	100 (3.7)
Chromium-49	24	1000 (37)
Chromium-49	24	1000 (37)
Chromium-51	24	1000 (37)
Cobalt-55	27	10 (.37)
Cobalt-56	27	10 (.37)
Cobalt-57	27	100 (3.7)
Cobalt-58	27	10 (.37)
Cobalt-58m	27	1000 (37)
Cobalt-60	27	10 (.37)
Cobalt-60m	27	1000 (37)

TABLE 2

(1)-Radionuclides	(2)-Atomic number	(3)-Reportable quantity (RQ) Ci (TBq)
Cobalt-61	27	1000 (37)
Cobalt-62m	27	1000 (37)
Copper-60	29	100 (3.7)
Copper-61	29	100 (3.7)
Copper-64	29	1000 (37)
Copper-67	29	100 (3.7)
Curium-238	96	1000 (37)
Curium-240	96	1 (.037)
Curium-241	96	10 (.37)
Curium-242	96	1 (.037)
Curium-243	96	0.01 (.00037)
Curium-244	96	0.01 (.00037)
Curium-245	96	0.01 (.00037)
Curium-246	96	0.01 (.00037)
Curium-247	96	0.01 (.00037)
Curium-248	96	0.001 (.000037)
Curium-249	96	1000 (37)
Dysprosium-155	66	100 (3.7)
Dysprosium-157	66	100 (3.7)
Dysprosium-159	66	100 (3.7)
Dysprosium-165	66	1000 (37)
Dysprosium-166	66	10 (.37)
Einsteinium-250	99	10 (.37)
Einsteinium-251	99	1000 (37)
Einsteinium-253	99	10 (.37)
Einsteinium-254	99	0.1 (.0037)
Einsteinium-254m	99	1 (.037)
Erbium-161	68	100 (3.7)
Erbium-165	68	1000 (37)
Erbium-169	68	100 (3.7)

TABLE 2

(1)-Radionuclides	(2)-Atomic number	(3)-Reportable quantity (RQ) Ci (TBq)
Erbium-171	68	100 (3.7)
Erbium-172	68	10 (.37)
Europium-145	63	10 (.37)
Europium-146	63	10 (.37)
Europium-147	63	10 (.37)
Europium-148	63	10 (.37)
Europium-149	63	100 (3.7)
Europium-150 (12.6 hr)	63	1000 (37)
Europium-150 (34.2 yr)	63	10 (.37)
Europium-152	63	10 (.37)
Europium-152m	63	100 (3.7)
Europium-154	63	10 (.37)
Europium-155	63	10 (.37)
Europium-156	63	10 (.37)
Europium-157	63	10 (.37)
Europium-158	63	1000 (37)
Fermium-252	100	10 (.37)
Fermium-253	100	10 (.37)
Fermium-254	100	100 (3.7)
Fermium-255	100	100 (3.7)
Fermium-257	100	1 (.037)
Fluorine-18	9	1000 (37)
Francium-222	87	100 (3.7)
Francium-223	87	100 (3.7)
Gadolinium-145	64	100 (3.7)
Gadolinium-146	64	10 (.37)
Gadolinium-147	64	10 (.37)
Gadolinium-148	64	0.001 (.000037)
Gadolinium-149	64	100 (3.7)
Gadolinium-151	64	100 (3.7)

TABLE 2

(1)-Radionuclides	(2)-Atomic number	(3)-Reportable quantity (RQ) Ci (TBq)
Gadolinium-152	64	0.001 (.000037)
Gadolinium-153	64	10 (.37)
Gadolinium-159	64	1000 (37)
Gallium-65	31	1000 (37)
Gallium-66	31	10 (.37)
Gallium-67	31	100 (3.7)
Gallium-68	31	1000 (37)
Gallium-70	31	1000 (37)
Gallium-72	31	10 (.37)
Gallium-73	31	100 (3.7)
Germanium-66	32	100 (3.7)
Germanium-67	32	1000 (37)
Germanium-68	32	10 (.37)
Germanium-69	32	10 (.37)
Germanium-71	32	1000 (37)
Germanium-75	32	1000 (37)
Germanium-77	32	10 (.37)
Germanium-78	32	1000 (37)
Gold-193	79	100 (3.7)
Gold-194	79	10 (.37)
Gold-195	79	100 (3.7)
Gold-198	79	100 (3.7)
Gold-198m	79	10 (.37)
Gold-199	79	100 (3.7)
Gold-200	79	1000 (37)
Gold-200m	79	10 (.37)
Gold-201	79	1000 (37)
Hafnium-170	72	100 (3.7)
Hafnium-172	72	1 (.037)
Hafnium-173	72	100 (3.7)

TABLE 2

(1)-Radionuclides	(2)-Atomic number	(3)-Reportable quantity (RQ) Ci (TBq)
Hafnium-175	72	100 (3.7)
Hafnium-177m	72	1000 (37)
Hafnium-178m	72	0.1 (.0037)
Hafnium-179m	72	100 (3.7)
Hafnium-180m	72	100 (3.7)
Hafnium-181	72	10 (.37)
Hafnium-182	72	0.1 (.0037)
Hafnium-182m	72	100 (3.7)
Hafnium-183	72	100 (3.7)
Hafnium-184	72	100 (3.7)
Holmium-155	67	1000 (37)
Holmium-157	67	1000 (37)
Holmium-159	67	1000 (37)
Holmium-161	67	1000 (37)
Holmium-162	67	1000 (37)
Holmium-162m	67	1000 (37)
Holmium-164	67	1000 (37)
Holmium-164m	67	1000 (37)
Holmium-166	67	100 (3.7)
Holmium-166m	67	1 (.037)
Holmium-167	67	100 (3.7)
Hydrogen-3	1	100 (3.7)
Indium-109	49	100 (3.7)
Indium-110 (69.1 min)	49	100 (3.7)
Indium-110 (4.9 hr)	49	10 (.37)
Indium-111	49	100 (3.7)
Indium-112	49	1000 (37)
Indium-113m	49	1000 (37)
Indium-114m	49	10 (.37)
Indium-115	49	0.1 (.0037)

TABLE 2

(1)-Radionuclides	(2)-Atomic number	(3)-Reportable quantity (RQ) Ci (TBq)
Indium-115m	49	100 (3.7)
Indium-116m	49	100 (3.7)
Indium-117	49	1000 (37)
Indium-117m	49	100 (3.7)
Indium-119m	49	1000 (37)
Iodine-120	53	10 (.37)
Iodine-120m	53	100 (3.7)
Iodine-121	53	100 (3.7)
Iodine-123	53	10 (.37)
Iodine-124	53	0.1 (.0037)
Iodine-125	53	0.01 (.00037)
Iodine-126	53	0.01 (.00037)
Iodine-128	53	1000 (37)
Iodine-129	53	0.001 (.000037)
Iodine-130	53	1 (.037)
Iodine-131	53	0.01 (.00037)
Iodine-132	53	10 (.37)
Iodine-132m	53	10 (.37)
Iodine-133	53	0.1 (.0037)
Iodine-134	53	100 (3.7)
Iodine-135	53	10 (.37)
Iridium-182	77	1000 (37)
Iridium-184	77	100 (3.7)
Iridium-185	77	100 (3.7)
Iridium-186	77	10 (.37)
Iridium-187	77	100 (3.7)
Iridium-188	77	10 (.37)
Iridium-189	77	100 (3.7)
Iridium-190	77	10 (.37)
Iridium-190m	77	1000 (37)

TABLE 2

(1)-Radionuclides	(2)-Atomic number	(3)-Reportable quantity (RQ) Ci (TBq)
Iridium-192	77	10 (.37)
Iridium-192m	77	100 (3.7)
Iridium-194	77	100 (3.7)
Iridium-194m	77	10 (.37)
Iridium-195	77	1000 (37)
Iridium-195m	77	100 (3.7)
Iron-52	26	100 (3.7)
Iron-55	26	100 (3.7)
Iron-59	26	10 (.37)
Iron-60	26	0.1 (.0037)
Krypton-74	36	10 (.37)
Krypton-76	36	10 (.37)
Krypton-77	36	10 (.37)
Krypton-79	36	100 (3.7)
Krypton-81	36	1000 (37)
Krypton-83m	36	1000 (37)
Krypton-85	36	1000 (37)
Krypton-85m	36	100 (3.7)
Krypton-87	36	10 (.37)
Krypton-88	36	10 (.37)
Lanthanum-131	57	1000 (37)
Lanthanum-132	57	100 (3.7)
Lanthanum-135	57	1000 (37)
Lanthanum-137	57	10 (.37)
Lanthanum-138	57	1 (.037)
Lanthanum-140	57	10 (.37)
Lanthanum-141	57	1000 (37)
Lanthanum-142	57	100 (3.7)
Lanthanum-143	57	1000 (37)
Lead-195m	82	1000 (37)

TABLE 2

(1)-Radionuclides	(2)-Atomic number	(3)-Reportable quantity (RQ) Ci (TBq)
Lead-198	82	100 (3.7)
Lead-199	82	100 (3.7)
Lead-200	82	100 (3.7)
Lead-201	82	100 (3.7)
Lead-202	82	1 (.037)
Lead-202m	82	10 (.37)
Lead-203	82	100 (3.7)
Lead-205	82	100 (3.7)
Lead-209	82	1000 (37)
Lead-210	82	0.01 (.00037)
Lead-211	82	100 (3.7)
Lead-212	82	10 (.37)
Lead-214	82	100 (3.7)
Lutetium-169	71	10 (.37)
Lutetium-170	71	10 (.37)
Lutetium-171	71	10 (.37)
Lutetium-172	71	10 (.37)
Lutetium-173	71	100 (3.7)
Lutetium-174	71	10 (.37)
Lutetium-174m	71	10 (.37)
Lutetium-176	71	1 (.037)
Lutetium-176m	71	1000 (37)
Lutetium-177	71	100 (3.7)
Lutetium-177m	71	10 (.37)
Lutetium-178	71	1000 (37)
Lutetium-178m	71	1000 (37)
Lutetium-179	71	1000 (37)
Magnesium-28	12	10 (.37)
Manganese-51	25	1000 (37)
Manganese-52	25	10 (.37)

TABLE 2

(1)-Radionuclides	(2)-Atomic number	(3)-Reportable quantity (RQ) Ci (TBq)
Manganese-52m	25	1000 (37)
Manganese-53	25	1000 (37)
Manganese-54	25	10 (.37)
Manganese-56	25	100 (3.7)
Mendelevium-257	101	100 (3.7)
Mendelevium-258	101	1 (.037)
Mercury-193	80	100 (3.7)
Mercury-193m	80	10 (.37)
Mercury-194	80	0.1 (.0037)
Mercury-195	80	100 (3.7)
Mercury-195m	80	100 (3.7)
Mercury-197	80	1000 (37)
Mercury-197m	80	1000 (37)
Mercury-199m	80	1000 (37)
Mercury-203	80	10 (.37)
Molybdenum-90	42	100 (3.7)
Molybdenum-93	42	100 (3.7)
Molybdenum-93m	42	10 (.37)
Molybdenum-99	42	100 (3.7)
Molybdenum-101	42	1000 (37)
Neodymium-136	60	1000 (37)
Neodymium-138	60	1000 (37)
Neodymium-139	60	1000 (37)
Neodymium-139m	60	100 (3.7)
Neodymium-141	60	1000 (37)
Neodymium-147	60	10 (.37)
Neodymium-149	60	100 (3.7)
Neodymium-151	60	1000 (37)
Neptunium-232	93	1000 (37)
Neptunium-233	93	1000 (37)

TABLE 2

(1)-Radionuclides	(2)-Atomic number	(3)-Reportable quantity (RQ) Ci (TBq)
Neptunium-234	93	10 (.37)
Neptunium-235	93	1000 (37)
Neptunium-236 (1.2 E 5 yr)	93	0.1 (.0037)
Neptunium-236 (22.5 hr)	93	100 (3.7)
Neptunium-237	93	0.01 (.00037)
Neptunium-238	93	10 (.37)
Neptunium-239	93	100 (3.7)
Neptunium-240	93	100 (3.7)
Nickel-56	28	10 (.37)
Nickel-57	28	10 (.37)
Nickel-59	28	100 (3.7)
Nickel-63	28	100 (3.7)
Nickel-65	28	100 (3.7)
Nickel-66	28	10 (.37)
Niobium-88	41	100 (3.7)
Niobium-89 (66 min)	41	100 (3.7)
Niobium-89 (122 min)	41	100 (3.7)
Niobium-90	41	10 (.37)
Niobium-93m	41	100 (3.7)
Niobium-94	41	10 (.37)
Niobium-95	41	10 (.37)
Niobium-95m	41	100 (3.7)
Niobium-96	41	10 (.37)
Niobium-97	41	100 (3.7)
Niobium-98	41	1000 (37)
Osmium-180	76	1000 (37)
Osmium-181	76	100 (3.7)
Osmium-182	76	100 (3.7)
Osmium-185	76	10 (.37)
Osmium-189m	76	1000 (37)

TABLE 2

(1)-Radionuclides	(2)-Atomic number	(3)-Reportable quantity (RQ) Ci (TBq)
Osmium-191	76	100 (3.7)
Osmium-191m	76	1000 (37)
Osmium-193	76	100 (3.7)
Osmium-194	76	1 (.037)
Palladium-100	46	100 (3.7)
Palladium-101	46	100 (3.7)
Palladium-103	46	100 (3.7)
Palladium-107	46	100 (3.7)
Palladium-109	46	1000 (37)
Phosphorus-32	15	0.1 (.0037)
Phosphorus-33	15	1 (.037)
Platinum-186	78	100 (3.7)
Platinum-188	78	100 (3.7)
Platinum-189	78	100 (3.7)
Platinum-191	78	100 (3.7)
Platinum-193	78	1000 (37)
Platinum-193m	78	100 (3.7)
Platinum-195m	78	100 (3.7)
Platinum-197	78	1000 (37)
Platinum-197m	78	1000 (37)
Platinum-199	78	1000 (37)
Platinum-200	78	100 (3.7)
Plutonium-234	94	1000 (37)
Plutonium-235	94	1000 (37)
Plutonium-236	94	0.1 (.0037)
Plutonium-237	94	1000 (37)
Plutonium-238	94	0.01 (.00037)
Plutonium-239	94	0.01 (.00037)
Plutonium-240	94	0.01 (.00037)
Plutonium-241	94	1 (.037)

TABLE 2

(1)-Radionuclides	(2)-Atomic number	(3)-Reportable quantity (RQ) Ci (TBq)
Plutonium-242	94	0.01 (.00037)
Plutonium-243	94	1000 (37)
Plutonium-244	94	0.01 (.00037)
Plutonium-245	94	100 (3.7)
Polonium-203	84	100 (3.7)
Polonium-205	84	100 (3.7)
Polonium-207	84	10 (.37)
Polonium-210	84	0.01 (.00037)
Potassium-40	19	1 (.037)
Potassium-42	19	100 (3.7)
Potassium-43	19	10 (.37)
Potassium-44	19	100 (3.7)
Potassium-45	19	1000 (37)
Praseodymium-136	59	1000 (37)
Praseodymium-137	59	1000 (37)
Praseodymium-138m	59	100 (3.7)
Praseodymium-139	59	1000 (37)
Praseodymium-142	59	100 (3.7)
Praseodymium-142m	59	1000 (37)
Praseodymium-143	59	10 (.37)
Praseodymium-144	59	1000 (37)
Praseodymium-145	59	1000 (37)
Praseodymium-147	59	1000 (37)
Promethium-141	61	1000 (37)
Promethium-143	61	100 (3.7)
Promethium-144	61	10 (.37)
Promethium-145	61	100 (3.7)
Promethium-146	61	10 (.37)
Promethium-147	61	10 (.37)
Promethium-148	61	10 (.37)

TABLE 2

(1)-Radionuclides	(2)-Atomic number	(3)-Reportable quantity (RQ) Ci (TBq)
Promethium-148m	61	10 (.37)
Promethium-149	61	100 (3.7)
Promethium-150	61	100 (3.7)
Promethium-151	61	100 (3.7)
Protactinium-227	91	100 (3.7)
Protactinium-228	91	10 (.37)
Protactinium-230	91	10 (.37)
Protactinium-231	91	0.01 (.00037)
Protactinium-232	91	10 (.37)
Protactinium-233	91	100 (3.7)
Protactinium-234	91	10 (.37)
RADIONUCLIDES§†	....	1 (.037)
Radium-223	88	1 (.037)
Radium-224	88	10 (.37)
Radium-225	88	1 (.037)
Radium-226**	88	0.1 (.0037)
Radium-227	88	1000 (37)
Radium-228	88	0.1 (.0037)
Radon-220	86	0.1 (.0037)
Radon-222	86	0.1 (.0037)
Rhenium-177	75	1000 (37)
Rhenium-178	75	1000 (37)
Rhenium-181	75	100 (3.7)
Rhenium-182 (12.7 hr)	75	10 (.37)
Rhenium-182 (64.0 hr)	75	10 (.37)
Rhenium-184	75	10 (.37)
Rhenium-184m	75	10 (.37)
Rhenium-186	75	100 (3.7)
Rhenium-186m	75	10 (.37)
Rhenium-187	75	1000 (37)

TABLE 2

(1)-Radionuclides	(2)-Atomic number	(3)-Reportable quantity (RQ) Ci (TBq)
Rhenium-188	75	1000 (37)
Rhenium-188m	75	1000 (37)
Rhenium-189	75	1000 (37)
Rhodium-99	45	10 (.37)
Rhodium-99m	45	100 (3.7)
Rhodium-100	45	10 (.37)
Rhodium-101	45	10 (.37)
Rhodium-101m	45	100 (3.7)
Rhodium-102	45	10 (.37)
Rhodium-102m	45	10 (.37)
Rhodium-103m	45	1000 (37)
Rhodium-105	45	100 (3.7)
Rhodium-106m	45	10 (.37)
Rhodium-107	45	1000 (37)
Rubidium-79	37	1000 (37)
Rubidium-81	37	100 (3.7)
Rubidium-81m	37	1000 (37)
Rubidium-82m	37	10 (.37)
Rubidium-83	37	10 (.37)
Rubidium-84	37	10 (.37)
Rubidium-86	37	10 (.37)
Rubidium-88	37	1000 (37)
Rubidium-89	37	1000 (37)
Rubidium-87	37	10 (.37)
Ruthenium-94	44	1000 (37)
Ruthenium-97	44	100 (3.7)
Ruthenium-103	44	10 (.37)
Ruthenium-105	44	100 (3.7)
Ruthenium-106	44	1 (.037)
Samarium-141	62	1000 (37)

TABLE 2

(1)-Radionuclides	(2)-Atomic number	(3)-Reportable quantity (RQ) Ci (TBq)
Samarium-141m	62	1000 (37)
Samarium-142	62	1000 (37)
Samarium-145	62	100 (3.7)
Samarium-146	62	0.01 (.00037)
Samarium-147	62	0.01 (.00037)
Samarium-151	62	10 (.37)
Samarium-153	62	100 (3.7)
Samarium-155	62	1000 (37)
Samarium-156	62	100 (3.7)
Scandium-43	21	1000 (37)
Scandium-44	21	100 (3.7)
Scandium-44m	21	10 (.37)
Scandium-46	21	10 (.37)
Scandium-47	21	100 (3.7)
Scandium-48	21	10 (.37)
Scandium-49	21	1000 (37)
Selenium-70	34	1000 (37)
Selenium-73	34	10 (.37)
Selenium-73m	34	100 (3.7)
Selenium-75	34	10 (.37)
Selenium-79	34	10 (.37)
Selenium-81	34	1000 (37)
Selenium-81m	34	1000 (37)
Selenium-83	34	1000 (37)
Silicon-31	14	1000 (37)
Silicon-32	14	1 (.037)
Silver-102	47	100 (3.7)
Silver-103	47	1000 (37)
Silver-104	47	1000 (37)
Silver-104m	47	1000 (37)

TABLE 2

(1)-Radionuclides	(2)-Atomic number	(3)-Reportable quantity (RQ) Ci (TBq)
Silver-105	47	10 (.37)
Silver-106	47	1000 (.37)
Silver-106m	47	10 (.37)
Silver-108m	47	10 (.37)
Silver-110m	47	10 (.37)
Silver-111	47	10 (.37)
Silver-112	47	100 (3.7)
Silver-115	47	1000 (37)
Sodium-22	11	10 (.37)
Sodium-24	11	10 (.37)
Strontium-80	38	100 (3.7)
Strontium-81	38	1000 (37)
Strontium-83	38	100 (3.7)
Strontium-85	38	10 (.37)
Strontium-85m	38	1000 (37)
Strontium-87m	38	100 (3.7)
Strontium-89	38	10 (.37)
Strontium-90	38	0.1 (.0037)
Strontium-91	38	10 (.37)
Strontium-92	38	100 (3.7)
Sulfur-35	16	1 (.037)
Tantalum-172	73	100 (3.7)
Tantalum-173	73	100 (3.7)
Tantalum-174	73	100 (3.7)
Tantalum-175	73	100 (3.7)
Tantalum-176	73	10 (.37)
Tantalum-177	73	1000 (37)
Tantalum-178	73	1000 (37)
Tantalum-179	73	1000 (37)
Tantalum-180	73	100 (3.7)

TABLE 2

(1)-Radionuclides	(2)-Atomic number	(3)-Reportable quantity (RQ) Ci (TBq)
Tantalum-180m	73	1000 (37)
Tantalum-182	73	10 (.37)
Tantalum-182m	73	1000 (37)
Tantalum-183	73	100 (3.7)
Tantalum-184	73	10 (.37)
Tantalum-185	73	1000 (37)
Tantalum-186	73	1000 (37)
Technetium-93	43	100 (3.7)
Technetium-93m	43	1000 (37)
Technetium-94	43	10 (.37)
Technetium-94m	43	100 (3.7)
Technetium-96	43	10 (.37)
Technetium-96m	43	1000 (37)
Technetium-97	43	100 (3.7)
Technetium-97m	43	100 (3.7)
Technetium-98	43	10 (.37)
Technetium-99	43	10 (.37)
Technetium-99m	43	100 (3.7)
Technetium-101	43	1000 (37)
Technetium-104	43	1000 (37)
Tellurium-116	52	1000 (37)
Tellurium-121	52	10 (.37)
Tellurium-121m	52	10 (.37)
Tellurium-123	52	10 (.37)
Tellurium-123m	52	10 (.37)
Tellurium-125m	52	10 (.37)
Tellurium-127	52	1000 (37)
Tellurium-127m	52	10 (.37)
Tellurium-129	52	1000 (37)
Tellurium-129m	52	10 (.37)

TABLE 2

(1)-Radionuclides	(2)-Atomic number	(3)-Reportable quantity (RQ) Ci (TBq)
Tellurium-131	52	1000 (37)
Tellurium-131m	52	10 (.37)
Tellurium-132	52	10 (.37)
Tellurium-133	52	1000 (37)
Tellurium-133m	52	1000 (37)
Tellurium-134	52	1000 (37)
Terbium-147	65	100 (3.7)
Terbium-149	65	100 (3.7)
Terbium-150	65	100 (3.7)
Terbium-151	65	10 (.37)
Terbium-153	65	100 (3.7)
Terbium-154	65	10 (.37)
Terbium-155	65	100 (3.7)
Terbium-156m (5.0 hr)	65	1000 (37)
Terbium-156m (24.4 hr)	65	1000 (37)
Terbium-156	65	10 (.37)
Terbium-157	65	100 (3.7)
Terbium-158	65	10 (.37)
Terbium-160	65	10 (.37)
Terbium-161	65	100 (3.7)
Thallium-194	81	1000 (37)
Thallium-194m	81	100 (3.7)
Thallium-195	81	100 (3.7)
Thallium-197	81	100 (3.7)
Thallium-198	81	10 (.37)
Thallium-198m	81	100 (3.7)
Thallium-199	81	100 (3.7)
Thallium-200	81	10 (.37)
Thallium-201	81	1000 (37)
Thallium-202	81	10 (.37)

TABLE 2

(1)-Radionuclides	(2)-Atomic number	(3)-Reportable quantity (RQ) Ci (TBq)
Thallium-204	81	10 (.37)
Thorium (Irradiated)	90	***
Thorium (Natural)	90	**
Thorium-226	90	100 (3.7)
Thorium-227	90	1 (.037)
Thorium-228	90	0.01 (.00037)
Thorium-229	90	0.001 (.000037)
Thorium-230	90	0.01 (.00037)
Thorium-231	90	100 (3.7)
Thorium-232**	90	0.001 (.000037)
Thorium-234	90	100 (3.7)
Thulium-162	69	1000 (37)
Thulium-166	69	10 (.37)
Thulium-167	69	100 (3.7)
Thulium-170	69	10 (.37)
Thulium-171	69	100 (3.7)
Thulium-172	69	100 (3.7)
Thulium-173	69	100 (3.7)
Thulium-175	69	1000 (37)
Tin-110	50	100 (3.7)
Tin-111	50	1000 (37)
Tin-113	50	10 (.37)
Tin-117m	50	100 (3.7)
Tin-119m	50	10 (.37)
Tin-121	50	1000 (37)
Tin-121m	50	10 (.37)
Tin-123	50	10 (.37)
Tin-123m	50	1000 (37)
Tin-125	50	10 (.37)
Tin-126	50	1 (.037)

TABLE 2

(1)-Radionuclides	(2)-Atomic number	(3)-Reportable quantity (RQ) Ci (TBq)
Tin-127	50	100 (3.7)
Tin-128	50	1000 (37)
Titanium-44	22	1 (.037)
Titanium-45	22	1000 (37)
Tungsten-176	74	1000 (37)
Tungsten-177	74	100 (3.7)
Tungsten-178	74	100 (3.7)
Tungsten-179	74	1000 (37)
Tungsten-181	74	100 (3.7)
Tungsten-185	74	10 (.37)
Tungsten-187	74	100 (3.7)
Tungsten-188	74	10 (.37)
Uranium (Depleted)	92	***
Uranium (Irradiated)	92	***
Uranium (Natural)	92	**
Uranium Enriched 20% or greater	92	***
Uranium Enriched less than 20%	92	***
Uranium-230	92	1 (.037)
Uranium-231	92	1000 (37)
Uranium-232	92	0.01 (.00037)
Uranium-233	92	0.1 (.0037)
Uranium-234**	92	0.1 (.0037)
Uranium-235**	92	0.1 (.0037)
Uranium-236	92	0.1 (.0037)
Uranium-237	92	100 (3.7)
Uranium-238**	92	0.1 (.0037)
Uranium-239	92	1000 (37)
Uranium-240	92	1000 (37)

TABLE 2

(1)-Radionuclides	(2)-Atomic number	(3)-Reportable quantity (RQ) Ci (TBq)
Vanadium-47	23	1000 (37)
Vanadium-48	23	10 (.37)
Vanadium-49	23	1000 (37)
Xenon-120	54	100 (3.7)
Xenon-121	54	10 (.37)
Xenon-122	54	100 (3.7)
Xenon-123	54	10 (.37)
Xenon-125	54	100 (3.7)
Xenon-127	54	100 (3.7)
Xenon-129m	54	1000 (37)
Xenon-131m	54	1000 (37)
Xenon-133	54	1000 (37)
Xenon-133m	54	1000 (37)
Xenon-135	54	100 (3.7)
Xenon-135m	54	10 (.37)
Xenon-138	54	10 (.37)
Ytterbium-162	70	1000 (37)
Ytterbium-166	70	10 (.37)
Ytterbium-167	70	1000 (37)
Ytterbium-169	70	10 (.37)
Ytterbium-175	70	100 (3.7)
Ytterbium-177	70	1000 (37)
Ytterbium-178	70	1000 (37)
Yttrium-86	39	10 (.37)
Yttrium-86m	39	1000 (37)
Yttrium-87	39	10 (.37)
Yttrium-88	39	10 (.37)
Yttrium-90	39	10 (.37)
Yttrium-90m	39	100 (3.7)
Yttrium-91	39	10 (.37)

TABLE 2

(1)-Radionuclides	(2)-Atomic number	(3)-Reportable quantity (RQ) Ci (TBq)
Yttrium-91m	39	1000 (37)
Yttrium-92	39	100 (3.7)
Yttrium-93	39	100 (3.7)
Yttrium-94	39	1000 (37)
Yttrium-95	39	1000 (37)
Zinc-62	30	100 (3.7)
Zinc-63	30	1000 (37)
Zinc-65	30	10 (.37)
Zinc-69	30	1000 (37)
Zinc-69m	30	100 (3.7)
Zinc-71m	30	100 (3.7)
Zinc-72	30	100 (3.7)
Zirconium-86	40	100 (3.7)
Zirconium-88	40	10 (.37)
Zirconium-89	40	100 (3.7)
Zirconium-93	40	1 (.037)
Zirconium-95	40	10 (.37)
Zirconium-97	40	10 (.37)

§ The RQs for all radionuclides apply to chemical compounds containing the radionuclides and elemental forms regardless of the diameter of pieces of solid material.

† The RQ of one curie applies to all radionuclides not otherwise listed. Whenever the RQs in TABLE 1—HAZARDOUS SUBSTANCES OTHER THAN RADIONUCLIDES and this table conflict, the lowest RQ shall apply. For example, uranyl acetate and uranyl nitrate have RQs shown in TABLE 1 of 100 pounds, equivalent to about one-tenth the RQ level for uranium-238 in this table.

** The method to determine the RQs for mixtures or solutions of radionuclides can be found in paragraph 6 of the note preceding TABLE 1 of this Appendix. RQs for the following four common radionuclide mixtures are provided: radium-226 in secular equilibrium with its daughters (0.053 curie); natural uranium (0.1 curie); natural uranium in secular equilibrium with its daughters (0.052 curie); and natural thorium in secular equilibrium with its daughters (0.011 curie).

*** Indicates that the name was added by RSPA because it appears in the list of radionuclides in 49 CFR 173.435. The reportable quantity (RQ), if not specifically listed elsewhere in this Appendix, shall be determined in accordance with the procedures in Paragraph 6 of this Appendix.

Appendix B to §172.101 - List of Marine Pollutants

1. This appendix lists potential marine pollutants as defined in §171.8 of this subchapter.

2. If a marine pollutant meets the definition of any hazard class or division as defined in this subchapter, other than Class 9, the class of the material must be determined in accordance with §173.2a of this subchapter.

3. This appendix contains two columns. The first column, entitled "S.M.P." (for severe marine pollutants), identifies whether a material is a severe marine pollutant. If the letters "PP" appear in this column for a material, the material is a severe marine pollutant, otherwise it is not. The second column, entitled "Marine Pollutant", lists the marine pollutants.

4. If a material not listed in this appendix meets the criteria for a marine pollutant, as provided in the General Introduction of the IMDG Code, Guidelines for the Identification of Harmful Substances in Packaged Form, the material may be transported as a marine pollutant in accordance with the applicable requirements of this subchapter.

5. If approved by the Associate Administrator for Hazardous Materials Safety, a material listed in this appendix which does not meet the criteria for a marine pollutant, as provided in the General Introduction of the IMDG Code, Guidelines for the Identification of Harmful Substances in Packaged Form, is excepted from the requirements of this subchapter as a marine pollutant.

APPENDIX B TO §172.101

S.M.P.	Marine Pollutant
(1)	(2)
.....	Acetal
.....	Acetone cyanohydrin, stabilized
.....	Acetylene dibromide
.....	Acetylene tetrabromide
.....	Acetylene tetrachloride
.....	Acraldehyde, inhibited
.....	Acrolein, inhibited
.....	Acrylic aldehyde, inhibited
.....	Alcohol C-12 - C-15 poly(1-3) ethoxylate
.....	Alcohol C-13 - C-15 poly (1-6) ethoxylate
.....	Alcohol C-6 C-17 (secondary)poly(3-6) ethoxylate
.....	Aldicarb
PP ..	Aldrin
.....	Alkyl (C12-C14) dimethylamine
.....	Alkyl (C7-C9) nitrates
.....	Alkylphenols, liquid, n.o.s. (*including C2-C8 homologues*)
.....	Alkylphenols, solid, n.o.s. (*including C2-C8 homologues*)
.....	Allyl bromide
.....	ortho-Aminoanisole
.....	Aminocarb
.....	Ammonium dinitro-o-cresolate
.....	n-Amylbenzene
.....	Amyl mercaptans
.....	ortho-Anisidines
.....	Arsenates, liquid, n.o.s.
.....	Arsenates, solid, n.o.s.
.....	Arsenic bromide
.....	Arsenic chloride
.....	Arsenical pesticides, liquid, toxic, flammable, n.o.s.
PP ..	Azenphos-methyl
PP ..	Azinphos-ethyl

S.M.P.	Marine Pollutant
(1)	(2)
....	Barium compounds, soluble, n.o.s.
....	Barium cyanide
....	Bendiocarb
....	Benomyl
....	Benquinox
....	Benzyl chlorocarbonate
....	Benzyl chloroformate
PP..	Binapacryl
.....	Biphenyl phenyl ether and diphenyl oxide, mixtures
PP..	Brodifacoum
....	Bromine cyanide
....	Bromoacetone
....	Bromoallylene
....	ortho-Bromobenzyl cyanide
....	Bromocyane
....	Bromoform
PP..	Bromophos-ethyl
....	3-Bromopropene
....	Bromoxynil
....	1-Butanethiol
....	2-Butenal, inhibited
....	Butyl benzenes
....	Butyl benzyl phthalate
....	n-Butyl butyrate
....	Butylphenols, liquid
....	Butylphenols, solid
....	para-tertiary-butyltoluene
PP..	Cadmium compounds
....	Cadmium sulphide
....	Calcium arsenate
....	Calcium arsenate and calcium arsenite, mixtures, solid

APPENDIX B TO §172.101

S.M.P.	Marine Pollutant
(1)	(2)
.....	Calcium cyanide
.....	Calcium naphthenate
PP ..	Camphechlor
.....	Carbaryl
.....	Carbendazim
.....	Carbofuran
.....	Carbon bisulphide
.....	Carbon tetrabromide
.....	Carbon tetrachloride
PP ..	Carbophenothion
.....	Cartap hydrochloride
PP ..	Chlordane
.....	Chlorfenvinphos
.....	Chlorinated paraffins (C-10 — C-13)
.....	Chlorine
.....	Chlorine cyanide, inhibited
.....	Chlomephos
.....	Chloroacetone, stabilized
.....	2-Chloro-6-nitrotoluene
.....	4-Chloro-2-nitrotoluene
.....	Chloro-ortho-nitrotoluene
.....	2-Chloro-5-trifluoromethylnitrobenzene
.....	para-Chlorobenzyl chloride, liquid or solid
.....	Chlorobenzylchlorides
.....	Chlorodinitrobenzenes
.....	1-Chloroheptane
.....	1-Chlorohexane
.....	Chloronitroanilines
.....	Chloronitrotoluenes, *liquid*
.....	Chloronitrotoluenes, *solid*
.....	1-Chlorooctane

S.M.P.	Marine Pollutant
(1)	(2)
PP..	Chlorophenates, liquid
PP..	Chlorophenates, solid
.....	Chlorophenols, liquid
.....	Chlorophenols, solid
.....	Chlorophenyltrichlorosilane
.....	alpha-Chloropropylene
.....	1-Chloropropylene
.....	2-Chloropropylene
.....	Chlorotoluenes
PP..	Chlorpyriphos
PP..	Chlorthiophos
.....	Chromyl chloride
.....	Coal tar
.....	Coal tar naphtha
.....	Cocculus
.....	Copper acetoarsenite
.....	Copper arsenite
.....	Copper arsenate
.....	Copper chloride
.....	Copper chloride (solution)
PP..	Copper cyanide
PP..	Copper metal powder
.....	Coumachlor
PP..	Coumaphos
.....	Creosote (coal tar)
.....	Creosote (wood tar)
.....	Cresols (*o-; m-; p-*)
PP..	Cresyl diphenyl phosphate
.....	Cresylic acid
.....	Cresylic acid sodium salt
.....	Crotonaldehyde, inhibited

APPENDIX B TO §172.101

S.M.P.	Marine Pollutant
(1)	(2)
.....	Crotonic aldehyde
.....	Crotoxyphos
.....	Cumene
.....	Cupric arsenite
.....	Cupric chloride
PP ..	Cupric cyanide
.....	Cupric sulfate
.....	Cupriethylenediamine solution
.....	Cuprous chloride
.....	Cyanide mixtures
.....	Cyanide solutions
.....	Cyanides, inorganic, n.o.s.
.....	Cyanogen bromide
.....	Cyanogen chloride, inhibited
.....	Cyanophos
PP ..	1,5,9-Cyclododecatriene
PP ..	Cyhexatin
PP ..	Cypermethrin
.....	2,4-D
PP ..	DDT
.....	Decyl acrylate
.....	Decyloxytetrahydrothiophene dioxide
.....	DEF
.....	Di-allate
.....	Di-n-Butyl phthalate
.....	1,4-Di-tert-butylbenzene
PP ..	Dialifos
.....	4,4'-Diaminodiphenylmethane
PP ..	Diazinon
.....	1,2-Dibromethene
.....	1,2-Dibromoethane

APPENDIX B TO §172.101

S.M.P.	Marine Pollutant
(1)	(2)
PP..	Dichlofenthion
.....	Dichloroanilines
.....	o-Dichlorobenzene
.....	p-Dichlorobenzene
.....	1,3-Dichlorobenzene
.....	1,2-Dichlorobenzene
.....	1,4-Dichlorobenzene
.....	Dichlorobenzene (meta; ortho; para)
.....	2,2-Dichlorodiethyl ether
.....	Dichloroether
.....	Dichloroethyl ether
.....	Dichloroethyl oxide
.....	1,1-Dichloroethylene, inhibited
.....	1,6-Dichlorohexane
.....	Dichlorophenols, liquid
.....	Dichlorophenols, solid
.....	2,4-Dichlorophenoxyacetic acid (see also 2,4D)
.....	2,4-Dichlorophenoxyacetic acid diethanolamine salt
.....	2,4-Dichlorophenoxyacetic acid dimethylamine salt
.....	2,4-Dichlorophenoxyacetic acid triisopropylamine salt
.....	Dichlorophenyltrichlorosilane
PP..	Dichlorvos
.....	Dicrotophos
PP..	Dieldrin
.....	Diethylbenzenes (mixed isomers)
.....	Diisopropylbenzenes
.....	Diisopropylnaphthalene
PP..	Dimethoate
.....	Dimethyl glyoxal (butanedione)
.....	Dimethyl sulphide
.....	Dimethylarsinic acid

APPENDIX B TO §172.101

S.M.P.	Marine Pollutant
(1)	(2)
.....	Dimethylphenols, liquid or solid
.....	Dinitro-o-cresol, *solid*
.....	Dinitro-o-cresol, *solution*
.....	Dinitrochlorobenzenes, liquid or solid
.....	Dinitrophenol, *dry or wetted with less than 15 per cent water, by mass*
.....	Dinitrophenol solutions
.....	Dinitrophenol, *wetted with not less than 15 per cent water, by mass*
.....	Dinitrophenolates *alkali metals, dry or wetted with less than 15 per cent water, by mass*
.....	Dinitrophenolates, *wetted with not less than 15 per cent water, by mass*
.....	Dinobuton
.....	Dinoseb
.....	Dinoseb acetate
.....	Dioxacarb
.....	Dioxathion
.....	Diphacinone
.....	Diphenyl
.....	Diphenyl ether
.....	Diphenyl oxide
.....	Diphenyl oxide and biphenyl phenyl ether mixtures
PP ..	Diphenylamine chloroarsine
PP ..	Diphenylchloroarsine, solid *or* liquid
.....	Disulfoton
.....	DNOC
.....	DNOC (pesticide)
.....	Dodecyl diphenyl oxide disulphonate
.....	Dodceyl hydroxypropyl sulfide
.....	1-Dodecylamine
PP ..	Dodecylphenol
.....	Drazoxolon

S.M.P.	Marine Pollutant
(1)	(2)
.....	Edifenphos
PP..	Endosulfan
PP..	Endrin
.....	Epibromohydrin
.....	Epichlorohydrin
PP..	EPN
.....	Esfenvalerate
PP..	Ethion
.....	Ethoprophos
.....	Ethyl acrylate, inhibited
.....	Ethyl chlorothioformate
.....	Ethyl fluid
.....	Ethyl mercaptan
.....	1-Ethyl-2-methylbenzene
.....	5-Ethyl-2-picoline
.....	Ethyl propenoate, inhibited
.....	2-Ethyl-3-propylacrolein
.....	Ethyl tetraphosphate
.....	Ethyldichloroarsine
.....	Ethylene chloride
.....	Ethylene dibromide and methyl bromide mixtures, liquid
.....	Ethylene dichloride
.....	2-Ethylhexenal
.....	2-Ethylhexyl nitrate
.....	Ethylidene dichloride
.....	Fenaminphos
.....	Fenbutatin oxide
PP..	Fenitrothion
PP..	Fenpropathrin
.....	Fensulfothion
PP..	Fenthion

APPENDIX B TO §172.101

S.M.P.	Marine Pollutant
(1)	(2)
PP ..	Fentin acetate
PP ..	Fentin hydroxide
.....	Ferric arsenate
.....	Ferric arsenite
.....	Ferrous arsenate
PP ..	Fonofos
.....	Formetanate
PP ..	gamma-BHC
.....	Gasoline, leaded
PP ..	Heptachlor
.....	Heptenophos
.....	normal-Heptyl chloride
.....	n-Heptylbenzene
PP ..	Hexachlorobutadiene
.....	1,3-Hexachlorobutadiene
.....	Hexaethyl tetraphosphate, *liquid*
.....	Hexaethyl tetraphosphate, *solid*
.....	normal-Hexyl chloride
.....	n-Hexylbenzene
.....	Hydrocyanic acid, anhydrous, stabilized
.....	Hydrocyanic acid, anhydrous, stabilized, absorbed in a porous inert material
.....	Hydrocyanic acid, aqueous solutions *not more than 20% hydrocyanic acid*
.....	Hydrogen cyanide, anhydrous, stabilized
.....	Hydrogen cyanide, anhydrous, stabilized, *absorbed in a porous inert material*
.....	Hydroxydimethylbenzenes, liquid or solid
.....	Ioxynil
.....	Iron oxide, spent
.....	Isoamyl mercaptan
.....	Isobenzan

APPENDIX B TO §172.101

S.M.P.	Marine Pollutant
(1)	(2)
.....	Isobutyl butyrate
.....	Isobutyl isobutyrate
.....	Isobutyl propionate
.....	Isobutylbenzene
.....	Isodecyl acrylate
.....	Isodecyl diphenyl phosphate
.....	Isofenphos
.....	Isooctyl nitrate
.....	Isoprocarb
.....	Isopropenyl chloride
.....	Isopropenylbenzene
.....	Isopropyl chloride
.....	Isopropylbenzene
.....	Isopropyltoluene
PP ..	Isoxathion
.....	Lead acetate
.....	Lead arsenates
.....	Lead arsenites
.....	Lead compounds, soluble, n.o.s.
.....	Lead cyanide
.....	Lead nitrate
.....	Lead perchlorate, solid or solution
.....	Lead tetraethyl
.....	Lead tetramethyl
PP ..	Lindane
.....	London Purple
.....	Magnesium arsenate
.....	Malathion
.....	Maneb or Maneb preparations with not less than 60% maneb
.....	Maneb *or* Maneb preparations *with not less than 60 per cent maneb*

APPENDIX B TO §172.101

S.M.P.	Marine Pollutant
(1)	(2)
.....	Maneb stabilized *or* Maneb preparations, stabilized *against self-heating*
.....	Manganese ethylene-1,2-bisdithiocarbamate
.....	Manganeseethylene-1,2-bisdithiocarbamate, stabilized against self-heating
.....	Mephosfolan
.....	Mercaptodimethur
.....	Mercarbam
PP ..	Mercuric acetate
PP ..	Mercuric ammonium chloride
PP ..	Mercuric arsenate
PP ..	Mercuric benzoate
PP ..	Mercuric bisulphate
PP ..	Mercuric bromide
PP ..	Mercuric chloride
PP ..	Mercuric cyanide
PP ..	Mercuric gluconate
.....	Mercuric iodide
PP ..	Mercuric nitrate
PP ..	Mercuric oleate
PP ..	Mercuric oxide
PP ..	Mercuric oxycyanide, desensitized
PP ..	Mercuric potassium cyanide
PP ..	Mercuric Sulphate
.....	Mercuric sulphide
PP ..	Mercuric thiocyanate
PP ..	Mercurol
PP ..	Mercurous acetate
PP ..	Mercurous bisuphate
PP ..	Mercurous bromide
PP ..	Mercurous chloride
PP ..	Mercurous nitrate

S.M.P.	Marine Pollutant
(1)	(2)
PP..	Mercurous salicylate
PP..	Mercurous sulphate
PP..	Mercury acetates
PP..	Mercury ammonium chloride
PP..	Mercury based pesticides, liquid, flam-mable, toxic, n.o.s.
PP..	Mercury based pesticides, liquid, toxic, flammable, n.o.s.
PP..	Mercury based pesticides, liquid, toxic, n.o.s.
PP..	Mercury based pesticides, solid, toxic, n.o.s.
PP..	Mercury benzoate
PP..	Mercury bichloride
PP..	Mercury bisulphates
PP..	Mercury bromides
PP..	Mercury compounds, liquid, n.o.s.
PP..	Mercury compounds, solid, n.o.s.
PP..	Mercury cyanide
PP..	Mercury gluconate
PP..	NMercury (I) (mercurous) compounds (pesticides)
PP..	NMercury (II) (mercuric) compounds (pesticides)
.....	Mercury iodide
.....	Mercury iodide, *solution*
PP..	Mercury nucleate
PP..	Mercury oleate
PP..	Mercury oxide
PP..	Mercury oxycyanide, desensitized
PP..	Mercury potassium cyanide
PP..	Mercury potassium iodide
PP..	Mercury salicylate
PP..	Mercury sulfates
PP..	Mercury thiocyanate
.....	Metaarsenic acid
.....	Metam-sodium

APPENDIX B TO §172.101

S.M.P.	Marine Pollutant
(1)	(2)
.....	Methamidophos
.....	Methanethiol
.....	Methidathion
.....	Methomyl
.....	ortho-Methoxyaniline
.....	Methyl bromide and ethylene dibromide mixtures, liquid
.....	1-Methyl-2-ethylbenzene
.....	1-Methyl-4-ethylbenzene
.....	2-Methyl-5-ethylpyridine
.....	Methyl mercaptan
.....	2-Methyl-2-phenylpropane
.....	3-Methyl pyridine
.....	Methyl salicylate
.....	3-Methylacroleine, Inhibited
.....	Methylchlorobenzenes
.....	Methylchloroform
.....	Methylene bromide
.....	Methylene dibromide
.....	Methylnaphthalenes, liquid
.....	Methylnaphthalenes, solid
.....	Methylnitrophenols
.....	3-Methylpyradine
.....	alpha-Methylstyrene
.....	Methylstyrenes, inhibited
.....	Methyltrithion
.....	Methylvinylbenzenes, inhibited
PP..	Mevinphos
.....	Mexacarbate
.....	Mirex
.....	Monocratophos
.....	Mononitrobenzene (nitro benzene)

APPENDIX B TO §172.101

S.M.P.	Marine Pollutant
(1)	(2)
.....	Motor fuel anti-knock mixtures
.....	Motor fuel anti-knock mixtures or compunds
.....	Nabam
.....	Naled
.....	Naptha, coal tar
.....	Naphthalene, crude *or* refined
.....	Naphtanlene, molten
.....	Naphthenic acids, liquid
.....	Naphthenic acids, solid
PP..	Nickel carbonyl
PP..	Nickel cyanide
PP..	Nickel tetracarbonyl
.....	Nitrates, inorganic, n.o.s.
.....	Nitrites, inorganic, n.o.s.
.....	3-Nitro-4-chlorobenzotrifluoride
.....	Nitrobenzotrifluorides
.....	Nitrocresols
.....	Nitrotoluenes (o-; m-; p-)
.....	Nitroxylenes, (o-; m-; p-)
.....	Nonylphenol
.....	Oleylamine
PP..	Organotin compounds, liquid, n.o.s.
PP..	Organotin compounds, (pesticides)
PP..	Organotin compounds, solid, n.o.s.
PP..	Organotin pesticides, liquid, flammable, toxic, n.o.s. *flash point less than 23 deg.C*
PP..	Organotin pesticides, liquid, toxic, flam-mable, n.o.s.
PP..	Organotin pesticides, liquid, toxic, n.o.s.
PP..	Organotin pesticides, solid, toxic, n.o.s.
.....	Orthoarsenic acid
.....	Osmium tetroxide
....	Oxamyl

APPENDIX B TO §172.101

S.M.P.	Marine Pollutant
(1)	(2)
.....	Oxydisulfoton
.....	Paraoxon
PP..	Parathion
PP..	Parathion-methyl
PP..	PCBs***
.....	Pentachloroethane
PP..	Pentachlorophenol
.....	Pentalin
.....	Pentanethiols
.....	n-Pentylbenzene
.....	Perchloroethylene
.....	Perchloromethylmercaptan
.....	Petrol, leaded
PP..	Phenarsazine chloride
.....	d-Phenothrin
PP..	Phenthoate
.....	1-Phenylbutane
.....	2-Phenylbutane
.....	Phenylethylene, inhibited
PP..	Phenylmercuric acetate
PP..	Phenylmercuric compounds, n.o.s.
PP..	Phenylmercuric hydroxide
PP..	Phenylmercuric nitrate
.....	2-Phenylpropene
PP..	Phorate
PP..	Phosaione
.....	Phosmet
PP..	Phosphamidon
PP..	Phosphorus, white *or* yellow dry *or* under water *or* in solution
PP..	Phosphorus white, or yellow, molten
.....	Pindone (and salts of)

APPENDIX B TO §172.101

S.M.P.	Marine Pollutant
(1)	(2)
.....	alpha-Pinene
.....	Pirimicarb
PP ..	Pirimiphos-ethyl
PP ..	Polychlorinated biphenyls
PP ..	Polyhalogenated biphenyls, liquid *or* Ter-phenyls liquid
PP ..	Polyhalogenated biphenyls, solid *or* Ter-phenyls solid
PP ..	Potassium cuprocyanide
.....	Potassium cyanide
PP ..	Potassium cyanocuprate I
PP ..	Potassium cyanomercurate
.....	Potassium dihydrogen arsenate
PP ..	Potassium mercuric iodide
.....	Promecarb
.....	Propachlor
.....	Propanethiols
.....	Propaphos
.....	Propenal, inhibited
.....	Propenyl chloride (cis-; trans-)
.....	Propoxur
.....	n-Propylbenzene
.....	Propylene dichloride
.....	Propylidene dichloride
.....	Prothoate
.....	Prussic acid, anhydrous, stabilized
.....	Prussic acid, anhydrous, stabilized, absorbed in a porous inert material
PP ..	Pyrazophos
.....	Quinalphos
.....	Quizalofop
.....	Quizalofop-p-ethyl
.....	Rotenone
.....	Salithion

APPENDIX B TO §172.101

S.M.P.	Marine Pollutant
(1)	(2)
.....	Silver arsenite
.....	Silver cyanide
PP ..	Silver orthoarsenite
PP ..	Sodium copper cyanide, solid
PP ..	Sodium copper cyanide solution
PP ..	Sodium cuprocyanide, solid
PP ..	Sodium cuprocyanide, solution
.....	Sodium cyanide
.....	Sodium dinitro-o-cresolate, *dry or wetted* with less than 15 per cent water, by mass
.....	Sodium dinitro-ortho-cresolate, wetted *with not less than 15 per cent water, by mass*
.....	Sodium metaarsenite
.....	Sodium orthoarsenate
PP ..	Sodium pentachlorophenate
.....	Strontium orthoarsenite
.....	Strychnine *or* Strychnine salts
.....	Styrene monomer, inhibited
.....	Sulfotep
PP ..	Sulprophos
.....	Sym-Dichloroethyl ether
.....	Temephos
.....	TEPP
PP ..	Terbufos
.....	1,1,2,2-Tetrabromoethane
.....	Tetrabromomethane
.....	Tetrachloroethane
.....	1,1,2,2-Tetrachloroethylene
.....	Tetrachloromethane
.....	Tetrachlorophenol
.....	Tetrachlorvinphos
.....	Tetraethyl dithiopyrophosphate

S.M.P.	Marine Pollutant
(1)	(2)
.....	Tetraethyl lead, liquid
.....	Tetramethrin
.....	Tetramethylbenzenes
.....	Tetramethyllead
.....	Thallium chlorate
.....	Thallium compounds, n.o.s.
.....	Thallium compounds (pesticides)
.....	Thallium nitrate
.....	Thallium sulfate
.....	Thallous chlorate
.....	Thiocarbonyl tetrachloride
.....	Triaryl phosphates, isopropylated
PP ..	Triaryl phosphates, n.o.s.
.....	Triazophos
.....	Tribromomethane
PP ..	Tributyltin compounds
.....	Trichlorfon
.....	Trichlorobenzenes, liquid
.....	Trichlorobutene
.....	Trichlorobutylene
.....	Trichloromethane sulphuryl chloride
.....	Trichloromethyl sulphochloride
.....	Trichloronat
.....	Tricresyl phosphate (less than 1% ortho-isomer)
PP ..	Tricresyl phosphate (not less than 1% ortho-isomer)
.....	Tricresyl phosphate *with more than 3 per cent ortho isomer*
.....	Triethylbenzene
.....	Triisopropylated phenyl phosphates
.....	1,2,3-Trimethylbenzene
.....	1,2,4-Trimethylbenzene
.....	1,3,5-Trimethylbenzene

APPENDIX B TO §172.101

S.M.P.	Marine Pollutant
(1)	(2)
.....	Trimethylene dichloride
PP ..	Triphenyltin compounds
.....	Tritolyl phosphate (less than 1% ortho-isomer)
PP ..	Tritolyl phosphate (not less than 1% ortho-isomer)
.....	Trixylenyl phosphate
.....	Turpentine
.....	Turpentine substitute
.....	Vinylbenzene, inhibited
.....	Vinylidene chloride, inhibited
.....	Vinyltoluenes, inhibited *mixed isomers*
.....	Warfarin (and salts of)
.....	White arsenic
PP ..	White phosphorus, dry
PP ..	White phosphorus, molten
PP ..	White phosphorus, wet
.....	White spirit, low (15-20%) aromatic
.....	Xylenols
PP ..	Yellow phosphorus, dry
PP ..	Yellow phosphorus, molten
PP ..	Yellow phosphorus, wet
.....	Zinc bromide
.....	Zinc cyanide

Special Provisions

§172.102 Special provisions.

(a) *General.* When Column 7 of the §172.101 Table refers to a special provision for a hazardous material, the meaning and requirements of that provision are as set forth in this section. When a special provision specifies packaging or packaging requirements—

(1) The special provision is in addition to the standard requirements for all packagings prescribed in §173.24 of this subchapter and any other applicable packaging requirements in subparts A and B of part 173 of this subchapter; and

(2) To the extent a special provision imposes limitations or additional requirements on the packaging provisions set forth in Column 8 of the §172.101 Table, packagings must conform to the requirements of the special provision.

(b) *Description of codes for special provisions.* Special provisions contain packaging provisions, prohibitions, exceptions from requirements for particular quantities or forms of materials and requirements or prohibitions applicable to specific modes of transportation, as follows:

(1) A code consisting only of numbers (for example, "11") is multi-modal in application and may apply to bulk and non-bulk packagings.

(2) A code containing the letter "A" refers to a special provision which applies only to transportation by aircraft.

(3) A code containing the letter "B" refers to a special provision which applies only to bulk packagings requirements. Unless otherwise provided in this sub-

chapter, these special provisions do not apply to IM portable tanks.

(4) A code containing the letter "H" refers to a special provision which applies only to transportation by highway.

(5) A code containing the letter "N" refers to a special provision which applies only to non-bulk packaging requirements.

(6) A code containing the letter "R" refers to a special provision which applies only to transportation by rail.

(7) A code containing the letter "T" refers to a special provision which applies only to transportation in IM portable tanks.

(8) A code containing the letter "W" refers to a special provision which applies only to transportation by water.

(c) *Tables of special provisions.* The following tables list, and set forth the requirements of, the special provisions referred to in Column 7 of the §172.101 Table.

(1) *Numeric provisions.* These provisions are multimodal and apply to bulk and non-bulk packagings:

Code/Special Provisions

1   This material is poisonous by inhalation (see §171.8 of this subchapter) in Hazard Zone A (see §173.116(a) or §173.133(a) of this subchapter), and must be described as an inhalation hazard under the provisions of this subchapter.

2   This material is poisonous by inhalation (see §171.8 of this subchapter) in Hazard Zone B (see §173.116(a) or §173.133(a) of this subchapter), and must be described as an inhalation hazard under the provisions of this subchapter.

3   This material is poisonous by inhalation (see §171.8 of this subchapter) in Hazard Zone C (see §173.116(a) of

this subchapter), and must be described as an inhalation hazard under the provisions of this subchapter.

4 This material is poisonous by inhalation (see §171.8 of this subchapter) in Hazard Zone D (see §173.116(a) of this subchapter), and must be described as an inhalation hazard under the provisions of this subchapter.

5 If this material meets the definition for a material poisonous by inhalation (see §171.8 of this subchapter), a shipping name must be selected which identifies the inhalation hazard, in Division 2.3 or Division 6.1, as appropriate.

6 This material is poisonous-by-inhalation and must be described as an inhalation hazard under the provisions of this subchapter.

7 An ammonium nitrate fertilizer is a fertilizer formulation, containing 90% or more ammonium nitrate and no more than 0.2% organic combustible material (calculated as carbon), which does not meet the definition and criteria of a Class 1 (explosive) material (See §173.50 of this subchapter).

8 A hazardous substance that is not a hazardous waste may be shipped under the shipping description "Other regulated substances, liquid *or* solid, n.o.s.", as appropriate. In addition, for solid materials, special provision B54 applies.

9 Packaging for certain PCBs for disposal and storage is prescribed by EPA in 40 CFR 761.60 and 761.65.

10 An ammonium nitrate mixed fertilizer is a fertilizer formulation, containing less than 90% ammonium nitrate and other ingredients, which does not meet the definition and criteria of a Class 1 (explosive) material (See §173.50 of this subchapter).

11 The hazardous material must be packaged as either a liquid or a solid, as appropriate, depending on its physical form at 55°C (131° F) at atmospheric pressure.

12 In concentrations greater than 40 percent, this material has strong oxidizing properties and is capable of starting fires in contact with combustible materials. If appropriate, a package containing this material must conform to the additional labeling requirements of §172.402 of this subchapter.

13 The words "Inhalation Hazard" shall be entered on each shipping paper in association with the shipping description, shall be marked on each non-bulk package in association with the proper shipping name and identification number, and shall be marked on two opposing sides of each bulk package. Size of marking on bulk package must conform to §172.302(b) of this subchapter. The requirements of §§172.203(m) and 172.505 of this subchapter do not apply.

14 Motor fuel antiknock mixtures are:

  a. Mixtures of one or more organic lead mixtures (such as tetraethyl lead, triethylmethyl lead, diethyldimethyl lead, ethyltrimethyl lead, and tetramethyl lead) with one or more halogen compounds (such as ethylene dibromide and ethylene dichloride), hydrocarbon solvents or other equally efficient stabilizers; or

  b. tetraethyl lead.

15 Chemical kits include boxes, cases, etc., containing small amounts of various compatible dangerous goods which are used for medical, analytical or testing purposes and for which exceptions are provided in this subchapter. Inner packagings must not exceed 250 mL for liquids or 250 g for solids and must be protected from other materials in the kit. For transportation by aircraft, any dangerous goods forbidden in passenger aircraft may not be included in these kits. Kits must be packed in wooden boxes (4C1, 4C2), plywood boxes (4D), reconstituted wood boxes (4F), fiberboard boxes (4G) or plastic boxes (4H1, 4H2) which must be marked "Chemical kits". Each package shall be classified for the material contained therein; if two or more different classes of material are present the class of the package shall be determined in accordance with §173.2a of this subchapter. Each package must be labeled according to each substance contained in the package; this includes the primary hazard label and any subsidiary risk labels applicable to each individual substance within the kit. The total quantity of dangerous goods in any one kit must not exceed either 1 L or 1 kg. The total quantity of dangerous goods in any one package must not exceed either 10 L or 10 kg. The packing group assigned to the kit as a whole must be the most stringent packing group assigned to any individual substance contained in the kit.

16   This description applies to smokeless powder and other solid propellants that are used as powder for small arms and have been classed as Division 1.3 and 4.1 in accordance with §173.56 of this subchapter.

17   Aqueous solutions of hydrogen peroxide containing less than 8 percent hydrogen peroxide are not subject to the requirements of this subchapter.

19   For domestic transportation only, the identification number "UN1075" may be used in place of the identification number specified in Column (4) of the §172.101 Table. The identification number used must be consistent on package markings, shipping papers and emergency response information.

20   The transport of this substance, when in concentrations of greater than 10% nitroglycerin, is prohibited. Concentrations of below 5% nitroglycerin may be transported as a Class 3 material; see UN 1204 and UN 3064.

21   This material must be stabilized by appropriate means (e.g., addition of chemical inhibitor, purging to remove oxygen) to prevent dangerous polymerization (see §173.21(f) of this subchapter).

22 If the hazardous material is in dispersion in organic liquid, the organic liquid must have a flash point above 50° C (122° F).

23 This material may be transported under the provisions of Division 4.1 only if it is so packed that the percentage of diluent will not fall below that stated in the shipping description at any time during transport.

24 Alcoholic beverages containing more than 70 percent alcohol by volume must be transported as materials in Packing Group II. Alcoholic beverages containing more than 24 percent but not more than 70 percent alcohol by volume must be transported as materials in Packing Group III.

26 This entry does not include ammonium permanganate, the transport of which is prohibited except when approved by the Associate Administrator for Hazardous Materials Safety.

27 Sodium carbonate peroxhydrate is considered nonhazardous.

28 The dihydrated sodium salt of dichloroisocyanuric acid is not subject to the requirements of this subchapter.

29 Lithium cells and batteries and equipment containing or packed with lithium cells and batteries which do not comply with the provisions of §173.185 of this subchapter may be transported only if they are approved by the Associate Administrator for Hazardous Materials Safety.

30 Sulfur which is transported domestically is not subject to the requirements of this subchapter if transported in a non-bulk packaging or is formed to a specific shape (e.g., prills, granules, pellets, pastilles, or flakes).

31 Materials which have undergone sufficient heat treatment to render them non-hazardous are not subject to the requirements of this subchapter.

32 These beads are made from polystyrene, poly(methyl methacrylate) or other polymeric material.

33 Ammonium nitrates and mixtures of an inorganic nitrite with an ammonium salt are prohibited.

34 The commercial grade of calcium nitrate fertilizer, when consisting mainly of a double salt (calcium nitrate and ammonium nitrate) containing not more than 10 percent ammonium nitrate and at least 12 percent water of crystallization, is not subject to the requirements of this subchapter.

35 Antimony sulphides and oxides which do not contain more than 0.5 percent of arsenic calculated on the total mass do not meet the definition of Division 6.1.

36 The maximum net quantity per package is 5 liters (1 gallon) or 5 kg (11 pounds).

37 Unless it can be demonstrated by testing that the sensitivity of the substance in its frozen state is no greater than in its liquid state, the substance must remain liquid during normal transport conditions. It must not freeze at temperatures above −15°C(5°F).

38 If this material shows a violent effect in laboratory tests involving heating under confinement, the labeling requirements of Special Provision 53 apply, and the material must be packaged in accordance with packing method OP6B in §173.225 of this subchapter. If the SADT is higher than 75° C, the technically pure substance and formulations derived from it are not self-reactive materials.

39 This substance may be carried under provisions other than those of Class 1 only if it is so packed that the percentage of water will not fall below that stated at any time during transport. When phlegmatized with water and inorganic inert material, the content of urea nitrate must not exceed 75 percent by mass and the mixture should not be capable of being detonated by test 1(a)(i) or test 1(a) (ii) in the UN Recommendations Tests and Criteria.

40 Polyester resin kits consist of two components: a base material (Class 3, Packing Group II or III) and an activator (organic peroxide), each separately packed in an inner packaging. The organic peroxide must be type D, E, or F, not requiring temperature control, and be limited to a quantity of 125 ml (4.22 ounces) per inner packaging if liquid, and 500 g (1 pound) if solid. The components may be placed in the same outer packaging provided they will not interact dangerously in the event of leakage. Packing group will be II or III, according to the criteria for Class 3, applied to the base material.

43 The nitrogen content of the nitrocellulose must not exceed 11.5 percent. Each single filter sheet must be packed between sheets of glazed paper. The portion of glazed paper between the filter sheets must not be less than 65 percent, by mass. The membrane filters/paper arrangement must not be liable to propagate a detonation as tested by one of the tests described in the UN Recommendations, Tests and Criteria, Part I, Test series 1 (a).

44 The formulation must be prepared so that it remains homogeneous and does not separate during transport. Formulations with low nitrocellulose contents and neither showing dangerous properties when tested for their ability to detonate, deflagrate or explode when heated under defined confinement by the appropriate test methods and criteria in the UN Recommendations, Tests and Criteria, nor being a flammable solid when tested in accordance with Appendix E to Part 173 of this subchapter (chips, if necessary, crushed and sieved to a particle size of less than 1.25 mm) are not subject to this subchapter.

45 Temperature should be maintained between 18° C (64.4° F) and 40° C (104° F). Tanks containing solidified methacrylic acid must not be reheated during transport.

46 This material must be packed in accordance with packing method OP6B (see §173.225 of this subchapter). During transport, it must be protected from direct sunshine and stored (or kept) in a cool and well-ventilated place, away from all sources of heat.

47 Mixtures of solids which are not subject to this subchapter and flammable liquids may be transported under this entry without first applying the classification criteria of Division 4.1, provided there is no free liquid visible at the time the material is loaded or at the time the packaging or transport unit is closed. Each packaging must correspond to a design type that has passed a leakproofness test at the Packing Group II level.

48 Mixtures of solids which are not subject to this subchapter and toxic liquids may be transported under this entry without first applying the classification criteria of Division 6.1, provided there is no free liquid visible at the time the material is loaded or at the time the packaging or transport unit is closed. Each packaging must correspond to a design type that has passed a leakproofness test at the Packing Group II level. This entry may not be used for solids containing a Packing Group I liquid.

49 Mixtures of solids which are not subject to this subchapter and corrosive liquids may be transported under this entry without first applying the classification criteria of Class 8, provided there is no free liquid visible at the time the material is loaded or at the time the packaging or transport unit is closed. Each packaging must correspond to a design type that has passed a leakproofness test at the Packing Group II level.

50 Cases, cartridge, empty with primer which are made of metallic or plastic casings and meeting the classification criteria of Division 1.4 are not regulated for domestic transportation.

51 This description applies to items previously described as "Toy propellant devices, Class C" and includes reloadable kits. Model rocket motors containing 30 grams or less propellant are classed as Division 1.4S and items containing more than 30 grams of propellant but not more than 62.5 grams of propellant are classed as Division 1.4C.

52 Ammonium nitrate fertilizers may not meet the definition and criteria of Class 1 (explosive) material (see §173.50 of this subchapter).

53 Packages of these materials must bear the subsidiary risk label, "EXPLOSIVE", unless otherwise provided in this subchapter or through an approval issued by the Associate Administrator for Hazardous Materials Safety, or the competent authority of the country of origin. A copy of the approval shall accompany the shipping papers.

54 Maneb or maneb preparations not meeting the definition of Division 4.3 or any other hazard class are not subject to the requirements of this subchapter when transported by motor vehicle, rail car, or aircraft.

55 This device must be approved in accordance with §173.56 of this subchapter by the Associate Administrator for Hazardous Materials Safety.

56 A means to interrupt and prevent detonation of the detonator from initiating the detonating cord must be installed between each electric detonator and the detonating cord ends of the jet perforating guns before the charged jet perforating guns are offered for transportation.

81 Polychlorinated biphenyl items, as defined in 40 CFR 761.3, for which specification packagings are impractical, may be packaged in non-specification packagings meeting the general packaging requirements of subparts A and B of part 173 of this subchapter. Alternatively, the item itself may be used as a packaging if it meets the general packaging requirements of subparts A and B of part 173 of this subchapter.

101 The name of the particular substance or article must be specified.

102 The articles may be transported as in Division 1.4 Compatibility Group D (1.4D) if all of the conditions specified in §173.63(a) of this subchapter are met.

103 Detonators which will not mass detonate and undergo only limited propagation in the shipping package may be assigned to 1.4B classification code. Mass detonate means that more than 90 percent of the devices tested in a package explode practically simultaneously. Limited propagation means that if one detonator near the center of a shipping package is exploded, the aggregate weight of explosives, excluding ignition and delay charges, in this and all additional detonators in the outside packaging that explode may not exceed 25 grams.

104 Detonators which meet the following conditions may be assigned to 1.4S classification code: Each detonator may contain no more than 1 g of explosive, excluding ignition and delay charges, and if one detonator near the center of a package detonates it will not cause functioning of any other device in the same or adjacent packages.

105 The word "Agents" may be used instead of "Explosives" when approved by the Associate Administrator for Hazardous Materials Safety.

106 The recognized name of the particular explosive may be specified in addition to the type.

107 The classification of the substance is expected to vary especially with the particle size and packaging but the border lines have not been experimentally determined; appropriate classifications should be verified following the test procedures in §§ 173.57 and 173.58 of this subchapter.

108 Fireworks must be so constructed and packaged that loose pyrotechnic composition will not be present in packages during transportation.

109 Rocket motors must be nonpropulsive in transportation unless approved in accordance with §173.56 of this subchapter. A rocket motor to be considered "nonpropulsive" must be capable of unrestrained burning and must not appreciably move in any direction when ignited by any means.

110 Cartridges containing 3.2 grams or less of deflagrating (propellant) explosives installed in a fire extinguisher are not subject to the requirements of this subchapter.

111 Explosive substances of Division 1.1 Compatibility Group A (1.1A) are forbidden for transportation if dry or not desensitized, unless incorporated in a device.

112 "Cartridges, small arms" and "Cartridges, power devices" (which are used to project fastening devices)" which have been classed in Division 1.4 Compatibility Group S (1.4S) may be reclassed and offered for domestic transportation as ORM-D material if they are offered for transportation and transported in accordance with the limitations and packaging requirements of §173.230 of this subchapter.

113 The sample must be given a tentative approval by an agency or laboratory in accordance with §173.56 of this subchapter.

114 Jet perforating guns, charged, oil well, without detonator may be reclassed to Division 1.4 Compatibility Group D (1.4D) if the following conditions are met:
   a. The total weight of the explosive contents of the shaped charges assembled in the guns does not exceed 90.5 kg (200 pounds) per vehicle; and
   b. The guns are packaged in accordance with Packing Method US006 as specified in §173.62 of this subchapter.

115 Boosters with detonator (detonating primers) in which the total explosive charge per unit does not exceed 25 g, and which will not mass detonate and undergo only limited propagation in the shipping package may be assigned to 1.4B classification code. Mass detonate means more than 90 percent of the devices tested in a package explode practically simultaneously. Limited propagation means that if one booster near the center of the package is exploded, the aggregate weight of explosives, excluding ignition and delay charges, in this and all additional boosters in the outside packaging that explode may not exceed 25 g.

116 Fuzes, detonating may be classed in Division 1.4 if the fuzes do not contain more than 25 g of explosive per fuze and are made and packaged so that they will not cause functioning of other fuzes, explosives or other explosive devices if one of the fuzes detonates in a shipping packaging or in adjacent packages.

117 If shipment of the explosive substance is to take place at a time that freezing weather is anticipated, the water contained in the explosive substance must be mixed with denatured alcohol so that freezing will not occur.

(2) *"A" codes.* These provisions apply only to transportation by aircraft:

Code/Special Provisions

A1 Single packagings are not permitted on passenger aircraft.

A2 Single packagings are not permitted on aircraft.

A3 For combination packagings, if glass inner packagings (including ampoules) are used, they must be packed with absorbent material in tightly closed metal receptacles before packing in outer packagings.

A4 Liquids having an inhalation toxicity of Packing Group I are not permitted on aircraft.

A5 Solids having an inhalation toxicity of Packing Group I are not permitted on passenger aircraft and may not exceed a maximum net quantity per package of 15 kg (33 pounds) on cargo aircraft.

A6 For combination packagings, if plastic inner packagings are used, they must be packed in tightly closed metal receptacles before packing in outer packagings.

A7 Steel packagings must be corrosion-resistant or have protection against corrosion.

A8 For combination packagings, if glass inner packagings (including ampoules) are used, they must be packed with cushioning material in tightly closed metal receptacles before packing in outer packagings.

A9 For combination packagings, if plastic bags are used, they must be packed in tightly closed metal receptacles before packing in outer packagings.

A10 When aluminum or aluminum alloy construction materials are used, they must be resistant to corrosion.

A11 For combination packagings, when metal inner packagings are permitted, only specification cylinders constructed of metals which are compatible with the hazardous material may be used.

A13 Non-bulk packagings conforming to §173.197 of this subchapter not exceeding 16 kilograms (35 pounds) gross mass containing only used sharps are permitted for transportation by aircraft. Maximum liquid content in each inner packaging may not exceed 50 milliliters (1.7 ounces).

A14 Non-bulk packagings of regulated medical waste conforming to §173.197 of this subchapter not exceeding 16 kilograms (35 pounds) gross mass for solid waste or 12 liters (3 gallons) total volume for liquid waste may be transported by passenger and cargo aircraft when means of transportation other than air are impracticable or not available.

A19 Combination packagings consisting of outer fiber drums or plywood drums, with inner plastic packagings, are not authorized for transportation by aircraft

A20 Plastic bags as inner receptacles of combination packagings are not authorized for transportation by aircraft.

A29 Combination packagings consisting of outer expanded plastic boxes with inner plastic bags are not authorized for transportation by aircraft.

A30 Ammonium permanganate is not authorized for transportation on aircraft.

A34 Aerosols containing a corrosive liquid in Packing Group II charged with a gas are not permitted for transportation by aircraft.

(3) *"B" codes.* These provisions apply only to bulk packagings:

Code/Special Provisions

B1 If the material has a flash point at or above 38°C (100°F) and below 93°C (200°F), then the bulk packaging requirements of §173.241 of this subchapter are applicable. If the material has a flash point of less than 38°C (100°F), then the bulk packaging requirements of §173.242 of this subchapter are applicable.

B2 MC 300, MC 301, MC 302, MC 303, MC 305, and MC 306, and DOT 406 cargo tanks are not authorized.

B3 MC 300, MC 301, MC 302, MC 303, MC 305, and MC 306, and DOT 406 cargo tanks and DOT 57 portable tanks are not authorized.

B4 AAR 206 tank car tanks and MC 300, MC 301, MC 302, MC 303, MC 305, and MC 306, and DOT 406 cargo tanks are not authorized.

B5 Only ammonium nitrate solutions with 35 percent or less water that will remain completely in solution at a maximum lading temperature of 116°C (240°F) are authorized for transport in the following bulk packagings:

DOT 103 ALW, 111A60 ALW tank car tanks and MC 307, MC 312, DOT 407 and DOT 412 cargo tanks with at least 172 kPa (25 psig) design pressure. The packaging shall be designed for a working temperature of at least 121°C (250°F). Only Specifications MC 304, MC 307 or DOT 407 cargo tank motor vehicles are authorized for transportation by vessel.

B6 Packagings shall be made of steel.

B7 Safety relief devices are not authorized on multi-unit tank car tanks. Openings for safety relief devices shall be plugged or blank flanged.

B8 Packagings shall be made of nickel, stainless steel, or steel with nickel, stainless steel, lead or other suitable corrosion resistant metallic lining.

B9 Bottom outlets are not authorized.

B10 AAR 206 tank car tanks, MC 300, MC 301, MC 302, MC 303, MC 305 and MC 306 and DOT 406 cargo tanks, and DOT 57 portable tanks are not authorized.

B11 Tank car tanks must have a test pressure of at least 2,068.5 kPa (300 psi). Cargo and portable tanks must have a design pressure of at least 1,207 kPa (175 psig).

B12 Tank car tanks must be marked with the name of the lading in accordance with the requirements of §172.330 of this subchapter.

B13 A nonspecification cargo tank motor vehicle authorized in §173.247 of this subchapter must be at least equivalent in design and in construction to a DOT 406 cargo tank or MC 306 cargo tank (if constructed before August 31, 1995), except as follows:

a. Packagings equivalent to MC 306 cargo tanks are excepted from §§178.340-10, certification; 178.341-4, vents; and 178.341-5, emergency flow control.

b. Packagings equivalent to DOT 406 cargo tanks are excepted from §§178.345-7(d)(5), circumferential reinforcements, 178.345-14, marking; 178.345-15, certification; 178.346-10, pressure relief; and 178.346-11, outlets.

c. Packagings are excepted from the design stress limits at elevated temperatures, as described in the ASME Code. However, the design stress limits may not exceed 25 percent of the stress, as specified in the Aluminum Association's "Aluminum Standards and Data" (7th Edition June 1982), for 0 temper at the maximum design temperature of the cargo tank.

B14 Each bulk packaging, except a tank car or a multi-unit-tank car tank, must be insulated with an insulating material so that the overall thermal conductance at 15.5° C (60° F) is no more than 1.5333 kilojoules per hour per square meter per degree Celsius (0.075 Btu per hour per square foot per degree Fahrenheit) temperature differential. Insulating materials must not promote corrosion to steel when wet. Notwithstanding the requirements in §171.14(b)(4)(ii) of this subchapter, compliance with this provision is delayed until October 1, 1994, for a bulk packaging containing a material poisonous by inhalation which, when in contact with moisture, becomes highly corrosive to the tank and could cause a degree of corrosion under an insulation blanket that would have an adverse effect on tank integrity.

B15 Packagings must be protected with non-metallic linings impervious to the lading or have a suitable corrosion allowance.

B16 The lading must be completely covered with nitrogen, inert gas or other inert materials.

B18 Open steel hoppers or bins are authorized.

B23 Tanks must be made of steel that is rubber lined or unlined. Unlined tanks must be passivated before being placed in service. If unlined tanks are washed out with water, they must be repassivated prior to return to service. Lading in unlined tanks must be inhibited so that the corrosive effect on steel is not greater than that of hydrofluoric acid of 65 percent concentration.

B25 Packagings must be made from monel or nickel or monel-lined or nickel-lined steel.

B26 Tanks must be insulated. Insulation must be at least 100 mm (3.9 inches) except that the insulation thickness may be reduced to 51 mm (2 inches) over the exterior heater coils. Interior heating coils are not authorized. The packaging may not be loaded with a material outside of the packaging's design temperature range. In addition, the material also must be covered with an inert gas or the container must be filled with water to the tank's capacity. After unloading, the residual material also must be covered with an inert gas or the container must be filled with water to the tank's capacity.

B27 Tanks must have a service pressure of 1,034 kPa (150 psig). Tank car tanks must have a test pressure rating of 1,379 kPa (200 psi). Lading must be blanketed at all times with a dry inert gas at a pressure not to exceed 103 kPa (15 psig).

B28 Packagings must be made of stainless steel.

B30 MC 312, MC 330, MC 331 and DOT 412 cargo tanks and DOT 51 portable tanks must be made of stainless steel, except that steel other than stainless steel may be used in accordance with the provisions of §173.24b(b) of this subchapter. Thickness of stainless steel for tank shell and heads for cargo tanks and portable tanks must be the greater of 7.62 mm (0.300 inch) or the thickness required for a tank with a design pressure at least equal to 1.5 times the vapor pressure of the lading at 46°C (115° F). In addition, MC 312 and DOT 412 cargo tank motor vehicles must:

a. Be ASME Code (U) stamped for 100% radiography of all pressure-retaining welds;

b. Have accident damage protection which conforms with §178.345–8 of this subchapter;

c. Have a MAWP or design pressure of at least 87 psig: and

d. Have a bolted ASA manway cover.

B32 MC 312, MC 330, MC 331, DOT 412 cargo tanks and DOT 51 portable tanks must be made of stainless steel, except that steel other than stainless steel may be used in accordance with the provisions of §173.24b(b) of this subchapter. Thickness of stainless steel for tank shell and heads for cargo tanks and portable tanks must be the greater of 6.35 mm (0.250 inch) or the thickness required for a tank with a design pressure at least equal to 1.3 times the vapor pressure of the lading at 46°C (115° F). In addition, MC 312 and DOT 412 cargo tank motor vehicles must:

a. Be ASME Code (U) stamped for 100% radiography of all pressure-retaining welds;

b. Have accident damage protection which conforms with §178.345-8 of this subchapter;

c. Have a MAWP or design pressure of at least 87 psig; and

d. Have a bolted ASA manway cover.

B33 MC 300, MC 301, MC 302, MC 303, MC 305, MC 306, and DOT 406 cargo tanks equipped with a 1 psig normal vent used to transport gasoline must conform to Table I of this Special Provision. Based on the volatility class determined by using ASTM D439 and the Reid vapor pressure (RVP) of the particular gasoline, the maximum lading pressure and maximum ambient temperature permitted during the loading of gasoline may not exceed that listed in Table I.

TABLE I—MAXIMUM AMBIENT TEMPERATURE—GASOLINE

ASTM D439 volatility class	Maximum lading and ambient temperature (see note 1)
A (RVP<=9.0 psia)	131° F

ASTM D439 volatility class	Maximum lading and ambient temperature (see note 1)
B (RVP<=10.0 psia)	124° F
C (RVP<=11.5 psia)	116° F
D (RVP<=13.5 psia)	107° F
E (RVP<=15.0 psia)	100° F

Note 1: Based on maximum lading pressure of 1 psig at top of cargo tank.

B35 Tank cars containing hydrogen cyanide may be alternatively marked "Hydrocyanic acid, liquefied" if otherwise conforming to marking requirements in subpart D of this part. Tank cars marked "HYDROCYANIC ACID" prior to October 1, 1991 do not need to be remarked.

B37 The amount of nitric oxide charged into any tank car tank may not exceed 1,379 kPa (200 psig) at 21°C (70°F).

B42 Each 105J500W tank car must be marked as 105J200W. Each tank car must have a safety relief valve with a start-to-discharge pressure of 1,034 kPa (150 psig).

B44 All parts of valves and safety relief devices in contact with lading must be of a material which will not cause formation of acetylides.

B45 Safety relief valves must be equipped with stainless steel or platinum frangible discs approved by the AAR Committee on Tank Cars.

B46 The detachable protective housing for the loading and unloading valves of multi-unit tank car tanks must withstand tank test pressure and must be approved by the Associate Administrator for Hazardous Materials Safety.

B47 A safety relief device with a start-to-discharge pressure setting of 310 kPa (45 psig) is permitted.

B48 Portable tanks in sodium metal service may be visually inspected at least once every 5 years instead of being retested hydrostatically. Date of the visual inspection must be stenciled on the tank near the other required markings.

B49 Tanks equipped with interior heater coils are not authorized. Single unit tank car tanks must have a safety relief valve set at no more than 1551 kPa (225 psig).

B50 Each valve outlet of a multi-unit tank car tank must be sealed by a threaded solid plug or a threaded cap with inert luting or gasket material. Valves must be of stainless steel and the caps, plugs, and valve seats must be of a material that will not deteriorate as a result of contact with the lading.

B52 Notwithstanding the provisions of §173.24b of this subchapter, non-reclosing pressure relief devices are authorized on DOT 57 portable tanks.

B53 Except for IBCs, packagings must be made of either aluminum or steel.

B54 Open-top, sift-proof rail cars are also authorized.

B55 Water-tight, sift-proof, closed-top, metal-covered hopper cars, equipped with a venting arrangement (including flame arrestors) approved by the Associate Administrator for Hazardous Materials Safety are also authorized.

B56 Water-tight, sift-proof, closed-top, metal-covered hopper cars also authorized if the particle size of the hazardous material is not less than 149 microns.

B57 Class DOT 115A tank car tanks must be equipped with a safety vent of a diameter not less than 305mm (12 inches) complying with §179.221-1 of this subchapter an the outer shell must be stenciled "CHLOROPRENE" on both sides in letters not less than 102 mm (4 inches) high.

B59 AAR Specification 207A40W, 207A40W6, 207A48W, 207A60W, 207A80W tank car tanks are also authorized provided that the lading is covered in a nitrogen blanket.

B60 DOT Specification 106A500X multi-unit tank car tanks that are not equipped with a safety relief device of any type are authorized. For the transportation of phosgene, the outage must be sufficient to prevent tanks from becoming liquid full at 55°C (130°F).

B61 Written procedures covering details of tank car appurtenances, dome fittings, safety devices, and marking, loading, handling, inspection, and testing practices must be approved by the Associate Administrator for Hazardous Materials Safety before any single unit tank car tank is offered for transportation.

B64 Each single unit tank car tank built after December 31, 1990 must be equipped with a tank head puncture resistance system that conforms to §179.16 of this subchapter.

B65 Notwithstanding the provisions of §173.244 of this subchapter, only DOT 105A500W tank cars are authorized. Each DOT 105A500W tank car must be marked as DOT 105A300W. Each tank car must have a safety relief valve with a start-to-discharge pressure of 1,551 kPa (225 psig)

B66 Each tank must be equipped with gas tight valve protection caps. Outage must be sufficient to prevent tanks from becoming liquid full at 55°C (130°F). Specification 110A500W tanks must be stainless steel.

B67 All valves and fittings must be protected by a securely attached cover made of metal not subject to deterioration by the lading, and all valve openings, except safety valve, must be fitted with screw plugs or caps to prevent leakage in the event of valve failure.

B68 Sodium must be in a molten condition when loaded and allowed to solidify before shipment. Outage must be at least 5 percent at 98°C (208°F). Bulk packagings must have exterior heating coils fusion welded to the tank shell which have been properly stress relieved. The only tank car tanks authorized are Class DOT 105 tank cars having a test pressure of 2,069 kPa (300 psig) or greater.

B69 Dry sodium cyanide or potassium cyanide may be shipped in sift-proof weather-resistant metal covered hopper cars, covered motor vehicles, portable tanks or non-specification bins. Bins must be approved by the Associate Administrator for Hazardous Materials Safety. Flexible intermediate bulk containers (FIBCs) may also be used under conditions approved by the Associate Administrator for Hazardous Materials Safety.

B70 If DOT 103ANW tank car tank is used: All cast metal in contact with the lading must have 96.7 percent nickel content; and the lading must be anhydrous and free from any impurities.

B71 The only tank cars authorized are Class DOT 105, 112, and 114 tank car tanks with a test pressure of 2069 kPa (300 psig) or greater.

B72 Notwithstanding the provisions of §173.244(a) of this subchapter, only the following tank car tanks are authorized: DOT 105J500W tank car tanks and Class DOT 106 and 110 multi-unit tank car tanks.

B74 Notwithstanding the requirements of §173.244 of this subchapter, only the following are authorized: DOT 105S300W, 105S300ALW, 112J340W, and 114J340W tank cars; and Class DOT 106 and 110 multi-unit-tank car tanks.

B76 Notwithstanding the requirements of §173.244 of this subchapter, only the following are authorized: DOT 105S300W, 105S300ALW, 112J340W, and 114J340W tank cars. Each tank car must be marked DOT 105S200W, 105S200ALW, 112J200W, or 114J200 respectively. Each tank car must have a safety relief valve with a start-to-discharge pressure of 1,034 kPa (150 psig).

B77 Other packaging are authorized when approved by the Associate Administrator for Hazardous Materials Safety.

B78 Notwithstanding §173.240 of this subchapter, the only bulk packagings authorized for transportation by rail are Class DOT 103, 104, 105, 109, 111, 112, and 114 tank car tanks. Heater pipes must be of welded construction designed for a test pressure of 500 pounds per square inch. A 25 mm (1 inch) woven lining of asbestos or other approved material must be placed between the bolster slabbing and the bottom of the tank. If a tank car tank is equipped with a safety vent of the frangible disc type, the frangible disc must be perforated with a 3.2 mm (0.13 inch) diameter hole. If a tank car tank is equipped with a safety relief valve, the tank car tank must also be equipped with a vacuum relief valve.

B79 Tank car tanks shall have head puncture resistance and thermal protection in accordance with §§179.16 and 179.18 of this subchapter for tanks built before April 1, 1989.

B80 Each cargo tank must have a minimum design pressure of 276 kPa (40 psig).

B81 Venting and pressure relief devices for tank car tanks and cargo tanks must be approved by the Associate Administrator for Hazardous Materials Safety.

B82 Cargo tanks and portable tanks are not authorized.

B83 Bottom outlets are prohibited on tank car tanks transporting sulfuric acid in concentrations over 65.25 percent.

B84 Packagings must be protected with non-metallic linings impervious to the lading or have a suitable corrosion allowance for sulfuric acid or spent sulfuric acid in concentration up to 65.25 percent.

B85 Cargo tanks must be marked with the name of the lading accordance with the requirements of §172.302(b).

B90 Steel tanks conforming or equivalent to ASME specifications which contain solid or semisolid residual motor fuel antiknock mixture (including rust, scale, or other contaminants) may be shipped by rail freight or highway. The tank must have been designed and constructed to be capable of withstanding full vacuum. All openings must be closed with gasketed blank flanges or vapor tight threaded closures.

B100 Intermediate bulk containers are not authorized.

B101 Authorized only in metal intermediate bulk containers.

B103 If an intermediate bulk container is used, the package must be transported in a closed freight container or transport vehicle.

B104 Intermediate bulk containers must be provided with a device to allow venting during transport. The inlet to the pressure relief valve must communicate with the vapor space of the packaging and lading during transport.

B105 Authorized only in rigid intermediate bulk containers.

B106 Authorized in intermediate bulk containers that are vapor tight.

B108 Authorized in sift-proof, water-resistant flexible, fiberboard or wooden intermediate bulk containers; packed in a closed transport vehicle.

B109 Not authorized in flexible intermediate bulk containers.

B110 This material also may be packaged in IBCs authorized in §173.242(d) of this subchapter.

(4) *"H" codes.* These provisions apply only to transportation by highway.
(Reserved)

(5) *"N" codes.* These provisions apply only to non-bulk packagings:

Code/Special Provisions

N3 Glass inner packagings are permitted in combination or composite packagings only if the hazardous material is free from hydrofluoric acid.

N4 For combination or composite packagings, glass inner packagings, other than ampoules, are not permitted.

N5 Glass materials of construction are not authorized for any part of a packaging which is normally in contact with the hazardous material.

N6 Battery fluid packaged with electric storage batteries, wet or dry, must conform to the packaging provisions of §173.159(g) or (h) of this subchapter.

N7 The hazard class or division number of the material must be marked on the package in accordance with §172.302 of this subchapter. However, the hazard label corresponding to the hazard class or division may be substituted for the marking.

N8 Nitroglycerin solution in alcohol may be transported under this entry only when the solution is packed in metal cans of not more than 1 L capacity each, overpacked in a wooden box containing not more than 5

L. Metal cans must be completely surrounded with absorbent cushioning material. Wooden boxes must be completely lined with a suitable material impervious to water and nitroglycerin.

N9 If the substance is impregnated with less than 5% oil, it is excepted from the labeling requirements of subpart D of this part and the packaging tests of subpart M of part 178 of this subchapter.

N10 Lighters and their inner packagings, which have been approved by the Associate Administrator for Hazardous Materials Safety (see §173.21(i) of this subchapter), must be packaged in one of the following outer packagings at the Packing Group II level: 4C1 or 4C2 wooden boxes; 4D plywood boxes; 4F reconstituted wood boxes; 4G fiberboard boxes; or 4H1 or 4H2 plastic boxes.

N11 This material is excepted for the specification packaging requirements of this subchapter if the material is packaged in strong, tight non-bulk packaging meeting the requirements of subparts A and B of part 173 of this subchapter.

N12 Plastic packagings are not authorized.

N20 A 5M1 multi-wall paper bag is authorized if transported in a closed transport vehicle.

N25 Steel single packagings are not authorized.

N32 Aluminum materials of construction are not authorized for single packagings.

N33 Aluminum drums are not authorized.

N34 Aluminum construction materials are not authorized for any part of a packaging which is normally in contact with the hazardous material.

N36 Aluminum or aluminum alloy construction materials are permitted only for halogenated hydrocarbons that will not react with aluminum.

N37 This material may be shipped in an integrally-lined fiber drum (1G) which meets the general packaging requirements of subpart B of part 173 of this subchapter, the requirements of part 178 of this subchapter at the packing group assigned for the material and to any other special provisions of column 7 of the §172.101 table.

N40 This material is not authorized in the following packagings:
   a. A combination packaging consisting of a 4G fiberboard box with inner receptacles of glass or earthenware;
   b. A single packaging of a 4C2 sift-proof, natural wood box; or
   c. A composite packaging 6PG2 (glass, porcelain or stoneware receptacles within a fiberboard box).

N41 Metal construction materials are not authorized for any part of a packaging which is normally in contact with the hazardous material.

N43 Metal drums are permitted as single packagings only if constructed of nickel or monel.

N45 Copper cartridges are authorized as inner packagings if the hazardous material is not in dispersion.

N50 A Class 9 material that meets the definition of a marine pollutant, but does not meet the definition of a hazardous substance or a hazardous waste or the definition in §173.140(a) of this subchapter, is excepted from the labeling requirements of this part.

N65 Outage must be sufficient to prevent cylinders or spheres from becoming liquid full at 55°C (130°F). The vacant space (outage) may be charged with a nonflammable nonliquefied compressed gas if the pressure in the cylinder or sphere at 55°C (130°F) does not exceed 125 percent of the marked service pressure.

N71 Combination packagings consisting of inner glass packagings of not over 1.0 L (0.3 gallon) capacity each or inner metal packagings of not over 5.0 L (1 gallon) capacity each, placed in strong outer packagings, are authorized. Packagings are not subject to the requirements of part 178 of this subchapter.

N72 Packagings must be examined by the Bureau of Explosives and approved by the Associate Administrator for Hazardous Materials Safety.

N73 Packagings consisting of outer wooden or fiberboard boxes with inner glass, metal or other strong containers; metal or fiber drums; kegs or barrels; or strong metal cans are authorized and need not conform to the requirements of part 178 of this subchapter.

N74 Packages consisting of tightly closed inner containers of glass, earthenware, metal or polyethylene, capacity not over 0.5 kg (1.1 pounds) securely cushioned and packed in outer wooden barrels or wooden or fiberboard boxes, not over 15 kg (33 pounds) net weight, are authorized and need not conform to the requirements of part 178 of this subchapter.

N75 Packages consisting of tightly closed inner packagings of glass, earthenware or metal, securely cushioned and packed in outer wooden barrels or wooden or fiberboard boxes, capacity not over 2.5 kg (5.5 pounds) net weight, are authorized and need not conform to the requirements of part 178 of this subchapter.

N76 For materials of not more than 25 percent active ingredient by weight, packages consisting of inner metal packagings not greater than 250 ml (8 ounces) capacity each, packed in strong outer packagings together with sufficient absorbent material to completely absorb the liquid contents are authorized and need not conform to the requirements of part 178 of this subchapter.

N77 For materials of not more than two percent active ingredients by weight, packagings need not conform to the requirements of part 178 of this subchapter, if liquid contents are absorbed in an inert material.

N78 Packages consisting of inner glass, earthenware, or polyethylene or other nonfragile plastic bottles or jars not over 0.5 kg (1.1 pounds) capacity each, or metal cans not over five pounds capacity each, packed in outer wooden boxes, barrels or kegs, or fiberboard boxes are authorized and need not conform to the requirements of part 178 of this subchapter. Net weight of contents in fiberboard boxes may not exceed 29 kg (64 pounds). Net weight of contents in wooden boxes, barrels or kegs may not exceed 45 kg (99 pounds).

N79 Packages consisting of tightly closed metal inner packagings not over 0.5 kg (1.1 pounds) capacity each, packed in outer wooden or fiberboard boxes, or wooden barrels, are authorized and need not conform to the requirements of part 178 of this subchapter. Net weight of contents may not exceed 15 kg (33 pounds).

N80 Packages consisting of one inner metal can, not over 2.5 kg (5.5 pounds) capacity, packed in an outer wooden or fiberboard box, or a wooden barrel, are authorized and need not conform to the requirements of part 178 of this subchapter.

N82 See §173.306 of this subchapter for classification criteria for flammable aerosols.

(6) *"R" codes.* These provisions apply only to transportation by rail.
(Reserved)

(7) *"T" codes.* These provisions apply only to transportation in IM portable tanks. They are divided into two groupings, one of which appears as the IM Tank Configurations in paragraph (c)(7)(i) of this section, and the second of which imposes specific requirements and appears in paragraph (c)(7)(ii) of this section.

(i) *IM Tank Configurations.* Column 1 lists the code for the special provisions as specified in column 7 of the §172.101 table. Column 2 specifies the IM tank type, either IM 101 (§§178.270 and 178.271 of this subchapter) or IM 102 (§§178.270 and 178.272 of this subchapter). Column 3 specifies the minimum test pressure, in bars (1 bar = 14.5 psig), at which the periodic hydrostatic testing required by §173.32b of this subchapter must be conducted. Column 4 specifies either the section referenced for requirements for bottom openings or "Prohibited", which means bottom openings are prohibited. Column 5 specifies the section reference for requirements applicable to pressure relief devices.

IM TANK CONFIGURATIONS

Code (1)	IM tank type (2)	Minimum test Pressure (bars) (3)	Bottom Outlets (4)	Pressure relief devices (5)
T1	102	1.5	§173.32c(g)(1)	§178.270-11(a)(1),(2)
T2	102	1.5	§173.32c(g)(2)	§178.270-11(a)(1),(2)
T7	101	2.65	§173.32c(g)(1)	§178.270-11(a)(1),(2)
T8	101	2.65	§173.32c(g)(2)	§178.270-11(a)(1),(2)
T9	101	2.65	Prohibited	§178.270-11(a)(1),(2)
T11	101	2.65	§173.32c(g)(2)	§178.270-11(a)(3)
T12	101	2.65	Prohibited	§178.270-11(a)(3)
T13	101	4	§173.32c(g)(1)	§178.270-11(a)(1),(2)
T14	101	4	§173.32c(g)(2)	§178.270-11(a)(1),(2)
T15	101	4	Prohibited	§178.270-11(a)(1),(2)
T16	101	4	§173.32c(g)(1)	§178.270-11(a)(3)
T17	101	4	§173.32c(g)(2)	§178.270-11(a)(3)
T18	101	4	Prohibited	§178.270-11(a)(3)
T20	101	6	§173.32c(g)(2)	§178.270-11(a)(1),(2)
T21	101	6	Prohibited	§178.270-11(a)(1),(2)
T22	101	6	§173.32c(g)(1)	§178.270-11(a)(1)(2)
T23	101	6	§173.32c(g)(2)	§178.270-11(a)(3)
T24	101	6	Prohibited	§178.270-11(a)(3)
T28	101	10	Prohibited	§178.270-11(a)(1),(2)
T39	101	10	Prohibited	§178.270-11(a)(3)
T43	101	9	Prohibited	§178.270-11(a)(3)

(ii) *IM Tank special provisions.*

Code/Special Provisions

T25 This hazardous material is not permitted for transport in IM portable tanks.

T26 Each tank must have a minimum shell thickness of 6.35 mm (0.250 inch) mild steel.

T27 Each tank must have a minimum shell thickness of 8.0 mm (0.315 inch) mild steel.

T28 See entry for T28 in the IM Tank Configuration Table in paragraph (c)(7)(i) of this section.

T29 The lading must be completely covered with nitrogen, inert gas or other inert materials.

T30 IM 102 portable tanks without bottom openings or with bottom openings conforming to §173.32c(g)(1) of this subchapter are authorized for a hazardous material with a flash point of 0°C (32°F) or greater and a vapor pressure not greater than 65.5 kPa (9.5 psia) at 65.6°C (150°F).

T31 IM 102 portable tanks without bottom openings or with bottom openings conforming to §173.32c(g)(2) of this subchapter are authorized for a hazardous material with a flash point of 0°C (32°F) or greater and a vapor pressure not greater than 65 kPa (9.4 psia) at 65.6°C (150°F).

T32 Each tank must have a minimum shell thickness of 10.0 mm (0.394 inch) mild steel with at least 5.0 mm (0.197 inch) lead lining.

T33 Dry phosphorus is not permitted. For transport in a molten state, the tank must be insulated in accordance with Note T38. Air must be eliminated from the interior of the tank. The tank may be heated, however, interior heating coils are prohibited.

T34 The IM Tank authorization is limited to aqueous solutions containing not more than 40% dimethylamine.

T35 Each tank must be equipped with reclosing (spring loaded) pressure relief valves set to discharge at pressures determined according to the pressure characteristics of the organic peroxide lading.

T36 Each tank must be equipped with pressure relief devices with sufficient venting capacity to prevent the tank from bursting.

T37 IM portable tanks are only authorized for the shipment of hydrogen peroxide solutions in water containing 72 percent or less hydrogen peroxide by weight. Pressure relief devices shall be designed to prevent the entry of foreign matter, the leakage of liquid and the development of any dangerous excess pressure. In addition, the tank shall be designed so that internal surfaces may be effectively cleaned and passivated. Each tank must be equipped with pressure relief devices conforming to the following requirements:

Concentration of hydrogen peroxide solution	Total venting capacity in standard cubic feet per hour (S.C.F.H.) per pound of hydrogen peroxide solution
52 percent or less	11
Over 52 percent but not greater than 60 percent	22
Over 60 percent but not greater than 72 percent	32

T38 Each tank must be insulated with an insulating material so that the overall thermal conductance at 15.5° C (60° F) is no more than 1.5333 kilojoules per hour per square meter per degree Celsius (0.075 Btu per hour per square foot per degree Fahrenheit) temperature differential. Insulating materials must not promote corrosion to steel when wet. Notwithstanding the requirements in §171.14(b)(4)(ii) of this subchapter, compliance with this provision is delayed until October 1, 1994, for a bulk packaging containing a material poisonous by inhalation which, when in contact with moisture, becomes highly corrosive and could cause corrosion under an insulation blanket.

T39 See entry for T39 in the IM Tank Configuration Table in paragraph (c)(7)(i) of this section.

T40 Each tank must have a minimum shell thickness of 10.0 mm (0.39 inch) mild steel.

T41 Each tank must have a minimum shell thickness of 12.0 mm (0.47 inch) mild steel.

T42 Transport in IM portable tanks is permitted only under conditions approved by the Associate Administrator for Hazardous Materials Safety.

T43 See entry for T43 in the IM Tank Configuration Table in paragraph (c)(7)(i) of this section.

T44 DOT Specification IM 101 portable tanks shall be made of stainless steel except that steel other than stainless steel may be used in accordance with the provisions of §173.24b(b) of this subchapter. Thickness of stainless steel for tank shell and heads must be the greater of 7.62 mm (0.300 inch) or the thickness required for a tank with a design pressure at least equal to 1.5 times the vapor pressure of the lading at 46°C (115°F).

T45 DOT Specification IM 101 portable tanks shall be made of stainless steel except that steel other than stainless steel may be used in accordance with the provisions of §173.24b(b) of this subchapter. Thickness of stainless steel for tank shell and heads must be the greater of 6.35 mm (0.250 inch) or the thickness required for a tank with a design pressure at least equal to 1.3 times the vapor pressure of the lading at 46°C (115°F).

T46 IM portable tanks in sodium metal service are not required to be hydrostatically retested.

(8) *"W" codes.* These provisions apply only to transportation by water:

Code/Special Provisions

W41 When offered for transportation by water, this material must be packaged in bales and be securely and tightly bound with rope, wire or similar means.

EMERGENCY PHONE NUMBERS

DOT National Emergency Response Center: (800) 424-8802

Center for Disease Control: (800) 232-0124

Chemtrec: (800) 424-9300

PLACARDS

CLASS 1	CLASS 1
EXPLOSIVES 1.1, 1.2, & 1.3 ✳ The Division number 1.1, 1.2 or 1.3 and compatibility group (when required) are in black ink. Placard any quantity of Division number 1.1, 1.2, or 1.3 material.	**EXPLOSIVES 1.4** ✳ The compatibility group (when required) is in black ink. Placard 454 kg (1001 lbs.) or more of 1.4 Explosives.
CLASS 1	**CLASS 1**
EXPLOSIVES 1.5 ✳ The compatibility group (when required) is in black ink. Placard 454 kg (1001 lbs.) or more of 1.5 Blasting Agents.	**EXPLOSIVES 1.6** ✳ The compatibility group (when required) is in black ink. Placard 454 kg (1001 lbs.) or more of 1.6 Explosives.
CLASS 2 Division 2.1	**CLASS 2** Division 2.2
FLAMMABLE GAS Placard 454 kg (1001 lbs.) or more of flammable gas. See DANGEROUS.	**NON-FLAMMABLE GAS** Placard 454 kg (1001 lbs.) or more aggregate gross weight of non-flammable gas. See DANGEROUS.
CLASS 2	**CLASS 2** Division 2.3
OXYGEN Placard 454 kg (1001 lbs.) or more aggregate gross weight of either oxygen compressed and oxygen, refrigerated liquid. See 172.504(f)(7).	**POISON GAS** Placard any quantity of Division 2.3 material.

PLACARDS

CLASS 3

FLAMMABLE
Placard 454 kg (1001 lbs.) or more gross weight of flammable liquid. See DANGEROUS.

CLASS 3

GASOLINE
May be used in place of FLAMMABLE on a placard displayed on a cargo tank or a portable tank being used to transport gasoline by highway. See 172.542(c).

CLASS 3

COMBUSTIBLE
Placard a combustible liquid when transported in bulk.
A FLAMMABLE placard may be used in place of a COMBUSTIBLE placard on a cargo tank or portable tank or a compartmented tank car which contains both flammable and combustible liquids.
See 172.504(f)(2).

CLASS 3

FUEL OIL
May be used in place of COMBUSTIBLE on a placard displayed on a cargo tank or portable tank being used to transport, by highway, fuel oil not classed as a flammable liquid. See 172.544(c).

CLASS 4 — Division 4.1

FLAMMABLE SOLID
Placard 454 kg (1001 lbs.) or more gross weight of flammable solid. See DANGEROUS.

CLASS 4 — Division 4.2

SPONTANEOUSLY COMBUSTIBLE
Placard 454 kg (1001 lbs.) or more gross weight of spontaneously combustible material. See DANGEROUS.

CLASS 4 — Division 4.3

DANGEROUS WHEN WET MATERIAL
Placard any quantity of Division 4.3 material.

PLACARDS & MARKINGS

DANGEROUS
Placard 454 kg (1001 lbs.) gross weight of two or more categories of hazardous materials listed in Table 2 (172.504). A freight container, unit load device, transport vehicle of rail car which contains non-bulk packagings with two or more categories of hazardous materials that require different placards specified in Table 2 may be placarded with DANGEROUS placards instead of the separate placarding specified for each of the materials in Table 2. However, when 2,268 kg (5000 lbs.) or more of one category of material is loaded therein at one loading facility, the placard specified in Table 2 for that category must be applied.
Division 1.4 (explosives)
Division 1.5 (blasting agents)
Division 1.6 (explosives)
Division 2.1 (flammable gas)
Division 2.2 (non-flammable gas)
Class 3 (flammable liquid)
Combustible liquid
Division 4.1 (flammable solid)
Division 4.2 (spontaneously combustible)
Division 5.1 (oxidizer)
Division 5.2 (organic peroxide)
Division 6.1, PG I or II, other than PG I inhalation hazard (poison)
Division 6.1, PG III (keep away from food)
Class 8 (corrosive)
Class 9 (miscellaneous)

DISPLAY OF IDENTIFICATION NUMBER WHEN TRANSPORTING HAZARDOUS MATERIALS IN PORTABLE TANKS, CARGO TANKS, AND TANK CARS.

ORANGE PANEL

PLACARD

FUMIGATED

Placard motor vehicle, freight container or rail car on or near each door when fumigated with Division 6.1 (Poison) or Division 2.3 (Poison Gas).

RAIL

Placard empty tank cars for residue of material last contained.

The square background is required for the following placards when on rail cars: EXPLOSIVES 1.1 or 1.2; POISON GAS or POISON GAS-RESIDUE (Division 2.3, Hazard Zone A); POISON or POISON-RESIDUE (Division 6.1, PGI, Hazard Zone A). The square background is required for placards on motor vehicles transporting highway route controlled quantities of radioactive materials (Class 7).